THE

RESOURCES OF CALIFORNIA,

COMPRISING

AGRICULTURE, MINING, GEOGRAPHY, CLIMATE, COMMERCE, ETC. ETC.

AND THE

PAST AND FUTURE DEVELOPMENT OF THE STATE.

BY JOHN S. HITTEL.

SAN FRANCISCO:
A. ROMAN & COMPANY.
NEW YORK: W. J. WIDDLETON.
1863.

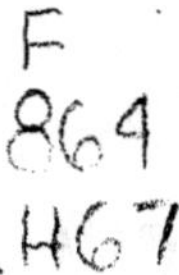

M'CREA & MILLER, STEREOTYPERS. C. A. ALVORD, PRINTER.

RESOURCES OF CALIFORNIA.

PREFACE.

I UNDERTAKE to write the resources of a state, which, though young in years, small in population, and remote from the chief centres of civilization, is yet known to the furthest corners of the earth, and, during the last twelve years, has had an influence upon the course of human life, and the prosperity and trade of nations, more powerful than that exerted during the same period by kingdoms whose subjects are numbered by millions, whose history dates back through thousands of years, and whose present stock of wealth began to accumulate before our continent was discovered, or our language was formed. I write of a land of wonders. I write of California, which has astonished the world by the great migration that suddenly built up the first large Caucasian community on the shores of the North Pacific; by her vast yield of gold, amounting within thirteen years to $700,000,000, which has sensibly affected the markets of labor and money in all the leading nations of Christendom; by the rapid development and great extent of her commerce; by the greatness of her chief port, which at one time had more large ships at her anchorage than were ever seen together in the harbor of either Liverpool, New York, or London; by the swift settlement of her remote districts; by the prompt organization of her government;

by the liberality with which the mines were thrown open and made free to all comers; by the rush of adventurers of every color and of every tongue; by the prices of her labor, and the rates of her interest for money, double those of the other American states, and quadruple those of Europe; by the vast extent of her gold-fields, and the facility with which they could be worked; by the auriferous rivers in which fortunes could be made in a week; by antediluvian streams richer than those of the present era; by beds of lava, which, after filling up the beds of antediluvian rivers, were left, by the washing away of the banks and adjacent plains, to stand as mountains, marking the position of great treasures beneath; by nuggets each worth a fortune; by the peculiar nature of her mining industry; by new and strange inventions; by the washing down of mountains; by filling the rivers of the Sacramento basin with thick mud throughout the year; by lifting a hundred mountains from their beds; by six thousand miles of mining ditches; by aqueducts less durable, but scarcely less wonderful than those of ancient Rome; by silver mines that promise to rival those of Peru; by quicksilver mines surpassing those of Spain; by great deposits of sulphur and asphaltum; by lakes of borax; by mud volcanoes, geysers, and natural bridges; by a valley of romantic and sublime beauty, shut in by walls nearly perpendicular and more than three-quarters of a mile high, with half a dozen great cascades, in one of which the water at two leaps falls more than the third of a mile; by a climate the most conducive to health, and the most favorable to mental and physical exertion—so temperate on the middle coast that ice is never seen and thin summer clothing never worn, and that January differs in average temperature only eight degrees of Fahrenheit from July; by a singular botany, including the most splendid known group of coniferous trees, of which half a dozen species grow to be more than two hundred and fifty feet high, and one species has reached a height of four hundred and fifty feet, and a diameter of forty feet in the trunk; by a peculiar zoology, composed almost of animals found only on this coast, and including the largest bird north of the equator, and the largest and most formidable quadruped of the continent; by the importation in early years of all articles of food, and then by the speedy development of agriculture, until her wheat and wine have gone to the furthest cities in search of buyers, and until her markets are unrivalled in the variety and magnificence of home-grown fruits; by the largest crop of grain, and the largest specimens of fruits and vegetables on record; by a society where for years there was not one woman to a score of men, and where all the men were in the bloom of manhood; by the first large migration of eastern Asiatics from their own continent; by the first settlement of Chinamen among white men; by the entire lack of mendicants, paupers, and alms-houses; by the rapid fluctuations of trade; by the accumulation of wealth in the hands of men, most of whom came to the country poor; by the practice, universal in early years, of going armed; by the multitude of deadly affrays, and by extra-constitutional courts, which sometimes punished villains with immediate execution, and sometimes proceeded with a gravity and slow moderation that might become the most august tribunals. I write of California while she is still youthful, and full of marvels; while her population is still unsettled; while her business is still fluctuating, her wages high, her gold abundant, and her birth still fresh in the memory of men and women who have scarcely reached their majority; and I write of her while she still offers a wide field for the adventurous, the enterprising, and the young, who have life before them, and wish to commence it where they may have the freest career, in full sight of the greatest rewards for success, and with the fewest chances of failure.

The general public are aware that California is a peculiar state, and their attention has often been called to certain prominent points of wonder, like those to which I have just referred; but hitherto there has been no careful attempt to sum up all that is known of her resources and natural history. I have undertaken that task, and the result of my undertaking is in this book placed before the reader. I have been a Californian since 1849, and expect to be as long as I may live. All the most interesting associations of my life are connected with this state. I arrived in the country while it was still under a territorial government, and more than a year before it was organized as a state under act of Congress. I saw the land in its original wildness, and saw society, order, trade, industry and polity developed; and I now see about me the beginnings and promises of science, art, literature, philosophy, and whatever can enrich or honor humanity. I have seen the state grow up, and its history is part of my life. The land-marks of its chorography, and the prominent events of its political, social, and industrial progress, mark epochs in my memory. Many of the happiest days of my life have been spent here, and here I hope to enjoy whatever blessings the future may have in store for me. If then I fail to do justice in my book to California, the failure will not be for any lack of love of her. Neither will it be for any lack of attention or industry. During the last nine years, I have assiduously collected every thing within my reach relative to the industry, resources, natural history and population of the state. I have looked through all the newspapers published between Crescent City and San Diego, and have examined all the books written about the country, Spanish, French and German, as well as English. I have been in the extreme north, and the extreme south; I have gone to both extremities by land and sea; I have travelled through the centre of her great basin; I am intimately acquainted with her richest agricultural districts; I know something of her mining and agriculture by experience and practice; and, finally, I have endeavored to compress into this book all the important attainable facts. Amidst so much information, there are undoubtedly some little errors; but the fair critic, before condemning and expatiating upon minor faults, will pass judgment upon the question whether the book is or is not more comprehensive and instructive than any other, or than all others relating to the same subject.

Of course, when I quote from the writings of others, I use quotation marks, and give credit according to the rules of honorable authors; but I have adopted, without quotation marks, various passages from articles written by myself, and published in different newspapers and magazines. Since the work is intended for popular use, and should be free from every thing not intelligible and interesting to the general reader, I have made no references to authorities; and, indeed, I have drawn my information from so great a variety of sources (in many instances newspapers), that it would have been very inconvenient for me, and cumbersome to the book, to cite the authority for every statement. In case, however, that the accuracy of any statement in the work should be called in question, I think that I can produce in every case credible evidence, and in most cases the conclusive proof. While I have drawn my material from many different sources, I claim as much originality as is possible for so comprehensive a collection of facts, in so many and so distinct branches of knowledge.

J. S. H.

SAN FRANCISCO, *March*, 1862.

INDEX OF CHAPTERS.

INDEX OF SECTIONS.

CHOROGRAPHY—CHAP. I.

CLIMATE—CHAP. II.

GEOLOGY—CHAP. III.

SCENERY—CHAP. IV.

BOTANY—CHAP. V.

TOPOGRAPHICAL NAMES—CHAP. XIII.

THE PAST AND FUTURE DEVELOPMENT OF THE STATE—CHAP. XIV.

RESOURCES OF CALIFORNIA.

CHAPTER I.

CHOROGRAPHY.

§ 1. *General Remarks.*—CALIFORNIA has a peculiar chorography. No other country comprises within so small a space, such various, so many, and such strongly-marked chorographical divisions. Mountains the most steep, barren, and rugged; valleys the most fertile and beautiful; deserts the most sterile; spacious bays, magnificent rivers, unparalleled waterfalls, picturesque lakes, extensive marshes, broad prairies, and dense forests—all these are hers.

In general shape, California is a long parallelogram, extending from latitude 32° 45′ to 42° north, seven hundred miles in length by one hundred and eighty in breadth, the course of the longitudinal axis being north-northwest by south-southeast. The first topographical division of the state may be into the *Coast* and *Interior* districts, separated from each other by the main ridge of the Coast Mountains, which runs the whole length of the state, nearly parallel with the ocean, and about fifty miles from it. The Coast district may be subdivided into the *Coast Mountains* and the *Coast Valleys.* The Interior district may be subdivided into the *Sierra Nevada*, the *Sac-*

ramento Basin, the *Plateau of the Sierra Nevada*, the *Klamath Basin*, the *Great Basin of Utah*, and the *Colorado Desert*.

§ 2. *Coast Mountains.*—The Coast range, though not so high or so wide as the Sierra Nevada, may be considered the main orographical feature of California, because it alone extends through the whole length of the state. The height of the range is from two thousand to six thousand feet; its width from twenty to forty miles. South of 34° 20′ the spurs are short and run at right angles to the course of the main divide, which is the easternmost ridge of the chain; nearly all the spurs, valleys, and streams, run to the westward. South of 34° 20′ a plain from twenty-five to forty miles wide lies between the mountains and the sea; north of that the spurs make up the greater part of the Coast line, and, where they enter the ocean, form the headlands and capes. The Santa Susanna spur starts from the main ridge in 34° 20′ and runs west by south, and is separated by the valley of the Santa Clara River from the Santa Inez ridge, which starts in 34° 30′ and runs west; then continuing our course northward, across the Santa Inez valley, we come to the Santa Barbara ridge, which starts from the main ridge in 34° 40′ and runs west-northwest. The Cuyama valley separates the Santa Barbara from the Santa Lucia ridge, which branches off at 35° in a northwestern direction, and forms the southern boundary of the Salinas valley, whose northern boundary is the Gabilan ridge, starting in 36° 10′ and running north-northwest; which is separated from the Contra Costa ridge, rising in 37° 10′ by the Santa Clara valley, and the Contra Costa ridge is separated from the main divide by the Amador and San Ramon valleys. The Gabilan ridge forms the back-bone of Santa Cruz, San Mateo, and San Francisco counties, each of which gives its name to that portion within its borders. The ridge is cut in two on the southern border of Santa Cruz county by the Pajaro River, and the Alameda Creek breaks through the Contra Costa ridge. North of the Golden Gate, the Gabilan ridge reappears, and is separated by Petaluma valley from the Sonoma ridge, that by Sonoma valley from the Carneros ridge, and that by Napa valley from the main Coast ridge. Farther north the spurs are so numerous, and connected so closely together, that they are scarcely distinguished by names; and a large portion of the coast, from the main ridge westward, is a mass of mountains. The Coast Mountains are steep, rocky, rugged, and brown: north of 38° they are covered with timber and brush; south of that the ridges nearest the ocean have some timber, those farther inland are nearly bare. The main ridge near the head of the Sacramento valley is called the Trinity ridge; near Mount Diablo it is called the Diablo ridge, or the Bolbones ridge; south of 34° it is called the San Bernardino ridge, and in one place the Cuyamaca Mountain.

§ 3. *Coast Peaks and Passes.*—The principal peaks of the main ridge are Mount Linn, in 40° 10′; Mount St. John, in 39° 25′; Mount Ripley, 7,500 feet high, in 39° 08′; Mount St. Helena, 3,700 feet high, in 38° 40′; Mount Diablo, 3,857 feet high, in 37° 50′; Pacheco's Peak, 2,700 feet high, in 36° 57′; Mount San Bernardino, 8,500 feet high, in 34° 20′; and Mount San Gorgonio, 7,000 feet high, in 33° 48′. In the Gabilan ridge are the following peaks: the Chupadero, in 36° 35′; the Gabilan, in 36° 50′; the Loma Prieta, 4,040 feet high, in 37° 08′; and Table Mountain, or Tamulpais, in 37° 53′. The principal passes in the main ridge are south of the outlet of the Sacramento basin, and are—Livermore's Pass, 686 feet high, in 37° 42′; Pacheco Pass, in 37°; the Pass de los Robles, in 35° 20′; the Cajon de Tenoco, in 34° 40′; the Pass of San Francisquito, 3,437 feet high, in 34° 35′; Williamson's Pass, 3,164 feet high, in 34° 30′; the Cajon Pass, 4,676 feet high, in 34° 10′; the San Gorgonio Pass, 2,808 feet high, in 33° 55′; and Warner's Pass, 3,780 feet high, in 33° 10′. The Santa Margarita Pass, with an altitude of 1,350 feet, leads across the Santa Lucia ridge, in 35° 20′; and the San Fernando Pass, 1,956 feet high, crosses the Santa Susanna ridge, in 34° 20′.

Having thus considered the mountains, let us look into the valleys of the coast. The flat land west of the San Bernardino Mountains, south of 34°, is rather composed of plains than of valleys, though watered by the San Gabriel, Los Angeles, Santa Ana, and other rivers. There are two of these plains: the lower one about two hundred and fifty feet above the sea, and skirting the coast; the other one thousand or twelve hundred feet high, nearer the mountains. On the lower plain are Los Angeles, Anaheim, and San Pedro; on the upper are San Fernando, San Bernardino, Cocomongo, Jurupa, Temescal, and Temecula. Northward of 34° we find long, flat, narrow, fertile valleys, shut in by steep, rugged hills. We have already mentioned the names of many of these valleys, as dividing certain ridges of the Coast Mountains from each other. South of the Salinas all these valleys open upon the ocean, save the Cuyama valley, the river of which runs in a cañon through mountains as it approaches its mouth. The Pajaro River breaks through the Gabilan Mountains, and makes a small but rich valley. The average width of these coast valleys is five miles at the mouth, with a length of from ten to forty miles, narrowing to a point near the head in the mountains. The Salinas valley, the largest of all the coast valleys, is ninety miles long, and from eight to fourteen wide. Three terraces are distinctly traceable on each side of the river. The first and lowest is about four miles wide, with a sort of a rich, sandy loam; the second rises with an abrupt edge, is eleven feet higher, has about two miles of width on each side, and has a coarser, poorer soil; the third terrace is less regular in height and width, and has a coarse, gravelly soil, scarcely fit for cultivation. This terraced formation, with its variations in richness of soil, is a strongly-marked feature of many valleys in the state. Ordinarily, the coast valleys are separated from each other by steep, rugged mountain-ridges, but there are occasional exceptions. Thus, there is a low plain between Russian River and Santa Rosa valley, which opens into Sonoma and Petulama valleys; and again, the Santa Clara and

Pajaro valleys are separated from each other by hills not more than two hundred and fifty feet high; and the valleys of the Pajaro and the Salinas open into each other. So also the divide between San Ramon and Amador valleys is so low as to be scarcely noticed by the traveller; and Amador valley is connected, by a level road through a cañon, with Suñol valley, and that by another cañon with the plain at San José Mission. North of San Francisco Bay, the valleys of Suisan, Vaca, Putah, and Cache Creek, lie eastward from Napa valley. The valley at the head of Putah Creek is sometimes called Berreyesa valley; and that at the head of Cache Creek, Clear Lake valley. North of Russian River there is little level land, and that little is found in Eel River valley, about the shores of Humboldt Bay, and about Crescent City.

§ 4. *Coast Rivers.*—The rivers of the Coast Mountains have necessarily but a short course. Those south of the bay of San Francisco are the San Lorenzo, Pajaro, Salinas, Cuyama, Santa Inez, Santa Clara, Los Angeles, San Gabriel, Santa Ana, Santa Margarita, San Luis Rey, San Diegnito, and San Diego. Some of these are large streams in wet winters; but, in the drought of autumn, all those south of the Salinas are swallowed up in the sands before reaching the ocean. Most of them are constant streams to within ten or fifteen miles of their mouths. The Santa Ana, the largest river on the southern coast, rises in Mount San Bernardino, and is in its meanderings nearly one hundred miles long, yet only in very wet seasons, once in six or eight years, succeeds in getting to the sea. The San Gabriel River sinks before reaching Monte, in Los Angeles county, and, after passing three miles under ground, rises again. The intervening space, where there is no river, is very moist, sandy ground, through which the water spreads and soaks.

W. H. Emory, in his report as member of the Mexican Boundary Commission, writes thus:

"The point at which water ceases to flow is quite variable; its more usual upward limit being marked at or near the passage of the stream from the first rocky ranges into the tertiary formation. The point, however, as before stated, is by no means a fixed one: thus, during the night it extends farther downward than in daytime; in cloudy weather, for the same reason, its course is more prolonged than under a clear sky. In the stream-beds themselves, however dry, water is generally found a short distance below the surface.

"The descent of these streams in the rainy season may be either a gradual process in the progressive saturation of their sandy beds, or, the saturation being accomplished by previous showers, the irruption may be sudden. A fine example of this sudden appearance was observed in the San Diego River, in December, 1849; when, after a rainy night, by which its sandy bed was completely saturated, the upper stream suddenly appeared in the form of a foaming body of water, moving onward at the rate of a fast walk, curling round the river-bends, absorbing the pools, and soon filling its bed with a brimming, swift current. An instance of the more gradual descent was seen in the following season (December, 1850), when, from the absence of local rain, its downward progress was slow and interrupted."

The only navigable stream south of San Francisco Bay is the Salinas, and that but for small vessels, and near its mouth.

North of San Francisco the main streams rising in the Coast Mountains are the Russian, Eel, Elk, Mad, and Smith Rivers, all permanent, but none navigable.

§ 5. *Coast Lakes.*—The only large lake in the Coast district is Clear Lake, which lies about eighty miles northward from San Francisco. It is twenty miles long, and varies in breadth from two to ten miles. Surrounded by a small valley of fertile land, it lies in a deep basin bounded by high mountains, with an outlet to the eastward, where its surplus waters are carried off by Cache Creek to the Sacramento. The water of Clear Lake is limpid; the vegetation on its banks abundant and vigorous; the scenery beautiful and romantic. In Amador valley, twenty-five miles eastward from San Francisco, there is a small lake, covering a couple of hundred acres. It lies in the course of the Alameda Creek. In the San Francis quito Pass, forty-five miles northward from Los Angeles, there was a lake called Lake Elizabeth, covering several hundred acres, but it has dried up of late.

These are the only lakes of note in the Coast district. Previous to 1860, there was a lake called the "Laguna Sal," six miles long and three wide, near Alamo, San Diego county; but it entirely dried up in that year. The water had a strong taste of alkali and sulphur. According to report, the lake was formed about the year 1820.

§ 6. *Capes.*—California has two capes: Cape Mendocino, in 40° 25′; and Point Conception, in 34° 25′. The former is reputed to be the stormiest place on our coast; the latter is the southern limit of the cold fogs and cool summers.

§ 7. *Islands.*—About forty miles westward from San Francisco are the Farallones, seven little islands of bare rocks, the largest with an extent of a couple of acres, and of no significance save as a danger to shipping, and as a point where a large lighthouse is maintained. All the other islands of California lie between 32° 50′ and 34° 10′, the farthest one being about sixty miles from the mainland. They are named Santa Cruz, Santa Catalina, San Clemente, Santa Rosa, San Nicolas, Anacapa, and Santa Barbara. They are all hilly, rocky, barren, and of little value. Santa Cruz, the largest and best of them, has good water and a few trees. It is twenty-one miles long, with an average width of about three miles. All these islands appear to be peaks of submerged mountain-ridges. Between them and the mainland lies the Santa Barbara channel.

§ 8. *Bays and Harbors.*—California has four land-locked bays—Humboldt, Tomales, San Francisco, and San Diego. All of them are comparatively long and narrow, and separated from the ocean by narrow peninsulas, their general course being parallel with the coast.

Humboldt Bay is twelve miles long, from two to five miles wide, and is separated from the ocean by two tongues of land, which are covered by high and dense timber, and offer an excellent protection against the strong winds of the coast. The mouth of the bay, in latitude 40° 44′, is a mile across, but has breakers on each side; and between them is a channel, a quarter of a mile wide, with about eighteen feet of water at low tide. The greater part of the bay is shallow, but there is an abundance of deep water, with good anchorage and perfect safety for shipping. The entrance is considered dangerous, and a steam-tug escorts nearly all sailing-vessels in and out.

Tomales Bay is fourteen miles long and two miles wide, separated from the ocean by a strip of land a mile and a half wide. Its mouth is in 38° 15′. Its course is southeastward, and it is open to the northwest winds. The water is about twelve feet deep. Tomales Bay is surrounded by hills, and is of little value for commerce.

San Francisco Bay, one of the finest bays in the world for the purposes of commerce, is about eight miles wide and fifty long, reaching from 37° 10′ to 38°. Its entrance, called the Golden Gate, or Chrysopylis, is a mile wide, between 37° 48′ and 37° 49′. The peninsulas which separate the bay from the ocean are from six to fifteen miles wide. The water on the bar is thirty feet deep at low water; inside much deeper, with excellent holding-ground, and room for all the shipping of the world.

Connected with this bay are those of San Pablo and Suisun, lying farther inland, on the course of the outlet of the waters of the Sacramento basin. San Pablo Bay is nearly round, about ten miles in diameter, and lies north of San Francisco Bay, with which it is connected by an unnamed strait, about three miles wide. Suisun Bay, about four miles wide by eight long, lies eastward of San Pablo Bay, with which it is connected by the strait of Carquinez, which is a mile wide. Both bays are deep, but the water in the strait is only sixteen feet deep at low tide, and large vessels cannot ascend beyond it. Benicia, on the bank of the strait, is the head of navigation, and aspires to be the main port of the coast, but in vain. Vallejo, seven miles from Benicia, still has hopes of that kind.

The harbor of Vallejo is excellent, lying between Mare Island and the mainland. It is half a mile wide, by three miles long, with four fathoms of water at low tide, excellent holding-ground, and perfect protection against all winds.

The bay of San Diego is twelve miles long, from one to two miles wide, and crescent-shaped, running from the entrance, and then turning to the southeastward. A channel, thirty feet deep and half a mile wide, extends more than half the length of the bay from the entrance. The holding-ground is good; the protection from the winds perfect. There is no difficulty in entering at any time, but it is not safe for sailing-vessels to go out during gales from the southeast.

In latitude 34° 38′, thirty-five miles southeastward from Los Angeles, is a land-locked estuary about eight miles long and from half a mile to a mile wide. It has not been surveyed, and its value for commerce is not known, but there has been some talk lately of using it as a port for some of the adjacent towns. The entrance is not more than ten feet deep, and probably not so deep as that.

Of the open harbors, that of Crescent City is the most northern, in latitude 41° 44′. It lies on the southern side of a rocky point that juts out about half a mile in a westward direction, at right angles to the general line of the coast. The harbor is small and shallow, with a bottom of sand and rocks. Vessels drawing twelve feet of water lie nearly half a mile from the shore. The harbor is safe while the wind blows from the north and northwest, but very dangerous when it blows from the southward. The harbor might be made much more safe by a breakwater, at a cost of one or two millions of dollars, but the trade of the place would never justify such an expenditure.

Trinidad, in 41° 03′, is a very small harbor, open to the south, with deep water and excellent holding-ground.

Bodega Bay, in 38° 18′, has nine feet of water, and opens to the southward, so that the anchorage is secure only while the wind blows from the north. Tomales Bay, just opposite, opens into the southern part of Bodega Bay, and is only five miles distant from the Bodega anchorage; and, as one is secure against northern and the other against southern winds, vessels are safe in all weathers, because they can easily run across to whichever may prove the sheltered side.

The bay of Sir Francis Drake, in latitude 38°, is small, open to the south, and of no value to commerce.

Half-Moon Bay is a small roadstead, eighteen miles south of the Golden Gate.

Santa Cruz Harbor, on the northern side of Monterey Bay, in 36° 57′, is small, has four fathoms of water, a sandy bottom, and is open to the south.

Twelve miles farther south is the mouth of the Salinas River, which is about two hundred yards wide, and has seven feet of water. It is entered by small schooners, with the help of a steam-tug.

Eight miles farther to the southward is the harbor of Monterey, which is large and deep, and has good holding-ground. It is open to the north.

San Simeon Harbor, in 35° 38′, has a good anchorage, and is safe while the wind blows from the north, but it offers no protection against storms from the southward. The bottom is sandy.

San Luis Obispo Harbor, in 35° 10′, has a good anchorage, safe at all times, except during storms from the southward.

Santa Barbara, in 34° 24′, has an open harbor, exposed to the south winds. The water is deep, and the bottom hard.

San Pedro, in 33° 43′, is open to the south, but probably might be made secure by a breakwater, to cost one million of dollars. The bottom is hard.

The difference between extreme high tide and extreme low tide is about nine feet at Crescent City, and seven feet at San Diego. At San Francisco, the establishment of the port is ten hours.

§ 9. *Sacramento Basin.*—The low land of the Sacramento basin, bounded on the west by the Coast Mountains and on the east by the Sierra Nevada, which ranges meet both at the north and the south, is the heart of the state, four hundred miles long by fifty wide, reaching from latitude 35° to 40° 30′. It is drained by two rivers: the Sacramento, running from the north; and the San Joaquin, from the south. They meet and unite in the centre of the basin, at 38°, and break through the Coast range to the Pacific, forming the bays of Suisun, San Pablo, and San Francisco, on their way. The mountains rise steeply from the edge of the valley, which is nearly level, about thirty feet above the level of the sea at the junction of the rivers, and two hundred feet higher where they issue from the mountains. Part of the Sacramento valley shows terraces, the farthest from the river being a coarse gravel. The richest soil is on the immediate bank. The great body of the valley is bare of trees. Its even surface is broken in only one place, by the "Buttes," a range of volcanic hills, six miles wide by twelve long, with three peaks, about two thousand feet high, which rise in lonely abruptness from the middle of the plain, in 39° 20′. The general course of the two main rivers of the basin lies nearly midway between the two mountain-chains, but almost all their tributaries come from the Sierra Nevada, which, like the Coast range, has most of its wealth on its western slope. In the four hundred miles from Tejon to Shasta there are a dozen creeks marked on the map as flowing eastward from the Coast range to the San Joaquin and Sacramento; but during the summer, three-fourths of them are swallowed up in the sands before reaching their mouths. Not one south of 38° is a permanent stream. From the Sierra Nevada a number of rivers run westward. Beginning at the north, we have the Pit, Feather, Yuba, American, Cosumnes, Mokelumne, Calaveras, Stanislaus, Tuolumne, Merced, San Joaquin, King's, White, and Kern Rivers—all of them considerable streams, though some of those in the southern part of the Sacramento basin are swallowed up in the sands, in the dry seasons, before reaching their mouths. The San Joaquin River does not rise at the extreme southern end of the basin, but one hundred miles northward from it, in the Sierra Nevada. After running westward to the middle of the valley, it turns northward. From its bend southward, the valley discharges no water to the ocean during the summer; but in wet winters there are continuous sloughs, or pieces of marsh-like ground, from the Tejon to the San Joaquin. In the dry season, no channel is visible for the escape of the waters of Tulare and Kern Lakes.

§ 10. *Rivers of the Sacramento Basin.*—The rivers flowing down from the Sierra Nevada are about one hundred and twenty miles long on an average, following their courses. The upper half of their length is in the mountains, where they are torrents, falling five thousand feet in fifty miles. Their beds are in deep cañons; after reaching the plain their currents are gentle, and they meander between low banks, fringed with oaks, sycamores, cottonwood, and willows. In the southern part of the Sacramento basin there are several large streams, which, soon after issuing from the mountains, divide into a number of channels, as do some large rivers which have deltas near their entrance to the sea. King's River, which is about eighty yards wide where it leaves the mountains, divides into seven or eight channels, which all unite again. The Cahuilla or Pipiyuma River, also a large stream, divides into a number of channels, which irrigate "the Four-Creek country," and render it one of the most fruitful parts of the state.

The Sacramento River is navigable for steamers drawing three feet of water, to Sacramento City, and to Red Bluffs for boats drawing fifteen inches. The Feather River is navigated by steamers drawing fifteen inches, to Marysville, seventy-five miles from Sacramento; and boats have ascended to Oroville, twenty-five miles farther. Steamers drawing five feet can run regularly to Stockton, on the San Joaquin, a distance of one hundred and thirty miles from San Francisco; and in times of high water, a boat drawing about fifteen inches ascends to Fresno City, one hundred and fifty miles farther. A number of sloughs or tide-water creeks, navigable for small vessels, open into the bays of San Francisco, San Pablo, and Suisun.

The most notable of these are the Alviso or Guadalupe slough, at the head of San Francisco Bay; the San Antonio slough, opposite San Francisco city; the Petaluma, Sonoma, and Napa sloughs, opening into San Pablo Bay; and Suisun and Pacheco sloughs, opening into Suisun Bay.

§ 11. *Tule-Land.*—Along the borders of these bays, and of the Tulare and Kern Lakes, and of the Sacramento and San Joaquin Rivers, there are extensive tracts of swamp-lands, usually called "tule-lands," from the *tule*, a species of rush which grows on them. Nearly all the tule-land west of Sacramento and Stockton, to which points the tides extend, are salt marshes; but north of Sacramento, and south of Stockton, the tule-lands are fresh-water swamps. The extent of this marshy land varies in different seasons; but at my estimate, there are eighty square miles on the borders of San Francisco Bay, eighty on San Pablo Bay, sixty on Suisun Bay, two hundred on the Sacramento River, one hundred on the San Joaquin, two hundred on the Tulare Lake, and the slough leading from it, and one hundred and twenty south of Tulare Lake—making eight hundred and forty square miles in all.

§ 12. *Sierra Nevada.*—The Sierra Nevada is four hundred and fifty miles long (in California) and seventy wide, with a height varying from five thousand to eight thousand feet above the sea-level. Nearly its whole width is occupied with its western slope, which descends to a level of three hundred feet above the ocean; whereas the slope on the eastern side is only five or six miles wide, and terminates in the Great Basin, which is itself from four thousand to five thousand feet above the sea. Nearly all the snows and rains that visit the Sierra Nevada fall on its western slope, which has all the large rivers. These rivers run westward, at right angles to the course of the chain, and cut it into steep hills and deep ravines, cañons, and chasms. The valleys are all small, and it is rare to see a hundred acres of level, tillable land, even on the banks of the largest mountain-streams. The greater part of the Sierra Nevada is covered with timber. The oak, manzanita, and nut-pine, grow to about twenty-five hundred feet above the sea; and then the coniferous trees appear, and are found in dense forests to a height of six thousand feet.

§ 13. *Peaks and Passes of the Sierra.*—Mount Shasta, which rises, in 41° 30′, high into the region of perpetual snow, the loftiest peak in the state, may be treated as belonging to the Sierra Nevada, though in fact it stands midway between that range and the Coast Mountains, and is connected by high mountain-ridges with both of them. Its height is given by Wilkes (Exploring Expedition, vol. v., p. 240) at 14,390 feet. Nearly a perpendicular mile of it is always covered with snow, and it is visible in every direction for more than a hundred miles, presenting to the traveller the most prominent landmark of the state. It is of volcanic origin, and still emits sulphureous vapors from its summit. Several parties have ascended to its top. The other most notable peaks in the Sierra Nevada are—Lassen's Peak, 9,000 feet high, in 40° 22′, also of volcanic origin; the Donnieville Buttes, 8,500 feet high; Pilot Peak, 7,300 feet high, in 39° 50; Castle Peak, 11,000 feet high, in 38° 10′; and Mount Breckenridge, 7,500 feet high, in 35° 20′. Mount Shasta is the only mountain which has snow on its southern and southwestern slopes throughout the year; the other-mentioned peaks lose all their snow in September and October, except in deep, shady ravines on their northern slopes.

The most notable passes in the Sierra Nevada are the following: Lassen's, 7,000 feet high, in latitude 41° 50′; Fredonyer's, 5,667 feet high, in 40° 47′; Beckworth's, in 39° 50′; Kenness's, in 39° 30′; Truckey, 5,636 feet high, in 39° 25′; Johnson's, 6,752 feet high, in 38° 50′; Carson's, 7,972 feet high, in 38° 43′; Sonora, 10,132 feet high, in 38° 15′; Walker's, 5,302 feet high, in 35° 40′; Hum-pa-ya-mup, 5,356 feet high, in 35° 35′; Tehachepa, 4,020 feet high, in 35° 10′; Tejon, 5,285 feet high, in 35°; and Cajon de las Uvas, 4,256 feet high, in 34° 50′. The last five passes are in the Sierra Nevada, south of its bend, where it turns westward to meet the Coast range.

The Johnson Pass is used by most of the travel and traffic between Sacramento and Utah; the Kenness Pass lies east of Marysville, and is used by the people of that neighborhood; and the Cajon de las Uvas is used by travellers between the San Joaquin valley and the Los Angeles district.

§ 14. *Lakes of the Sierra.*—The Sierra Nevada has few lakes. The most notable one is Lake Bigler, about twenty miles long and ten wide, and six thousand feet above the level of the sea, in latitude 39°, and on the eastern border of the state. Part of the lake is in Utah. Its waters flow eastward into Truckey River. In the eastern part of Nevada county there is a group of two dozen lakes, called the Eureka Lakes, the largest are three miles long and a mile wide.

§ 15. *Plateau of the Sierra Nevada.*—About latitude 40°, the Sierra Nevada seems to divide or to fork—one branch running northward, in the line of the main chain; the other northwestward, to Mount Shasta. Between these two branches, and between 40° and 42°, is a high table-land or plateau, about one hundred and twenty miles long, and five thousand feet above the ocean-level. This plateau is an independent basin, and its waters never leave it, but flow into a few lakes, where they are swallowed up in the sands. The district bears a strong resemblance in many of its features to the Great Basin of Utah, with which it should perhaps be classed. The main stream is Susan River, which, after a course of forty miles in an eastward direction, empties into Honey Lake, which is twelve miles long by five wide—or was, for in 1859 the lake dried up, and again dried up in 1860. The lake, when full, was shallow, with thick, yellowish water, of a saline taste. Northwestward from Honey Lake, and distant thirty miles from it, is Eagle Lake, about half the size of the other. The land is barren, and the vegetation scanty. Pit River starts in the northeastern corner of the state, and breaks through the plateau. North of the river are Wright Lake and Rhett Lake, within five miles of the Oregon line; and Goose Lake and Lower Klamath Lake, partly in Oregon and partly in California.

The largest is Goose Lake, ten miles long and five wide. All are destitute of large tributaries, sweet water, and valuable adjacent land.

§ 16. *Klamath Basin.*—North of latitude 41° lies the basin of the Klamath River, which rises in Oregon, crosses the Californian line about eighty miles from the sea, then turns southwestward, and, after a course of about one hundred and fifty miles, empties into the Pacific in 41° 33′. The basin of the Klamath is very rugged, particularly that part of it within forty miles of the ocean. Along the main river there is no valley, or bottom-land; its whole length is between steep hills and mountains, and through rocky cañons. Its largest tributaries, the Trinity and Salmon, run through a country almost as rugged as that bordering the main stream. Scott and Shasta Rivers, which are the only other notable tributaries of the Klamath—they all flow from the southward—have valleys of bottom-land, about five miles wide and forty long.

§ 17. *Utah Basin.*—A prominent feature of the North American continent is the Great Basin of Utah, a triangular district of country, bounded on the north by the basin of the Columbia, on the southeast by the basin of the Colorado, and on the southwest by the Sierra Nevada and San Bernardino Mountains. This Great Basin—an elevated tract of land, most of which is four thousand or five thousand feet above the sea-level, mountainous, barren, and cheerless, with no outlet for its waters—extends into California, including a district about two hundred miles long and one hundred wide, in the southeastern portion of the state. The Californian portion of the Great Basin is one of the driest and most sterile parts of the earth's surface, cut up by numerous irregular ridges of bare, rocky mountains, with intervening valleys of sand and volcanic scoriæ, and occasional springs and little streams which terminate in lakes, presenting a wide extent of muddy salt water after heavy rains, and, in the dry season, wide beds of dried and cracked mud, covered with a white alkaline efflorescence. The chief stream in the Californian portion of the Great Basin is the Mojave, which rises on the northern slope of Mount San Bernardino, and, after running about one hundred miles in a northeastward direction, sinks in the sand. The Mojave receives no tributaries after it leaves the side of Mount San Bernardino. After sinking, it rises again; or rather pools of water are found in the low places of its bed, the water evidently soaking through the sand and following the bed of the stream. The next stream in importance is Owen's River, which runs southward seventy-five miles along the foot of the Sierra Nevada, and terminates in Owen Lake, which lies in latitude 36° 25′, and is fifteen miles long by nine wide. Northward, one hundred miles from Owen Lake, is Mono Lake, eight miles long and six wide, sometimes called "the Dead Sea of California." No fish can live in the water, which is so heavy with saline substances, that the human body floats in it very lightly; though it is so strongly alkaline, that it scalds the skin. In the midst of the lake is an island, several miles long. While the greater part of the Utah Basin is high above the level of the sea, there is a portion of it, called "Death Valley," three hundred and seventy-seven feet below the sea-level; and, notwithstanding its great depth, it is one of the driest and most desolate parts of that basin of deserts.

§ 18. *Colorado Desert.*—A district, about seventy miles wide by one hundred and forty long, on the southeastern border of the state, belongs to the basin of the Colorado River. It is usually called the "Colorado Desert," because of its barren, sandy soil, and scanty vegetation. In some places the soil is composed of sand, packed together firmly, with a hard and smooth surface, which reflects light like a mirror; in other places are mountains of loose sand, which are continually shifting. In latitude 33° 20′, and longitude 115° 50′, is a district, thirty miles square or more, which is seventy feet below the level of the sea. It is supposed that at one time the Gulf of California extended several hundred miles farther north than it now does; and that the Colorado River, in long ages, deposited so much alluvium as to make banks down to the present head of the gulf, thus cutting off from its connection with the ocean that part of the gulf now dry. The evaporation in this desert far exceeds the fall of rain; so it was not long before this lake was dried up. When the Colorado River is very high, it breaks over its banks about forty miles southward from Fort Yuma, and sends a large stream, called New River, northwestward a distance of a hundred miles or more, to the lowest portion of the desert. A proposition has been made to cut a canal from the river to the low ground; so that the land, which is said to be of excellent quality, might be irrigated and cultivated: but no accurate survey has yet been made of a route for the canal, or of the district to be irrigated. The Colorado River is navigable to Fort Yuma, a distance of seventy-five miles from its mouth. The average depth is ten feet, but there are shoals which have not more than two feet at low water; the tide rises ten feet. The channel is crooked, and the bottom is of sand, which is constantly changing position. The banks of the river are low and muddy. The average current runs at the rate of two and a half miles per hour. The river is high in July, when the snows of the Rocky Mountains (in latitude 38°–44°) melt, and then the flood covers the low bottom-land along the river-banks.

§ 19. *Area of the State.*—The total area of the state amounts to about 155,000 square miles, of which there are, at my estimate, 42,000 in the mountains and valleys of the coast, 40,000 in the Sierra Nevada and its plateau, 20,000 in the low land of the Sacramento Basin, 30,000 in the Great Basin of Utah, 15,000 in the Colorado Desert, and 8,000 in the Klamath Basin. In the 42,000 square miles of the coast slope, 16,000 may be put down as valley and 26,000 as mountain.

CHAPTER II.

CLIMATE.

§ 20. *General Remarks.*—The climate of California is unlike that of every other country, and particularly dissimilar to that of the American states east of the Rocky Mountains. In general character it resembles the climate of western Europe. Its chief peculiarities, as distinguished from the Eastern states, are, that the winters are warmer; the summers—especially at night—cooler; the changes from heat to cold not so great nor so frequent; the quantity of rain less, and confined principally to the winter and spring months; the atmosphere drier; the cloudy days fewer; thunder, lightning, hail, snow and ice, and the aurora borealis, rarer; the winds more regular—blowing from the north for fair weather, and from the south for storms; and earthquakes more frequent.

The state reaches through nine and a quarter degrees of latitude, from 32° 45′ to 42°, San Diego being as far south as Charleston, and Crescent City as far north as Providence. Much of the Golden state has the winter of South Carolina, and the summer of Rhode Island. The orange, the lemon, the olive, the fig, the pomegranate, the vine, the peach, the apple, wheat, and barley, all find most congenial climes in California.

The state, indeed, has many climates: one for the western slope of the Coast range between Point Conception and Cape Mendocino; another for the low land of the Sacramento Basin;

another for the Sierra Nevada and Klamath Basin; another for the Great Basin of Utah; another for the coast south of Point Conception; and still another for the Colorado Desert.

The causes of these peculiarities of climate are chiefly to be found in the position of the country—a narrow strip on the western side of the continent, bounded on the east by a high range of mountains that shuts the coast off from all the influences of the interior; bordering on the wide Pacific Ocean, washed by a warm current flowing across from the China Sea; with a shore line that runs nearly north and south, and is exposed in all its length to the strong winds constantly blowing southeastward over the ocean.

§ 21. *Temperature of the Middle Coast.*—On the coast, between latitudes 35° and 40°, there is little difference in the temperatures of winter and summer. San Francisco is in the same latitude with Washington and St. Louis, but knows neither the cold winters nor the hot summers which afflict those places. Ice is rarely formed in the Californian metropolis, and never more than an inch in thickness; and the thermometer never stays at the freezing point twenty-four consecutive hours. The lowest point which the thermometer has ever reached in San Francisco, since observations have been taken, was 22° Fahrenheit in January, 1862; and previous to that time it had never fallen below 25°; while in St. Louis it goes down to 12° every winter, and remains near that figure for many consecutive days. The lowest figures which the mercury reached in the daytime at San Francisco, in January of the years 1851, '52, '53, '54, and '55, were respectively 30°, 35°, 41°, 25°, and 33°, showing that in three Januaries out of five no ice at all was formed in the daytime; and when the thermometer fell to 25° in 1854, the weather was declared to be colder than it had ever been before, "within the memory of the oldest inhabitant." During nine years' residence in the city, I never have seen ice formed here half an inch thick, and never saw the slightest film of it formed on water in a house. Snow sometimes falls, but I have never seen the streets dressed in white.

In St. Louis, the winter months rarely have a day which is really comfortable in the open air; while at least half the season is so in San Francisco, the sky being clear, the sun warm, and the breezes gentle, so that the weather bears a strong resemblance in temperature to the Indian summer in the upper Mississippi basin. Our coldest winter days, at noon, are as warm as the warmest in Philadelphia.

On the other hand, the summers are cool or cold. In November, 1854, the lowest figure reached by the thermometer in San Francisco, was 47°, while in July of the same year it was at 46°—showing that at no time in the former month was it so cold as at one time in the latter. The mean temperature of July is 57°, twenty-one degrees lower than in Washington city. There are not more than a dozen days in the year when the thermometer rises above 80°—at which figure heat first begins to be oppressive—while in St. Louis and at Washington there are in every year from sixty to ninety days that see that height. No matter how warm the day at noon, the evenings and mornings are always cool, and blankets are necessary —at least a pair of them—as a bed-covering, every night.

Although the mean temperature of summer differs little from that of winter, yet there are sometimes very warm days, which may be succeeded immediately by very cool nights. San Francisco never sees more than three hot days in succession. When the sun has had an opportunity to rage for so long a period, the air in the interior of the state becomes so hot, that it rises rapidly; and the ocean-winds, which must rush to supply the place, never fail to bring cool weather to the vicinity of the Golden Gate. Thus the mercury has risen (and that was its highest) to 97°, and it often falls in July to 46°; and such a change of fifty degrees might occur within twelve hours. The average range of the thermometer in July and August is about twenty degrees—from 50° to 70°. Yet, as the mornings and evenings are invariably cool, and the noons are not always warm, "summer clothing" is seldom worn by men, and never for twelve consecutive hours. The common custom is, to wear woollen coats and trousers of the same thickness in summer and winter. The persons who visit San Francisco during the summer, from the interior of the state, where the climate from May to October is much warmer, and where summer clothes are worn, are much bothered at having to bring their winter clothes with them. The editor of a Stockton paper, disgusted with the climate of the metropolis in July, expressed himself somewhat after this manner: "You go out in the morning shivering, notwithstanding the fact that you are dressed in heavy woollen clothing and under-clothing, and have a thick overcoat buttoned up to your throat. At 8.30 you unbutton two of the upper buttons; at 9 you unbutton the coat all the way down; at 9.30 you take it off; at 10 you take off your woollen coat, and put on a summer coat; at 11 you take off all your woollen and put on light summer clothing: at 4 it begins to grow cool, and you to put on your woollen again; and by 7 o'clock your overcoat is buttoned to the chin, and you shiver until bedtime."

The coolness of the summer is caused by the winds and fogs which blow in from the ocean, whose temperature at the Farallones never varies more than a degree or two from 42°. A strong wind blows along the coast from the north and north-west during almost the whole year; and it blows strongly upon the land for several hours after eleven o'clock in the morning and after five in the evening, and not unfrequently it continues the whole twenty-four hours. The common prevalence of this wind during the afternoon renders the mornings the pleasantest part of the summer weather in San Francisco; and the more delicate and fashionable ladies habitually make their calls and allow their children to go into the street only before mid-day. In June, July, and August, heavy, wet, cold mists come up from the sea at six in the evening, and continue until eight or nine in the morning. In the winter, fogs are rarer, and do not commence so early in the evenings, and the winds are not so strong; so that, in these respects, the winter is the pleasanter season of the year.

Dr. H. Gibbons, speaking of the mists and fogs at San Francisco, says:

"It is curious to observe the conflict between the absorbing power of the air and the supplying power of the ocean, in regard to moisture. Toward noon, when the wind rises, huge columns of mist may be seen piled along the coast, three or four miles west of the city, and pouring in like a deluge upon the land. But the air of the land, which is always thirsty, drinks it up with astonishing avidity; so that the impending wave, though in a current moving from thirty to fifty miles an hour, makes slow progress. By the middle of the afternoon it is within a mile or two of the city; and there it stands, like a solid mass of water several hundred feet in depth, rolling and tumbling toward you (not without grandeur and majesty), and threatening to overwhelm you in a few seconds. You await its coming, but it comes not; it even recedes, to return and recede again. Not until the sun has lost his calorific power does the atmosphere reach the point of saturation; and then, toward sunset or later, every.thing is submerged by the vapory flood. In the course of the evening the wind falls. During the night the mist is gradually dissolved and disappears from the lower stratum of air, while it forms a heavy cloud above. About the middle of the forenoon the cloud is dispersed by the rays of the sun. The dispersion is rapid, the sky often becoming entirely clear in less than half an hour.

"If it be possible to distinguish between fog and mist—regarding the former as impalpable, and the latter as composed of palpable particles of moisture—I may remark that mist belongs only to the summer and fog to the winter climate of San Francisco. There is no mist in winter, and no fog in summer. At all seasons the drying tendency of the atmosphere is observable. You notice none of those phenomena which in other climates depend on an excess of water in the air, and on sudden changes of temperature. The moisture does not condense on your windows, nor on the plastered walls; salt does not liquify, nor even exhibit the slightest dampness; and the housewife has no trouble in drying her clothes, provided it should not rain. In fact, the atmosphere of San Francisco, in spite of sea winds and mists, is a dry atmosphere."

The mean temperatures of spring, summer, autumn, and winter, are 54°, 57°, 56°, and 50° respectively, showing a difference of only seven degrees between the average of winter and summer; whereas a similar comparison in the climate of New York city shows a difference of thirty-nine degrees. There is a range of two degrees more in San Francisco by taking the months separately—January, the coldest month, having a mean temperature of 49°, and September, the warmest, a mean of 58°. October is as warm as July, and in some years it has been warmer. The mean of the whole year is 54°, a temperature that requires heavy woollen clothing for comfort. For vigorous, industrious men, the climate of San Francisco is the healthiest and most agreeable in the world. I prefer it to all others. But, to enjoy it, a man should have warm blood, full veins, and active habits; if he is weak or idle, he will find it too cool for him. It is a climate that allows a person to be out in the open air all the time; no hour is lost because of either excessive heat or excessive cold. Women do not like the climate so well as men; it is too cool for their less vigorous constitutions and sedentary habits.

San Francisco does not lie immediately on the ocean, but only six miles from it, and where there is a great gap to let in the winds and fogs. The nearer the Pacific, the denser and more frequent the fogs, the stronger the winds, the warmer the winters, and the cooler the summers. The great ocean is a powerful equalizer of climate: as you advance into the interior, the range of heat and cold becomes greater. In the coast valleys you can choose your distance. San Rafael is ten miles from the Pacific, Petaluma twenty, Sonoma thirty, Napa thirty-five, Suisun forty-five, and Vaca valley fifty. Sonoma valley has a delightful climate, free from fogs and cold winds, and yet blessed with a sea-breeze which tempers the heat of every summer day to the precise degree necessary to the perfect happiness of a man who wishes to take life easy, and do nothing. Indeed, all the valleys embosomed in the Coast Mountains, from Humboldt Bay to Santa Barbara, have beautiful climes, which will compare favorably, I think, with the best of Italy. The summer days are always warm, rarely hot; the mornings and evenings are clear and pleasant; in winter, ice never forms over two inches in thickness, and if snow falls, it never lies twenty-four hours.

§ 22. *Clear Days.*—On an average, there are two hundred and twenty perfectly clear days in a year, without a cloud; eighty-five days wherein clouds are seen, though in many of them the sun is visible; and sixty rainy. Italy cannot surpass that. New York has scarcely half so many perfectly clear days. From the first of April till the first of November, there are in ordinary seasons fifteen cloudy days; and from the first of November till the first of April, half the days are clear. It often happens that weeks upon weeks in winter, and months upon months in summer, pass without a cloud; that is, at a distance of thirty miles from the ocean. Near the shore, coast-clouds are frequently blown up from the sea, but they disappear after ten o'clock in the morning.

§ 23. *The Sirocco.*—One case, and only one, is on record, of a sirocco, or burning-hot wind, visiting the coast. This one was felt at the town of Santa Barbara, in latitude 34° 20′, on the ocean-shore, on the 17th of June, 1859. The *Gazette* newspaper of that place, published six days afterward, said:

"Friday, 17th June, will be long remembered by the inhabitants of Santa Barbara, from the burning, blasting heat experienced that day, and the effects thereof. Indeed, it is said that, for the space of thirty years, nothing in comparison has been felt in this county, and, we doubt, in any other. The sun rose like a ball of fire on that day; but though quite warm, no inconvenience was caused thereby until two o'clock, P. M., when suddenly a blast of heated air swept through our streets, followed quickly by others; and shortly afterward the atmosphere became so intensely heated, that no human being could withstand its force: all sought their dwellings, and had to shut doors and windows, and remain for hours confined to their houses. The effect of such intense and unparalleled heat was demonstrated by the death of calves, rabbits, birds, etc. The trees were all blasted; and the fruit, such as pears and apples, literally roasted on the trees ere they fell to the ground, and the same as if they had been cast on live coals. But, strange to say, they were only burned on one side—the direction whence came the wind. All kinds of metal became so heated, that for hours nothing of the kind could be touched with the naked hands. The thermometer rose to nearly fever-heat—in the shade. Near an open door, and during the prevalence of this properly-called sirocco, the streets were filled with impenetrable clouds of fine dust, or pulverized clay. Speculation has been rife since to ascertain the cause of such a terrible phenomenon; but, though we have heard of many plausible theories thereon, we have not been fully convinced yet: however that might be, we see its terrible effects all around us, in blighted trees, ruined gardens, blasted fruit, and almost a general destruction of the vegetable kingdom here. We hope we will never see the like again."

A correspondent of a San Francisco paper wrote thus: "At one o'clock in the afternoon of the 17th instant, a burning wind came upon us from the northwest, and smote us with terror. At two o'clock, the thermometer exposed to this wind rose to 133° of Fahrenheit; at five o'clock, it had fallen to 122°; and at seven o'clock, it stood at 77°, where it had been in the morning. During the whole time of this visitation, every one stayed in the house, taking good care to keep doors and windows closed. A fisherman who was out at sea came back with his arms all blistered. Many calves, rabbits, and birds, died of suffocation. The greatest losses are among the vegetables. The fruit-trees are all burned; the pears and apples have been literally cooked."

No similar case is mentioned in history or tradition, nor is there any explanation of this.

§ 24. *Temperature of the Southern Coast.*—The high mountain-spur which projects into the ocean at Point Conception seems to protect the coast south of it from the fogs, which are much rarer and warmer at Los Angeles than at San Francisco. But though the former is in latitude 34°, it is at times as cold in winter as the latter (in 37° 48′), because it is farther from the ocean, and is in sight of Mount San Bernardino and other high mountains, some of which wear snow-caps during a large part of the year. In summer, however, it is much warmer, even oppressively hot. The nights are sometimes so warm, that a sheet is as much covering as is necessary for comfort; but blankets are usually required.

The coast north of latitude 40° is much colder and cloudier in summer, and has more rain than any other part of the state.

§ 25. *Sacramento Basin.*—The climate of the Sacramento Basin differs from that of San Francisco in having no fogs, faint sea-breezes, winters four degrees colder, and summers from sixteen to twenty degrees warmer. The greater heat of summer is owing to the want of ocean winds and fogs; the greater cold of winter is caused by the distance from the Pacific, and the proximity of the snow-covered Sierra Nevada. While at San Francisco the thermometer usually stands at 70° at mid-day, it is at 86° in Sacramento city at the same moment; and these sixteen degrees make a vast difference, for they change comfort into oppression. And Sacramento city, lying near the great gap in the Coast Mountains, is cooler in summer than either end of the basin; for the upper portions of both the Sacramento and San Joaquin valleys, nearly every summer, see days when the thermometer stands at over 100° in the shade. The county assessor of Fresno county stated, in his annual report for 1857, that the mean temperature at Millerton during the three summer months was 106°. The Stockton *Argus* spoke thus of a great heat that was felt in Stanislaus county on the 23d of June, 1859:

"The thermometer was 113° in the shade. The wind was avoided, as it was heated so, that it felt as if actually burning the flesh—as if rushing from a hot oven. In one team of ten horses, three fell in the road, from heat; two died, but the other recovered by pouring sweet oil in its throat. The animal's throat was closed so that it could not drink, when the oil was used so as to soften the throat and open it, that it could swallow water, when it recovered. The two that died expired before such aid could be used with them. At Burton's public house, at Loving's Ferry, birds flew into the bar-room to the pitcher to get water, so tame were they made by the thirst caused by extreme heat. Birds were seen to fall dead off the limbs of trees in the middle of the day, from the heat, as if they were shot. The wind was of that burning heat never before witnessed by the settlers there since their arrival in the state."

In the Sierra Nevada, the heat of the summer at mid-day is about the same as in the Sacramento valley; but the winter is cold, and the amount of rain greater in proportion to the altitude above the sea. In places three thousand feet above the ocean-level, ice forms five and six inches thick, and snow deep enough for sleighing lies several weeks nearly every winter. In towns six thousand feet above the sea, the snow falls from five to ten feet deep, and covers the ground four or five months in the year.

In the Great Basin, the winters are cold and the summer days very hot; but there too the nights are always cool.

The Colorado Desert has exceedingly hot summer days and warm winters, but occasional frosts in the spring and fall as well as in the winter.

§ 26. *Comparative Tables.*—The following table shows the mean temperature of every month, and the average of the whole year, at San Francisco, Benicia, Sacramento, Fort Miller, Fort Reading, Fort Yuma, and also at various places in other parts of the world, some of them (such as Funchal, Naples, Honolulu, and Mexico) being famed for the beauty and equability of their climates. In addition to the temperature, the latitude of each place is given:

PLACES.	Jan.	Feb.	March.	April.	May.	June.	July.	Aug.	Sept.	Oct.	Nov.	Dec.	AVERAGE.	LATITUDE.
San Francisco ...	49	51	52	55	55	56	57	57	58	57	54	51	54	37° 48′
Benicia	47	52	53	57	59	67	67	66	64	62	54	47	58	58 03
Sacramento	45	48	51	59	67	71	73	73	66	64	52	45	59	38 34
Fort Miller	47	53	56	62	68	83	90	83	76	67	55	48	66	37
Fort Reading ...	44	49	54	59	65	77	82	79	71	62	52	44	62	40 28
Fort Yuma	56	58	66	73	76	87	92	90	86	76	64	55	73	32 43
New York	31	30	38	47	57	67	73	72	66	55	45	34	51	40 37
New Orleans	55	58	64	70	75	81	82	82	78	70	62	55	69	29 57
Steilacoom	38	40	42	48	55	60	64	63	57	52	45	39	50	47 10
London	37	40	42	46	53	58	62	62	57	50	44	40	49	51 29
City of Mexico...	52	54	61	63	66	65	65	64	64	60	55	52	60	19 26
Naples	46	47	51	56	64	70	76	76	69	61	53	49	60	40 52
Funchal	60	60	62	63	64	67	70	72	72	67	64	60	65	32 38
Honolulu	71	72	72	74	76	77	78	79	78	76	74	73	75	21 16
Jerusalem	47	53	60	54	66	71	77	72	72	60	58	47	62	31 47
Canton	52	55	62	70	77	81	83	82	80	73	65	57	69	23 08
Nagasaki	43	44	50	61	69	77	80	83	78	66	53	47	62	32 45

By the study of this table, we can form an excellent idea of the temperature of the different portions of the state, as compared with each other, and as compared with those of some other countries. So far as we know, San Francisco has the most equable and the mildest climate in the world. Within the tropics there are, no doubt, many places which have a more equable temperature, but it is the equability of intense heat.

Funchal, on the island of Madeira, has probably the mildest climate in the world, but in equability it is inferior to San Francisco. Benicia is thirty miles from the ocean, and has a warmer summer and a colder winter than the immediate coast. Sacramento has the climate of Naples and Jerusalem throughout the year: its summer being the same as that of New York, but its winter fourteen degrees warmer. Fort Reading and Nagasaki have nearly the same figures. Fort Yuma, in the Colorado Desert, in latitude 32° 45′, is warmer than New Orleans, in 29° 57′.

A railroad, one hundred and eighty miles long, running eastward from Oakland, a suburb of San Francisco, passing through Stockton and Sonora, near the Mammoth Grove of Mariposa and the Yosemite valley, to the summit of the Sierra Nevada, would enable the people near the line to place themselves, every summer's day, in any tolerable degree of heat or cold.

Fourteen miles west of Oakland is the ocean-beach, where a chilling wind blows without ceasing. Going from the coast, the traveller would gradually get into a warmer clime, until, in Stockton, he would find the thermometer indicating 100°, most of the summer noons; and proceeding up the sides of the Sierra, he would gradually rise into greater cold, to the eternal frost on the summit. A branch road running south to Fort Yuma would enable the traveller to enjoy almost as great a variety of temperature in the winter.

§ 27. *Rain.*—Nearly all the rain in California falls between the first of November and the first of June—the period called the "rainy season," as contradistinguished from the "dry season," which occupies the remainder of the year. Those names, however, when applied to any special season, do not signify an unchangeable number of months, but rather the term during which the rain falls or the dry weather lasts. Thus, we say that the rainy season of 1858–'59 began in October, because in that month the first heavy rains fell; the rainy season of 1855–'56 did not begin until December; the dry season of 1857 began in March; and so forth. The rainy season is so called, not because the rain falls then continuously, but because it does not fall at any other time. There are occasional showers in June, July, August, and September, but they are rare and light.

The following table gives the average amount of rain, in inches, which falls during the four seasons of spring, summer, autumn, and winter, at various places in California, as compared with the amount which falls in other places in the United States

PLACES.	SPRING.	SUMMER.	AUTUMN.	WINTER.	YEAR.
San Francisco	6.64	0.13	3.31	11.33	21.41
Sacramento	7.01	0.00	2.61	12.11	21.73
Fort Reading	11.30	0.39	4.89	12.44	29.02
Fort Humboldt	13.51	1.18	4.87	15.03	34.56
Fort Miller	9.57	0.02	2.80	9.79	22.18
Fort Yuma	0.27	1.30	0.86	0.72	3.15
San Diego	2.74	0.55	1.24	5.90	10.43
Astoria	16.43	4.00	21.77	44.15	86.35
Portland, Maine	12.11	10.28	11.93	10.93	45.25
New York city	11.69	11.64	9.93	10.39	43.65
New Orleans	11.29	17.28	9.62	12.71	50.90
St. Louis	12.86	14.09	8.71	6.29	41.95
Rome	7.27	3.39	10.89	9.31	30.86
Paris	5.53	5.92	6.51	4.68	22.64
Liverpool	6.19	9.78	10.81	7.32	34.10

From this table it appears that the amount of rain is about one-half as great in San Francisco as in any of the American states east of the Mississippi. Here, all the rain falls in the winter and spring; there, the amounts are nearly the same in the four seasons. They have as much rain in their summer and autumn as we in our winter and spring. We have less rain than Liverpool and Rome, and about the same amount with Paris. San Diego has only one-half and Fort Yuma one-seventh the rain-fall of San Francisco, which latter place is surpassed nearly seventy-five per cent. by Humboldt Bay. At Fort Yuma, and all through the Colorado Desert, the rain comes not in the rainy season of California, but chiefly in the summer and fall, synchronous with the wet season of North-western Mexico. Unfortunately, we have no statistics of the rain-fall in the Sierra Nevada, or in the Great Basin, within the limits of this state.

The least rain in San Francisco, during any rainy season since 1852, has been 19 inches; the largest amount, 24 inches. I obtain the following figures from statistics kept in this city by Mr. Thomas Tennant, from 1850 to the present time:

The average rain-fall in January is 3.52 inches. The most notable departures from that average were in 1858, when there were 9.40; in '59, only 1.28; in '60, 1.64; and in '62, 24.36 inches.

The average of February is 3.67 inches; but in 1853, the fall was 1.42; in '54, 8.04; in '56, 0.50; in '57, 8.59; in '58, 1.83; in '59, 6.32; in '60, 1.60; in '61, 3.72; and in '62, 7.53 inches. This shows a remarkable alternation. In only one year did the amount approach the average; in the others (excluding the last), the rains were very heavy and very light by turns. A dry February, after a wet December and January, is one of the pleasantest months of the year in California.

The average rain-fall of March is 3.38: in 1857, the amount was 1.62; in '58, 5.55.

The average of April is 2.63: in 1855, the amount was 5 inches; in '57, nothing; in '58, 1.35; and in '59, 0.27.

The average of May is 0.63: in 1857, the amount was 0.05; and in '60, 2.86.

The average of June is 0.08; of July, 0.02; of August, 0.03; and of September, 0.15. There are no large rains recorded in any of these months.

The average of October is 0.66: but in 1854, the amount was 2.41; and in '58, 2.74; in '55, nothing.

The average of November is 2.50: in 1854, the amount was 0.34; in '59, 7.28; and in '60, 0.58.

The average of December is 4.49: in 1854, the amount was 0.81; in '57, 8.08; in '59, 1.57; in '60, 6.16; and in '62, 9.54.

The rainy season of 1854–'55 did not commence, it may be said, until January; for although there were 2.41 inches of rain in October, yet the amount was only 0.34 in November, and 0.81 in December: so that the moisture from the October rain did no good to either the farmer or the miner, having been completely dried out from the earth before the rains of January came.

Let us now examine the rainy seasons since 1850, and see in what months more than three inches of rain fell. In 1851–'52, these months were December and March; in '52–'53, December, January, and March; in '53–'54, January, February,

March, and April; in '54–'55, January, February, March, and April; in '55–'56, December and January; in '56–'57, December and February; in '57–'58, December, January, and March; in '58–'59, December, February, and March; in '59–'60, November, March, and April; in '60–'61, December and February; and in '61–'62, from November to February, inclusive.

The rain of California usually comes with gentleness, and falls perpendicularly. The coast, above Humboldt Bay, receives a greater amount of rain than any other part of the immediate shore; and in this respect it resembles the humid clime of Western Oregon. At Fort Yuma the amount of rain is from one-fifth to one-seventh that at San Francisco, and it all falls during the spring and summer; for the rainy season of the Colorado Desert does not come at the same time with that of the remainder of the state, but is synchronous with the rainy season of Northwestern Mexico.

The rain along the middle coast of California usually comes slowly, and falls gently and perpendicularly. Here it is very seldom that two inches of rain fall in a day, and three inches have not fallen within twenty-four hours in ten years; while in the Eastern states the former figure is reached frequently, and the latter every year—where also the rain is generally accompanied with violent and long-continued storms of wind. The rains of the Sierra Nevada are far more abundant in quantity, and fiercer in the manner of their coming, than those about the bay of San Francisco. It is established that the amount of rain, and its equivalent snow, increases on the western slope of the Sierra Nevada with the elevation; but our statistics are not sufficiently extensive to enable us to determine whether the increase is in regular ratio to the altitude, or what the proportions are between the snow and rain at different heights. It is, however, an unquestioned fact that, in ordinary seasons, the amount of rain at Sonora, two thousand five hundred feet above the sea, is from twice to thrice as great as in Stockton, only seventy miles distant, at the sea-level;

and the same difference is observed between Nevada and Marysville, which bear similar relations of distance and elevation to each other.

The statistics given in the preceding part of this section relative to the amount of rain-fall at San Francisco, are intended to represent ordinary years, such as all those between 1847 and 1860. But the winter of 1861–'62 proved to be an extraordinary season, the amount of rain being double that which has fallen in any other winter since the American conquest. The average rain-fall during the winter months at San Francisco is about 12 inches; whereas, between the 1st of November, 1861, and the 1st of February, 1862, 37 inches fell in San Francisco, and during the same period 101 inches fell in Sonora, Tuolumne county. During the four months from the 1st of November, 1861, to the 28th of February, 1862, inclusive, 45.53 inches of rain fell in San Francisco, viz.: 4.10 in November; 9.54 in December; 24.36 in January; and 7.53 in February. This rain caused a great flood, which did much damage along most of the rivers, and especially in the Sacramento Basin, where Sacramento City, Stockton, Marysville, and numerous minor towns, were completely inundated, and the whole central part of the basin, including an area one hundred and fifty miles long and twenty miles wide, was converted into a great lake, which covered the land to a depth varying from two to ten feet for more than a month. The long duration of the flood, its great height, and the vast damage which it did, will render it an epoch in the history of the state, and make it well worthy of study, especially so far as relates to the Sacramento Basin, where the most serious injury was done —that basin extending north and south from Mount Shasta to the Tejon Pass, a distance of four hundred and fifty miles; and east and west from the summit of the Coast Range to that of the Sierra Nevada, a distance of one hundred miles. These two ranges unite at the two ends of the basin, which has its outlet in the middle, where the Sacramento from the north and the San Joaquin from the south, having united their waters in latitude 39°, break through the Coast Mountains to reach the Pacific. It may be said that the waters of these streams, after their union, pass through three straits: one at the Golden Gate, one hundred feet deep and a mile wide; one at the straits of Carquinez, fifteen feet deep and three-quarters of a mile wide, thirty-five miles from the ocean; and one near the head of Suisun Bay, half a mile wide and ten feet deep. The Golden Gate and the straits of Carquinez afford an abundant outlet for all the water from the interior, but not so with the pass at the head of Suisun Bay. The land at this place is low —not more than six or eight feet above low-water mark—for a width of three miles, beyond which there are hills, which prevent the spreading of the water to a greater distance. The river is shallow and crooked; the banks lined with bushes and covered with tules, which obstruct the passage of the water in time of flood.

During the flood of January, 1862, there was very little perceptible increase in the height of the water in San Francisco and Suisun Bays above the level of ordinary high tide. But there was no flow of the tide; a continual ebb of thick, muddy water, poured out at the Golden Gate for weeks together, discoloring the sea to a distance of forty miles from land. In the bays the water became almost fresh, and the planted oysters were killed by it in their beds near Oakland.

We may presume, since thirty-six inches of water fell at San Francisco from November to January inclusive, of 1861–'62, that the same amount fell in all the low lands of the Sacramento Basin, nearly one-half of its area. We may presume further that the amount which fell at Sonora is a fair representation of the amount which fell on the Sierra Nevada, one-half of the area of the basin. But possibly snow, which has not yet melted, formed one-third of the snow and rain which fell on the Sierra Nevada. It is not, therefore, necessary to take any account of that third, in this consideration of the flood of 1862—written, as it is, before the waters have gone down.

There was, then, a fall of three feet of water over an area of about twenty-two thousand five hundred square miles, and a fall of eight and a half feet over an area of fifteen thousand square miles. This would give us an average of five and one-fourth feet over an area of thirty-seven thousand five hundred square miles. The first foot was absorbed by the sand and earth, dried during a very arid summer and fall; and then there were four feet of water to escape through an outlet half a mile wide, from an area nearly as large as England, or the state of Ohio.

The outlet proved insufficient: the waters heaped themselves up in the lowest part of the Sacramento Basin, the size of which low portion I have already given as one hundred and fifty miles long and twenty wide, or an area of three thousand square miles. Now, four feet of water over an area of thirty-seven thousand five hundred square miles, will, if collected within three thousand square miles, form a body forty-eight feet deep; and that figure represents the amount of water that had to escape through the Sacramento River, below the mouth of the San Joaquin. It is to be observed that, as the outlet of the Sacramento Basin is in its centre, so the freshets come simultaneously from the north and from the south. The rains fall along the whole length of the Sierra Nevada at the same time; and as the mountain-streams are short and swift, they pour down their floods immediately and all together. Such are the circumstances which contributed to the great flood of 1862, and may contribute to other floods in the future.

During January, 1862, 24.36 inches of rain fell in San Francisco, according to records kept by Thomas Tennant, Esq.; 8.66 inches fell in Sacramento, according to Dr. T. M. Logan; 37.79 inches fell at Downieville, according to Dr. T. R. Kibbe; and 33.79 inches fell in Grass Valley, according to Mr. Atwood. I presume that all these figures are correct save those for San Francisco; and while I admit the care and accuracy of Mr. Tennant, I must suspect that somebody played tricks with his gauge, upon which he could not keep a constant watch.

Between the 1st of November, 1861, and the 1st of February, 1862, 37 inches of rain fell in San Francisco; 75.69 in Grass Valley; 79.28 in Downieville; 101 in Sonora; 42 inches of rain and 50 feet of snow (the snow probably equalling 60 inches of water) on the summit of the Sierra Nevada at Henness Pass; and 34 feet of snow and a great amount of rain (not measured) on the summit of the same range at the Big-Tree Road. The observations at the Henness Pass were kept by Mr. S. R. Dunham; those at the Big-Tree Road by Mr. Richey.

There have been "rainy seasons" in California which passed without rain; and the grass, receiving no moisture in winter, spring, or summer, has remained brown for a period of eighteen months. But no drought—more fearful than the worst of floods—has visited the country during the last twenty years, nor have we any accurate information about those that are reported to have happened before that time.

So long as the wind blows from the north, we expect fair weather; when it veers to the south, rain may be expected, usually within forty-eight hours. Sometimes, after a rain, the clouds near the earth move toward the south, while those higher up are going in the contrary direction: in such case, more rain may be expected. In no part of Europe or the Atlantic states can the state of the weather be predicted or guessed with so much reasonable confidence as in California. Here it is almost a certainty that nineteen days out of twenty in summer and fall, and that ten out of twenty on an average in winter and spring, will be clear and warm. Many circumstances of value, in furnishing grounds for predicting the state of the weather in other regions, are of no use here. In the Mississippi valley, for instance, three consecutive frosty mornings are considered as an almost certain indication of rain; but in California, frosts have no such significance: for a dozen may occur successively in the coast valleys or foot-hills of the Sierra Nevada, and nobody expects rain the more on that account.

§ 28. *Dryness of Climate.*—The small amount of rain during the winter, the entire want of it during the summer, the warmth of the sun, and the great number of cloudless days, render the climate a very dry one. As one conseqence or accompaniment of our dry climate and clear sky, it may be worth while to observe that near the ocean the clouds are rarely picturesque or sublimely beautiful. The magnificent sunsets, where the god of light goes down amid curtains of gold and crimson—those high-piled banks of clouds which adorn the heavens before and after thunder-showers, in the Mississippi valley—are never seen near the coast.

Dew is very rare or slight over a great part of the state. During the summer and autumn, many of the rivers sink in the sand soon after leaving the mountains in which they rise; the earth is dry and baked hard to a depth of many inches or even feet; the grass and herbage, except near springs, or on swampy land, are dried up, and as brown as the soil on which they grew.

It has been said that very hot days are less oppressive in California than equal heat in the Eastern states, because the cool nights serve to invigorate the system, and the extreme dryness of the climate favors the evaporation of sweat, and thus keeps the body cooler than in districts where the earth is always moist. Evaporation is so rapid, that a beefsteak hung up in the air will dry before it can commence to putrefy. A dead rat thrown into the street, where its body is crushed by wagon-wheels so that its viscera are exposed to the air, will "dry up," and its stiff hide and meat will lie during a whole summer in a mummy-like condition. In many places, steel may be exposed to the night air for weeks without getting a touch of rust.

It is common to ascribe the effects of the dryness of the atmosphere to the "purity" of the air; but it is rather the absence of moisture. I know no reason for supposing that, apart from its dryness, the air in California is purer than in any other part of the continent. It may be, however, that the constant decomposition of animal and vegetable manner, lying on wet ground, under a hot sun, causes the air in other states to be filled with such gases as are not set free to an equal extent here.

In May and June, all California "dries up"—the rivers, the brooks, the springs, the ditches, the vegetation—and, with them, many of the resources of the country.

§ 29. *Length of Days.*—The shortest day in the year, the 20th of December, measures nine hours and four minutes between sunrise and sunset at Crescent City, and ten hours at San Diego; while the longest day, the 20th of June, measures fifteen hours and seventeen minutes on the southern border, and fourteen hours and nineteen minutes on the northern border of the state—or, measuring from the beginning of twilight in the morning to the end of twilight at night, the day measures nineteen hours and forty-seven minutes on the Siskiyon Mountains, and seventeen hours and forty-three minutes at Fort Yuma.

§ 30. *Thunder-Storms.*—Thunder-storms are very rare in California. Lightning is not seen more than three or four times a year at San Francisco, and then it is never near, but far off, playing about the peak of Mount Diablo. Thunder is still more rare. Indeed, many persons have been here for years, and cannot say that they have ever seen the one or heard the other. During eleven years' residence in the state, I have never seen a brilliant flash of lightning or heard a loud clap of thunder. Thunder-storms are sometimes witnessed high up in the mountains, and in the Great Basin; very rarely in any of the low land of the state. In May, 1860, a house in Sonora was struck by lightning; and in February, 1861, three vessels in Humboldt Bay were struck in the same manner: and, though there were persons in the house and on all the vessels, no serious injury was done to either person or property in any case. On the 25th of May, 1860, a Chinaman was killed by lightning near the Lexington House, on the Coloma road, in Sacramento county.

The weather never has that peculiar condition which isolates everybody electrically, and then fills them with electricity. In New York, on a dry winter evening, a man dressed in woollen and shod in woollen slippers, after sliding along on the carpet a few steps, will accumulate so much electricity, that when he thrusts his finger at another person, a visible spark will fly off, and he can light gas with it! But this amusing experiment, common as it is in the Eastern states, never has been successful, and probably never will be often practised, here.

§ 31. *Hail.*—Hail is a rarity; and instead of falling in July and August, as is usual in the Eastern states and Europe, it is seen in California only between February and May. On the 10th of May, 1856, a storm of hail-stones, some of them weighing twelve pounds each, visited a small district at Butte Creek, in Shasta county. It has several times happened that hailstones more than an inch in diameter have fallen in the Sacramento valley.

The *Aurora Borealis* is seldom seen in California, perhaps not more than half a dozen times within the last ten years. The aurora of the 28th of August, 1859, seen over a great part of the world, was plainly visible in this state.

§ 32. *Earthquakes.*— Earthquakes are common in some parts of California, and especially at San Francisco, Los Angeles, and near the Tejon Pass, at the southern junction of the Sierra Nevada and Coast Mountains. They are rare at Sacramento, Marysville, Vallejo, and Napa. As a general rule, they are less frequent and less severe in the northern than in the southern part of the state. The vicinity of Humboldt is more often shaken than any other place north of the bay of San Francisco. About a dozen earthquakes are felt in a year at different places in the state; not so many at one place. Most of the shocks are so slight as to pass unnoticed by a great majority of the people; and there are persons who have resided six or eight years in San Francisco, and many who have resided ten years in other parts of the state, and say they have never felt an earthquake. No person has been hurt, nor has any strongly-built house been injured, by an earthquake in California, north of latitude 35°, since the American conquest. Several brick walls have been cracked in San Francisco, but they were weak structures, built on "made ground," and would perhaps have cracked by settling, of their own weight. On three or four occasions, large four-story houses have been so much shaken, that the inmates have run out in great alarm; but on examination it was found that the buildings were uninjured, even in the slightest perceptible manner.

On one such occasion, a friend of mine, who thought his life in great danger, and ran to save it, observed before he left his room that the water was splashed out of his basin by the movement of the house. The basin was of earthen ware, about fifteen inches in diameter at the top, six inches deep, half full of water, and it stood on an ordinary wash-stand. He supposed that, with another such a shock or two, the building must be in ruins; and he was very much astonished the next morning to find that there was not the slightest crack in the plastering. His room was in the fourth story of a brick hotel. It seems that the whole building had moved together.

The fear of earthquakes prevents the erection of high structures for show; and, for this reason, there is not a tall steeple in San Francisco. The largest churches have been commenced on such a plan that they might be crowned with lofty spires, but it was thought more prudent to leave them with low towers. The same motive induces many wealthy families to reside in wooden houses, which are considered better fitted to resist the shocks of earthquakes. These wooden houses, it must be kept in mind, are not "framed" with mortices and tenons, as large wooden houses are usually erected in the Atlantic states, but are "Chicago frames," held together with nails. This style of building, though introduced solely because of its cheapness and simplicity, is considered by far the most secure against earthquakes.

No earthquake felt at San Francisco since 1846 has been more severe than one which visited Buffalo, New York, in 1857, as described in the *American Journal of Science and Art* for September, 1858. On the 10th of July, 1855, an earthquake cracked the walls of twenty-six houses in Los Angeles; but no wall was thrown down, nor was any person injured.

Most of the earthquakes of California are confined to very small districts. Thus, not more than one in ten of those felt in San Francisco is perceived in Sacramento. The most extensive Californian earthquake of which we have any record was that of January 9th, 1857. It shook the earth from Fort Yuma to Sacramento, a distance of five hundred miles, being most severe at Fort Tejon, about half way between these two points. Loud noises, either rumbling or like explosions, were heard to accompany the shock at Tejon, San Bernardino, Visalia, and in the Mojave valley. The waters of the Mokelumne River were thrown upon the banks so as almost to leave the bed bare in one place. The current of Kern River was turned up-stream, and the water ran four feet deep over the bank. The water of Tulare Lake was thrown upon its shores; and the Los Angeles River was flung out of its bed. In Santa Clara valley the artesian wells were much affected: some ceased to run, and others had an increased supply of water. Near San Fernando a large stream of water was found running from the mountains, where there was no water before. In San Diego, and at San Fernando, several houses were thrown down; and at San Bueneventura the roof of the Mission Church fell in. Several new springs were formed near Santa Barbara by the shock. In the San Gabriel valley the earth opened in a gap several miles long; and in one place the river deserted its ancient bed, and followed this new opening. In the valley of the Santa Clara River there were large cracks in the earth. A large fissure was made in the western part of the town of San Bernardino. At Fort Tejon the shock threw down nearly all the buildings; snapped off large trees close to the ground, and overthrew others, tearing them up by the roots; and tore the earth apart in a fissure twenty feet wide and forty miles long, the sides of which rent then came together with so much violence, that the earth was forced up in a ridge ten feet wide and several feet high. At Reed's ranch, not far from Fort Tejon, a house was thrown down, and a woman in it killed.

In September, 1812, on a Sunday, an earthquake threw down the Mission Church at San Juan Capistrano, in latitude 33° 20′, and thirty persons were killed. The church at Santa Inez, in Santa Barbara county, was thrown down on the same day; but the shock, according to report, was an hour later than that at San Juan Capistrano, and there was nobody in the church when it fell. At the same time the sea receded a long distance from the ordinary place of the water's edge on the beach of Santa Barbara; and the people there, knowing that it would soon rush upon the shore, fled to the higher ground, and by that means alone saved their lives. These reports made about this earthquake of 1812, to Dr. J. B. Trask, by old residents, have never been contradicted, though published six or eight years ago.

The old Mission Church at Santa Clara was thrown down by an earthquake in 1818. On the 15th of May, 1851, a severe shock was felt in San Francisco. Windows were broken; merchandise was thrown down from shelves in stores; and vessels in the harbor rolled heavily. On the 26th of November, 1858, nearly every brick building in San José was injured by an earthquake. On the 3d of July, 1861, Amador valley, in Alameda county, was severely shaken. Adobe houses were seriously injured, chimneys toppled down, furniture was flung from side to side of the houses and much broken, and men in the fields were thrown down.

A severe shock of an earthquake was felt at Fort Yuma and

vicinity on the 29th of November, 1852. The low grounds near the Colorado cracked open with long, wide fissures, from which water, sand, and mud, spouted up. The fissures were in some places so large, that they turned the river from its course; and the change was so sudden, that great multitudes of fish were left to die in the mud. At the same time, the mud-volcanoes of Lower California, distant forty-five miles southwestward from Fort Yuma, resumed their activity; for, although there is no record of their previous action, yet they probably existed before. A pool of hot, sulphurous water had been observed at the place by Americans since 1849. Immediately after the shock of 1852, the officers at Fort Yuma saw a great body of steam shoot up at least one thousand feet in the desert to the southwest; and when, soon afterward, some of them went out to examine into the cause of it, they found the mud-volcanoes on the site of the old pool, throwing up steam, boiling water, and mud, very much like the *salses* farther north.

Earthquakes, according to the common theory of Californians, are electrical in their origin, or closely connected with electrical influences. Many of the strongest shocks have been preceded by a condition of the atmosphere very similar to that which precedes thunder-storms in other lands. When the weather is sultry and oppressive in San Francisco, people say, "Look out for an earthquake!" And it usually comes—perhaps so faint as to be barely perceptible, and sometimes not until several hours after a change in the weather.

The frequency of earthquakes in California has caused a number of persons, perhaps a hundred or more, to leave the state, and return to their former homes on the Atlantic side of the continent. And yet there they are in more danger from lightning than here from earthquakes, for there are fifty killed by lightning in the Mississippi valley for one killed by an earthquake in California. A year rarely passes that a dozen persons are not struck by thunderbolts within three hundred miles of St. Louis. [See *Appendix*, p. 464.]

§ 33. *Sand-Storms.*—In the Colorado Desert, and in some other districts in the southern part of the state, sand-storms, similar to the simooms of Africa, but not so dangerous, occasionally occur. The sand, which forms the greater portion of the soil, unprotected by sod, vegetation, or moisture, is swept away in dense clouds by every high wind, and carried many miles, a terror to man and beast. The storm stops the traveller, because he dare not open his eyes to the little flinty particles; nor can he eat, for the dust covers his food and fills his mouth: and even in the most tightly-built houses the sand penetrates and fills the air.

A newspaper correspondent speaks thus of a Colorado sand-storm:

"Should the traveller happen to encounter a sand-storm, however, he may not get along so smoothly. A huge, black cloud, rising from the western horizon, warns him of its approach. Rapidly it spreads over the sky, darkens the sun, and the fine particles of sand are swept before the gale in a dense and suffocating cloud; even the larger gravel and pebbles are sometimes lifted from the plain and carried like hail before the force of the blast. The horses are blinded, paralyzed with fear, and no urging can induce them to go forward. Were it otherwise, to go on would be folly; the road and sun are hid from view; no landmarks by which to be guided—safety bids you remain. The traces are unhitched, and the horses tethered to the wagon; the only course is to securely fasten down the sides to the wagon-top, and wait with what patience one can command until the storm has passed, which will be, doubtless, in from six to ten hours.

"Once the stage encountered a sand-storm while within three hundred yards of a station; the horses could not be induced to move, and there was no remedy but to stay by them till the gale had spent its force, though the station was even in sight.

"I have found such a storm sufficiently disagreeable while housed by the river-side, the fine sand penetrating everywhere, and have no ambition to encounter one upon the central desert. Luckily, they are not very common in the severest aspect; in summer, quite rare."

NOTE.—The chief writers upon the Meteorology of California have been, Dr. Henry Gibbons, of Alameda, and Dr. T. M. Logan, of Sacramento, whose writings have been published in the books of the Smithsonian Institute, and in the newspapers of San Francisco and Sacramento.

CHAPTER III.

GEOLOGY.

§ 34. *General Geological Character.*—California, geologically considered, belongs chiefly to the paleozoic and tertiary epochs. The carboniferous rocks are wanting, or their existence in the state is confined to a very small district, and has not been demonstrated even there. A tertiary sandstone, some of which is metamorphic, having lost its original stratification under the influence of intense heat, underlies the valleys of the Sacramento, the San Joaquin, and the coast, and is seen in the Coast Mountains, the Great Basin, and the Colorado Desert. Granite occupies the higher portions of all the mountainous districts, and considerable portions of the Great Basin and the borders of the Colorado Desert. The scarcity of stratified rocks is plainly discoverable by the traveller in the number and ruggedness of the mountains; only primary, eruptive, and metamorphic rocks make such steep hill-sides. The thinly-stratified rocks, with intervening layers of clay, are soon worn down by the water into gentle slopes, and covered with fertile soil, every foot of which may be turned over by the plough, and with profit. Such is not the character of California, nearly all of which is primary or metamorphic.

Many rocks besides granite and tertiary sandstone appear in irregularly-distributed patches. About Mounts Shasta and Lassen, Castle Peak, the Marysville Buttes, in the plateau of the Sierra Nevada, the Great Basin, and the Colorado Desert, there are considerable tracts of basalt, lava, trap, and trachyte; and in other places there are small tracts. Some very remarkable hills of basalt, called "Table Mountains," are found in the

Sierra Nevada. The largest of these is in Tuolumne county, about three thousand feet above the sea, and one hundred miles eastward from San Francisco. It is thirty miles long, three hundred to eight hundred feet high above the surrounding country, and about a quarter of a mile wide, in many places less. The basaltic formation is evident at a distance, from the perpendicularity of the sides near the top, and the flatness of its summit, gently descending toward the west. Along the sides of the Sierra Nevada, near the line of separation between the sandstone of the valley and the granite of the higher parts of the Sierra, are found various other rocks, among which slate, quartz, and limestone, are prominent. The slates are usually soft, their cleavage often perpendicular to the horizon. Limestone is abundant about two thousand feet above the sea, between latitudes 37° 30′ and 39°. It is all metamorphic, and some of it is a fine marble, which may prove of value for statuary. Most of it is gray in color. Metamorphic limestone is also found near Santa Cruz, near New Almaden, at Monte Diablo, and in Shasta and Siskiyou counties. It is said that some stratified secondary limestone has been found in Shasta county, but this is a matter of doubt. No secondary coal has been found in the state. Tertiary coal, much of it a lignite, has been found at various places in San Diego, Santa Clara, Alameda, San Francisco, San Mateo, Contra Costa, and Sonoma counties, and much money has been spent in opening veins.

§ 35. *Diluvium.*—The Sacramento and San Joaquin valleys are covered with a diluvium from four hundred to fifteen hundred feet deep. It is composed of alternate layers of sand, gravel, and clay. The most complete information which we have as to the nature of this diluvium is given in the report of the boring of an artesian well, one thousand feet deep, at Stockton. This is, I believe, the only artesian well in the Sacramento Basin; at least, it is the most notable one. An attempt was made to bore an artesian well in Sacramento, but the auger struck a stratum of boulders about four hundred feet from the surface, and was unable to pass through it. The following is a statement of the different strata encountered in the Stockton well, and the thickness of each in feet, beginning at the surface:

Black loam, 6; red clay, 6; dark-red clay and sand, 18; blue clay, mica, and sand, 10; blue clay, hard, highly stratified, 4; blue clay, mica, and sand, 3; blue clay, hard, and highly stratified, 4; green sandstone and clay, very hard, 29; blue clay, sand, and gravel, slightly impregnated with gold, 2; blue clay, sand, and gravel, 18 (100); green sandstone, clay, and mica, hard, 15; fine gravel, 5; gray quicksand, 15; blue clay, 8; gray sand and clay, 27; dark-blue clay and sand, 33 (203); coarse gravel and pebble-stone, 27; blue clay, 7; gray sand, 13; blue clay and sand conglomerated, 12; light-gray sand, 3; blue clay, 6; light sand, 9; blue clay, 1; fine gray sand, 12; dark clay, 2; fine gray sand, 7 (302); clay and sand, 10; coarse gray sand, 1; light clay, 19; coarse sand, 14. (NOTE.—At 340 feet, in this stratum of sand, a red-wood stump was found, and a stream of water ascended to within three feet of the surface.) Light clay, 8; fine gray sand, 1½; light clay, 20½; coarse gray sand, 20; clay, very hard, 4 (400); gray sand and clay, 5; clay, 20; coarse gray sand, 3; light clay, 15; fine gray sand, 4; light clay and sand, 1; coarse gray sand and clay, 1; light-blue clay, 11; gray sand and clay, 7; light-blue clay, 15; fine gravel, 1; light-blue clay and gravel, 13 (496); fine gravel, 25; clay and sand, 2; sand and clay, 5; coarse gray sand, 7; fine blue clay, 8; fine gray sand, 42. (NOTE.—At 560 feet, in this stratum of sand, obtained a stream of water, rising five feet above the surface.) Gray sand and clay, 15 (600); light clay and sand, 6; fine gray sand, 24; clay and sand, 3; fine sand, 9; fine gravel, 3; fine gray sand, 5; coarse sand, 2; blue clay and sand, 8; gray sand and clay, 8; clay and sand, 5; fine gray sand, 10; clay, 5; coarse gray sand, 2; fine light-blue clay, 4; hard, chocolate-colored clay, 2; blue clay, 2 (698); fine gray sand, 30; clay and sand, 8; gray sand, 4; light clay, 10; coarse sand, 6; blue clay, 4; dark clay, very hard, 2; gray sand, 4; blue clay, 14; light-drab clay, 3; very fine gray sand, 24 (807); light-drab clay, 1; light-gray sand, very fine, 27; dark-gray clay, 17; light-blue clay, very hard, 22; light clay, 11; dark, chocolate-colored clay, very hard, 10; light clay, very hard, 15 (910); fine gray sand—a good stream of water, 2; clay and sand, 11. (NOTE.—A large stream of water was obtained in this stratum, rising seven feet above the surface.) Fine sand and gravel, 10; blue clay, 20; sand and gravel, 6; blue clay, 27; clay, gravel, and mica, 14 (1,000); in sand, 2.

The depth of the well is 1,002 feet. The temperature of the water, as it issues from the well-surface, is 77°, the atmosphere being 60° Fahrenheit. The water rises eleven feet above the surface of the plain, and nine feet above the established grade of the city. The quantity of water discharged is about sixty thousand gallons in twenty-four hours.

The diluvium in the coast valleys bears a strong general resemblance, in its material and stratification, to that of the San Joaquin valley. In many places where artesian wells have been sunk, fossil wood and bone have been found three hundred and four hundred feet below the surface of the ground, and two hundred feet or more below the present level of the sea. In the mountains there are also large bodies of diluvium, but the material is coarser than in the valleys, being usually a gravelly clay, deposited in distinctly-marked layers, with intervening strata of sand and boulders.

§ 36. *Gold.*—Gold is found in nearly all parts of California, but is most abundant on the western slope of the Sierra Nevada, between two thousand and six thousand feet above the sea, from latitude 37° to 40°—a district two hundred and twenty miles long by forty wide. This may be called the Sacramento district. It is drained by the Feather, Yuba, American, Cosumnes, Calaveras, Stanislaus, and Tuolumne rivers. The next district in importance is in the northwestern corner of the state, including that part of the Sacramento Basin west of Shasta, and the lower portion of the Klamath valley. Next is the Kern River district, including White River between 35° and 36° on the western slope of the Sierra Nevada. There are gold diggings on the San Gabriel and Santa Anita Rivers, and in the San Francisquito cañon in Los Angeles county. In the Colorado valley, fifty miles above Fort Yuma, gold has been found. Nearly every one of the coast counties has more or less gold: it has been found in the valleys of Russian River, Putah Creek, Soquel Creek, Coyote Creek, the Salinas River, and in the earth in which the city of San Francisco is built.

§ 37. *Auriferous Lodes.*—Gold is found fastened in stony veins, and loose in earthy matter: the latter called placer diggings, the former auriferous quartz lodes.

It is the accepted theory among geologists that all gold was once enclosed in quartz lodes, and that the gold in the placers was obtained from the disintegration or breaking up of the lodes. The surface of the earth was once all rock; the earthy matter was formed by the action of air and water on this rock. The earthy matter was then deposited in diluvium, among which was the gold that had existed in the rock previous to its disintegration.

Gold is sometimes found in granite, syenite, limestone, slate, and other rocks; but the auriferous lodes, regularly worked, are all of quartz. Most of the quartz veins run parallel with the main divide of the Sierra—that is, north-northwest and south-southeast—are from a line to thirty feet thick, and are nearly perpendicular, dipping to the eastward. They are between two thousand and six thousand feet above the sea. The general color of the rock is white, occasionally bluish, frequently reddish-brown, the color of iron-rust, derived from the decomposition of iron pyrites. In some veins the rock is compact, and then it is usually very white; in others it is full of cracks and crevices, and ready to break into small pieces with a little pounding. Most of the veins have gold in them; only a few have enough to pay for working. The gold is in particles of irregular shape, but with some regularity of size, scattered through the rock. The particles are seldom larger than a pea, never weigh more than an ounce, and often are so fine as to be invisible to the naked eye.

A lode has usually a peculiar kind of particles, either large or small. Most of the gold in a lode is usually in a rich streak, near the "foot wall" or lower side, as if the metal had settled down by its gravity. The rock near the "hanging wall" or upper side of the lode is poorest. Occasionally several rich streaks will be found in a lode—one streak with coarse particles of gold, another with fine. All parts of a lode are not equally rich; but the gold is found in spots. A lode which is very rich in one place may be very poor in another not far off; indeed, there is no auriferous vein in the state known to be regularly rich for a long distance on the surface. The gold is found in streaks or pockets; the rich streak runs downward, and has a regular dip in the lode. It is a matter of very great importance to the miner to ascertain the direction of this dip, and here is the rule: Take out some of the vein stone, and examine the wall rock carefully. In most veins it will be found that the wall has little furrows, as though the lode had been pushed upward. These furrows indicate the direction of the dip of the rich streaks. Pockets may be considered as interrupted streaks; and when one rich pocket is discovered, others may usually be found by going down into the vein in the proper direction, and that is ascertained in the same manner as for continuous streaks. This is an important rule, and it is now published for the first time. I am indebted for it to J. E. Clayton, Esq., mining engineer.

§ 38. *Placers.*—The placers are of two kinds—*diluvial*, or those deposited under large bodies of water, as if in a deluge; and *alluvial*, or those deposited under the influence of streams of water, such as the present rivers and brooks of the country. It is evident, from an examination of the mining districts, that large tracts of auriferous ground have been deposited under diluvial influences. The same strata are found extending over wide areas, and the deposition is different from that made by a river. The gold, being nineteen times heavier than water, and from six to eight times heavier than the clay and stones with which it is found, has sunk to the bottom of the diluvium. The best diggings are therefore near the "bed-rock." The dirt in which the gold is found is usually a stiff clay, with gravel and large stones intermixed. The common phrase "golden sands" may mislead. Pure, fine sand rarely has any gold in it; and the richest deposits of the precious metal are in a clay so tough as to give the miner much trouble to dissolve it, with stones in it weighing from a pound to several hundred weight. The character of the pay-dirt varies greatly. A hill of diluvium may be three hundred feet deep, with a dozen strata of different material, and all of them auriferous in different degrees. In some places the pay-dirt is full of boulders, weighing several hundred pounds or more; in other places the stones are all about as large as a man's head; in others, as large as a hen's egg. In one stratum the dirt is red, in another blue, in another brown. In some places the dirt was deposited in basins of rock, four or five miles across, and from ten to fifty feet deeper at the centre than at the rim.

The placer-diggings are all found in a very rough, mountainous country. The gold has not been carried far; its weight has anchored it near its mother-vein. There may be much gold in the Sacramento valley; but if so, it is deposited beneath one thousand feet of diluvium, and nine hundred feet below the level of the sea, where it will never be disturbed.

The *diluvial* placers are in what are called hill and flat diggings; the *alluvial*, made by streams running through the diluvium, are in river-beds, bars, ravines, and gullies. The alluvial placers, as a general rule, are richer than the diluvial. The streams have carried away much of the dirt, and left nearly all the metal. Most of the gold of the rivers comes from gullies. There is a gully on a mountain: it is dry, except during heavy rains; it has steep sides. The rain comes; the water pours down its sides, fiercely sweeping clay, gravel, and gold along. The bed of the ravine is not so steep as the sides; most of the gold stops there; the dirt is carried away into the river, with a little gold. A thousand such gullies, contributing each a little gold, make a river rich. The heavier particles are deposited in the middle of the river-bed; many of the smaller particles are deposited at the sides. The richest spots in gullies are usually where the bed-rock is full of crevices, as where slate, with perpendicular strata, crosses the gully. In those parts of a gully where the bed is very steep, there is usually less gold than in spots nearly level. In rivers, the richest spots are usually just below cañons, where bars are formed. Wherever there are eddies at times of high water, there the gravel, clay, and gold will be deposited; and when the river falls, a bar is exposed. The richest bars of California have been found at the mouths of cañons.

§ 39. *Mineralogy of Gold.*—The particles of gold in quartz are usually very rough in shape, but sometimes they are in octahedral crystals, and at others in smooth, leaflike sheets. Very rarely specimens of crystals are found, clustered together so as to resemble pieces of coral.

Placer-gold, called "gold dust," is *fine* and *coarse.*

The fine is in *scales*, *grains*, *flour*, *shot*, and *wire.*

Flour-gold is very fine dust.

Grain-gold is in particles about as large as the end of a pin.

Scale-gold is in scales from a sixteenth to an eighth of an inch across, and as thick as heavy wrapping-paper.

Shot-gold is in roundish particles, the size of a pin-head.

Wire-gold is in small wires, from an eighth of an inch to an inch long, and as thick as a pin. Sometimes the wires are fluted, or hollow on one side; and sometimes they are knotted together.

Coarse gold is coarse-shot, pea, bean, moccasin, cucumber-seed, pumpkin-seed, large wire, and miscellaneous coarse.

The pea, bean, cucumber-seed, and pumpkin-seed varieties of gold, have particles resembling in shape the seeds whence they derive their names. The coarse-shot is in particles resembling coarse shot in shape and size. Moccasin-gold is in pieces resembling a low shoe or moccasin in shape, and about half an inch long. Large wire-gold is in wirelike pieces, about a sixteenth of an inch thick, and from a quarter of an inch to an inch long. Miscellaneous coarse gold is in pieces of very irregular shape and size.

Fine gold is often found without any admixture of coarse; coarse is rarely found without some admixture of fine.

The different varieties of gold are often found separate from each other. One gully will have scale gold, another fine wire-gold, another moccasin-gold, another pumpkin-seed gold, and so on. These different varieties of gold are frequently found very near to each other: a cucumber-seed gully will not be more than a hundred yards from a pea gully. There is a small hill in El Dorado county; all the gold on one side is fine, all on the other coarse. The gold as it originally comes from the quartz is rough, but by friction among the gravel and sand it becomes smooth. Where all the pieces of gold are rough, it has not moved far from its maternal lode; where all the particles are small and smooth, the presumption is that it has moved a considerable distance. The larger the stream, the finer and smoother its gold, as a general rule.

Most of the gold now obtained is miscellaneous coarse; the little gullies which yielded the delicate varieties are now nearly all exhausted.

Most of the placer-gold is coarse, in pieces worth half a dollar or more. Pieces worth five dollars are very common, and numberless nuggets worth one hundred dollars or more have been found in California. The largest nugget of gold on record was found at Ballaarat, Australia, on the 9th of June, 1858; it weighed two hundred and twenty-four pounds Troy, of nearly pure gold, and was called "The Welcome Nugget." The next, weighing one hundred and ninety-five pounds Troy, was found in Calaveras county, California, in November, 1854. The third, called "The Blanche Barkly Nugget," weighed one hundred and forty-five pounds Troy, and was also found in Australia. Smaller lumps are too numerous to mention. All placer-gold is called "dust," but the particles of the dust are sometimes pretty large. No satisfactory explanation has been given of the fact that placer-gold is usually in particles so much larger than that found in quartz.

Gold is fine and coarse mechanically—that is, in the size of its particles—and chemically in its composition. Most metals are found in ores, combined chemically with non-metallic substances which hide them. The ores have usually neither the color, the specific gravity, the strength, nor the other peculiar features of the metals. Native gold is never found as an ore; it is always in a metallic form. The reason of this is, that it does not rust on exposure to the air, nor is it dissolved by any of the simple acids. And yet it is never found pure, but always mixed with silver, in nearly all possible proportions. Frequently copper and lead are also found in native gold. The amount of other metal in gold is designated by figures of fineness, estimated according to thousandths. Perfectly pure gold is 1,000 fine; gold containing one-tenth of its weight in silver is 900 fine; that is, 900 parts in 1,000 are gold, and 100 are silver. In gold 600 fine, 400 parts in 1,000 are of other metal.

The native gold in California varies in fineness from 500 to 990, averaging about 880. One large piece, found at Downieville, was 992 fine. In Mariposa, Fresno, and Buena Vista counties, and at Mono Lake and Walker's River, east of the Sierra Nevada, the fineness is very low. The gold of the Colorado is very fine. In other districts there are great variations in the fineness within small distances. One has gold 900 fine; another, one hundred yards distant, has gold only 800 fine. Ordinarily, all the gold in a gully or in a river-bar is of the same fineness; so also all the gold in a quartz-lode is of the same fineness. But there are the same differences of fineness between the gold taken from different quartz-lodes as in that taken from different gullies. For these differences there is no satisfactory explanation.

Let us now run through the list of the principal mining districts of the state, giving the fineness of the placer-gold of each:

The gold found in the bars and beds of the Klamath and Salmon Rivers is in coarse particles, averaging 868 fine, and mixed with iridium.

The gold of Scott River is coarser and poorer. That of the South Fork of Scott River is about 875 fine, and coarse.

The gold on Trinity River, near Weaverville, is in small particles, from 885 to 940 fine.

Twenty miles below Weaverville, on the same river, the gold is poorer—from 865 to 870.

Still farther down, iridium becomes so abundant, that the gold is worth a dollar or a dollar and a half per ounce less than it would be if clear of that metal.

In the gullies and hills near Weaverville there are rich diggings of coarse gold, from 890 to 960 fine.

At Gold Bluff the gold is 950 fine.

The gold found at Yuba is coarse, in rough, flat particles, from 820 to 830 fine. In some lumps from Yuba little pebbles have been found hidden in their centre.

The gold of most of the creeks near Yuba is from 845 to 850 fine.

McAdam's Creek, near Shasta, yields gold from 875 to 885 fine.

Cottonwood Creek, near the Oregon line, yields gold of the same fineness.

At Oro Fino, near Yuba, the gold is in wires, and, like all wiry gold, is of poor quality, from 760 to 780 fine.

The Shasta gold is generally coarse, ranging from 865 to 925 fine, except at French Gulch, where it is only 830 fine.

The Pit River gold is coarse and poor, 830 fine.

The Feather River gold is all good, from 890 to 920 fine.

At Oroville it is from 920 to 940 fine.

At La Porte, Gibsonville, and Pine Grove, in Sierra county, the diggings are deep, and the gold coarse, and from 915 to 970 fine. At Poor Man's Creek, the gold is coated with something like an enamel, the color of iron-rust. When the gold is pounded, this enamel breaks off.

The Downieville gold is coarse, and from 895 to 925 fine.

At Goodyear's Bar the fineness varies from 880 to 885.

At Foster's Bar, and farther down, from 890 to 900.

At Slate Creek and Cañon Creek, from 900 to 910.

At Forbestown, in Yuba county, the gold is 890 fine.

At Camptonville, the fineness is from 920 to 970, except in one hill, which has wiry gold, 870 fine.

The gold of the Middle Fork of the Yuba River is fine, from 885 to 890.

That of the creeks and ravines tributary to that stream is still finer—from 900 to 920.

At North San Juan the fineness is from 960 to 965.

At French Corral it is 920 fine.

The gold of the Southern Fork of the Yuba River is from 890 to 900 fine.

Every hill about the town of Nevada has a quality of gold differing from that of other places in the vicinity. Some of it is very good, but the general character is poor, from 800 to 850. Most of it is rough.

The Grass Valley gold is poor, from 800 to 840.

At Rough and Ready the gold varies as at Nevada.

The gold of Yuba county is good, from 940 to 970.

The Timbuctoo gold in Yuba county is 950 fine.

The gold of Park's Bar is 910 fine.

The North Fork of the American River has gold from 915 to 920 fine; the Middle Fork from 865 to 900; the South Fork from 915 to 920.

At Iowa Hill, the gold is 900 fine; at Michigan Bluff, 930; at Forest Hill, 900; at Yankee Jim's, from 890 to 910; at Auburn, 860; below Auburn, and near Ophir, 820; and at Secret Ravine, 780. All these placers are in Placer county. At Gold Hill, in Placer county, iridium is found fastened in with the gold; the only place on the coast where the two metals are found fastened together.

Next we come to El Dorado county. The Coloma gold is good, about 915 fine; and some of the hill gold in the vicinity of the town is from 920 to 930 fine. The Georgetown gold is coarse, and from 890 to 920 fine. At Kelsey's Diggings the fineness is 860. At Placerville the range is from 870 to 970. Coon Hollow, near Placerville, produces gold 970 fine. At Diamond Springs the fineness varies from 870 to 900. The El Dorado or Mud Springs gold is 860 fine. The gold of the Cosumnes River varies from 865 to 875.

The gold along the American River, in Sacramento county, is from 915 to 920 fine.

Most of the gold in Amador county is poor, from 865 to 920. The gold about Volcano averages 870; Jackson, 865 to 868; the Buttes, near Jackson, 920.

The gold of the Mokelumne River, above Mokelumne Hill, varies from 850 to 910. At the hill, some of the gold has been 980 fine. At Campo Seco the fineness is 900. Along the Mokelumne River are many placers where the gold is in coarse, wiry pieces, and poor in quality. At Vallecito, the fineness is 915; at San Andres, 920; at Murphy's, 885; at Douglas Flat, 890; at Angel's, 890; at Jesus Maria, from 858 to 860; and at Carson, from 890 to 920.

The gold found along the Stanislaus River is from 875 to 880 fine.

Columbia, in Tuolumne county, is the best place for good gold in the state; it ranges from 930 to 970. There are many other places where gold equally fine is obtained, but none which produces so much gold of a high fineness, with so small a proportion of poor gold. The average of the Sonora gold is 900; Jamestown, 870; Montezuma Flat, 900; ravines at Chinese Camp, 950; the river and creek claims at Chinese Camp, 860; Don Pedro's Bar, 880; Big-Oak Flat, 800; and Garote, 810.

La Grange, in Stanislaus county, has gold 885 fine; that of the Merced River is from 845 to 860 fine.

The gold of Mariposa county is poor, and much of it wiry, ranging from 700 to 820 fine, averaging about 760. Some of the gold of Mariposa county is coarse, and resembles that of high fineness from other districts, but, on assaying, it proves to be poor.

The gold from the San Joaquin River is in small scales, and about 800 fine.

The Kern River gold is 630 fine.

The Colorado gold is 930.

The Walker's River gold is 560; the Mono gold 600.

The above figures of the fineness of gold from various districts refer only to the placer-gold. The gold from different quartz-lodes also varies. As a general rule, the quartz-gold is not so rich as the placer-gold; but some of the gold taken from quartz is nearly as rich as any from the placers.

§ 40. *Silver.*—A large amount of silver is found in California. One-tenth in weight of the gold-dust is silver. But it was not until 1860 that valuable veins of argentiferous ore were discovered in the state. As gold predominates on the western side of the main divide of the Sierra Nevada, so does silver on the eastern side. The Washoe mines are not in California; but the Esmeralda and Coso districts are.

Esmeralda is in latitude 38° 15′, about two hundred and fifty miles in a direct line from San Francisco in an eastward direction, and just within the limits of the state, as is generally supposed, though some persons assert that it is in Utah. The country is mountainous; the rock is porphyry and trap. The argentiferous veins are of a bluish-white quartz, containing sulphuret of silver. The principal vein of the district runs north and south, and this has very little gold; many other veins, containing considerable quantities of gold mixed with the silver, run east and west. The main vein is about twenty feet wide. The ores assay from thirty to fifteen hundred dollars per ton, but the latter figure was only obtained from picked specimens. Timber suitable for firewood is abundant, and there is sufficient water to supply the wants of a small town.

The Coso silver district is about one hundred and fifty miles northward from Los Angeles, one hundred miles eastward from Visalia, and twenty miles southeastward from Owen's Lake.

Wood and water are very scarce. Little is known about the district as yet. The Coso and the Esmerelda districts are both in the Great Basin of Utah, and about five thousand feet above the level of the sea.

It is said that rich silver-mines have been discovered in the valley of the Mojave River.

In the Coast Mountains, in Monterey county, a silver-mine called Allisal has long been known to exist, but it is not rich enough to pay. Similar veins are found in the Coast Mountains, in Santa Cruz county.

Veins of silver ore are also found in Nevada county.

§ 41. *Platinum.*—Platinum, iridium, and osmium, are three white metals resembling steel, often found in the placer mines of California. They usually occur together; and are found more abundantly in the lower part of the Klamath valley than in any other part of the state. In many districts they are entirely lacking. Platinum is found in lumps by itself; iridium and osmium are found united, and are then called irid-osmium. These metals are found in small particles, usually fine scales; the largest piece was of irid-osmium, found on the Lower Klamath, and weighed an ounce and a quarter. They are not found separate from the gold, nor are they ever the main object of search; they are obtained in small quantities only, and are rarely bought and sold in the state; they have no fixed market price. When mixed with gold-dust, they injure its value, and prevent its reception at the mint on deposit.

§ 42. *Quicksilver.*—There is probably no country in the world so rich in quicksilver as California. That metal is obtained only from its sulphuret or cinnabar, of which extensive deposits are found in Santa Clara county, about sixty miles southward from San Francisco, and fifteen miles from San José. There are three mines here—the New Almaden, the Enriqueta, and the Guadalupe. The ore is found between trap on one side, and metamorphic limestone on the other. The mines are about one thousand feet above the level of the sea.

The gangue or vein-stone is quartz. The ore varies in richness from 5 to 70 per cent.: the ore of the Almaden mine averages 18 per cent.; that of the Enriqueta mine is about the same; that of the New Idria mine is about 8 per cent. The lodes are extremely irregular; sometimes there will be a mere thread, which will widen out into a mass forty feet in breadth, twenty high, and seventy long, of rich ore, and then diminish again to a thread. The total annual production of the state will probably amount to three million seven hundred thousand pounds, of which two million four hundred thousand at least will come from New Almaden. The New Idria mines, in the Coast Mountains, about seventy-five miles southeastward from San José, furnish about three hundred and fifty thousand pounds per annum. In Napa county, on the slopes of Mount St. Helena, and in Sonoma county, in the Geyser Mountains, west of Clear Lake, cinnabar has been found, and companies are now at work opening the mines. Whether they will prove to be of value is as yet a matter of doubt. It is a singular feature of the cinnabar veins in these two last-mentioned places, that they are accompanied by a porous limestone which is full of pure quicksilver; and when the stone is shaken or struck, the liquid metal flies out in minute globules. There have been rumors of discoveries of cinnabar in other parts of the state, but they are not well authenticated.

§ 43. *Copper.*—Copper ore has been found near Crescent City; at Copper Cañon, in the southwestern corner of Calaveras county; in San Diego county; in Napa county; six miles below Grizzly Flat, in El Dorado county; near Sweetland, in Nevada county; and in Shasta county. Sulphuret of copper, or copper pyrites, is found in auriferous quartz-lodes in nearly all the mining counties. Near Sweetland, Nevada county, there is a claim in which so much copper is found with the gold, that the dust is worth only eleven dollars per ounce. No copper has been smelted in the state; the only attempts to mine for the ore have been at Copper Cañon, and on the bank of the Cosumnes River, below Grizzly Flat. Some of the Copper Cañon ore has been exported. There are rich veins of copper-ore near Crescent City, in Del Norte county; but, with the present high prices of labor and coal, they cannot be profitably wrought. Vitreous copper is found at Williamson's Pass, sixty miles from Los Angeles.

§ 44. *Coal.*—The old red sandstone and the "true carboniferous" rocks, as they are called, are wanting in California; and it was long supposed that no valuable coal would ever be discovered in the state; but within the last year some veins of a very good quality have been found near Mount Diablo. The mineral belongs to the tertiary epoch, but contains far more solid combustible matter and less incombustible material than most tertiary coal. In the strict geological meaning of the terms, it is not "coal," but "lignite," belonging to a later date than the true coal, and lying in a different formation. The rocks are sandstone and shale, of the upper tertiary or pliocene age, and were formed by alternating depositions in salt and fresh water. The coal-veins are situated on the north-eastern slope of Mount Diablo, are from two to nine feet in thickness, dip to the north at an average of 30°, and open on the southern declivities of the hills. A chemical analysis of some of the best specimens showed 50 per cent. of carbon, 46 per cent. of volatile bituminous substances, and 4 per cent. of ashes. The coal is bituminous in character, breaks readily, shows a bright surface where fractured, and burns with a brilliant flame. The quantity is large, and it can be profitably supplied in San Francisco at eight dollars per ton, whereas imported coal has hitherto cost twice as much.

§ 45. *Asphaltum.*—Bituminous springs are numerous near the coast, from the northern line of Monterey county to San Diego. They throw up a dark, pitch-like fluid, of a strong odor, which on exposure to the air grows thick, and finally solid. It collects in great masses about the springs, and in some places covers several acres of ground. After being exposed to the air for some time, it is called "asphaltum," which is very hard in cold weather, but grows soft at about 75°, and becomes liquid at 85°. Some springs of it rise in the sea, near San Diego, and others near Santa Barbara; and masses of the asphaltum are seen floating many miles from shore. The air at sea is even scented with it, and on several occasions frights on shipboard have been caused by its odor, which was supposed to come from some hidden fire.

The principal places in which these springs of asphaltum are found are the following:

1. In the Santa Cruz Mountains, in the southeastern part of Santa Clara county. A tract of twenty-five acres is here covered by the hardened asphaltum.

2. In San Luis Obispo valley. The asphaltum covers thirty acres.

3. The Napoma ranch, in San Luis Obispo county. The springs are small, and yield but little.

4. On the ranch of La Purissima, in Santa Barbara county.

5. A place six miles west of the town of Santa Barbara. The deposit of asphaltum covers three hundred acres from two to eight feet thick.

6. Rincon of San Buenaventura, Santa Barbara county.

7. A place near the San Buenaventura River, twelve miles from its mouth, in Santa Barbara county.

8. A place near the Santa Clara River, eighteen miles from its mouth, in Santa Barbara county.

9. A place in the Sierra Santa Susanna, in Los Angeles county.

10. In Los Angeles valley, Los Angeles county.

11. The San Pedro Hills, in Los Angeles county.

12. San Juan Capistrano, Los Angeles county.

One of the deposits in Santa Barbara is so near the sea, that the mineral might be thrown with a shovel into a shute which would carry it into the hold of a vessel at anchor.

A spring of mineral oil has been found in Mattole valley, Humboldt county. This is probably the same material with that of the asphaltum springs of the southern coast.

The asphaltum generally comes up through sandstone. The springs of Santa Barbara seem to have ceased to flow, while those in Los Angeles county are still active. It is supposed that the amount lying on the surface at the various deposits is not less than five thousand tons.

§ 46. *Other Minerals.*—Iron pyrites, or the sulphuret of iron, is found with gold in many of the quartz-veins. Iron is found also in a number of chalybeate springs. Iron-ore containing, it is said, 83 per cent. of metal, has been found near Auburn, in Placer county; and the assessor of Shasta county, in his report for 1857, said "rich iron-ore" had been found in that county. Magnetic iron-ore is found in the Cañada de las Uvas, and at Williamson's Pass.

Tin-ore, of the kind called "tin-stone," of a rich quality, has been found in a large vein at Temascal, in San Bernardino county; and it is reported that another lode of similar character has been found in the valley of White River, Buena Vista county.

Galena has been found in Humbug valley, Siskiyou county; in Tuolumne county; and on the banks of the Cosumnes River, in El Dorado county. Plumbago has been discovered near Columbia, Tuolumne county.

Cobalt is found, in various ores, in many counties in the state.

At San Emidio, about twenty miles westward from Tejon Pass, is a rich and large lode of sulphuret of antimony; the vein is from four to twelve feet thick, and is about six thousand feet above the level of the sea.

Arsenic exists in many of the lodes of auriferous quartz, in the argentiferous lodes at Esmeralda and Coso, and in the antimonial ore of San Emidio.

Sulphur is abundant in California. It exists in large beds near the Geysers, in Sonoma county; near Clear Lake, in Napa county; in San Diego county, thirty miles northward from the town of that name, and twelve miles from the sea; near the sea-shore, fifteen miles eastward from Santa Barbara;

in the valley of Santa Clara River, near its head; and near San Juan Bautista, Monterey county. The most abundant and most accessible supply is that near Clear Lake.

Alum is found in Santa Clara county, eastward of San José; near Lancha Plana and Campo Seco, in Calaveras county; at the Geysers, and at Owen's Lake. At the two last-named places there are hot alum-springs.

Three miles above the forks of Clear Creek, in Shasta county, there are twenty salt-springs.

Springs strong with sulphate of magnesia, or Epsom salts, are found at the Geysers.

Chalk is found in Amador county and in Sonoma county.

Suisun marble, the most beautiful form of sulphate of lime, is found in Solano county. Gypsum is found in Santa Cruz and Amador counties. Fine specimens of alabaster have been obtained in El Dorado and Monterey counties.

Fine varieties of porcelain clay exist in many of the mining counties, particularly in Tuolumne. Clay, suitable for making fire-brick, may be obtained near Richmond, in Honey Lake valley.

Chromium is found on the bank of Feather River, near the mouth of Nelson Creek; near the town of Nevada; on the bank of Bear River, above Anson's Ferry; and on the ridge between the North and Middle Forks of the American River. Chrome iron is found in Monterey county.

During the summer and fall, in many parts of the state, a saline efflorescence covers the earth, or those low parts of it where water collects during the rains of winter. This efflorescence is composed chiefly of carbonate and borate of soda, mixed with various other salts. The largest deposit of this kind is probably found at Soda Lake, at the sink of the Mojave River.

California is very rich in borax, and the day is probably not far distant when we shall supply a large amount of it to commerce. It is found in springs in Tehama county, and in springs and lakes in Napa county. One of these lakes covers a hundred and fifty acres of ground, and is strong with the solution. In the mud at the bottom of this lake, the borax is found crystallized in large quantities. Boracic acid has been discovered in the sea-water near the coast.

§ 47. *Artesian Wells.*—There are a great number of artesian wells in California. In Santa Clara county, within a district six miles wide by fifteen long, there are three hundred and eighteen—more than are to be found in any other district of equal size in the world. Their water is nearly all used to irrigate land; some for manufacturing purposes. They supply about two million gallons in twenty-four hours. The wells are from fifty to four hundred feet deep; the bore varies from six to nine inches. Only a small portion of Santa Clara valley yields artesian water; the artesian district lies north of a line commencing at Mountain View; thence running nine miles with the road through the town of Santa Clara to San José; and thence southeast to the mountains. South of this line no artesian water is found.

It is supposed that the water comes from certain subterranean streams. One well has abundant water at one hundred feet; another, not more than one hundred yards distant, has no water short of three hundred feet. The wells throw up living fish and shell-fish, which are of different species in different wells. Some wells throw up soft-shell clams good to eat, and of a kind not found in the superterrene waters of the state, before the opening of these artesian supplies. One well throws up a snail, with a long spiral shell; another has snails with flat shells; and others have blind fish, evidently of a species that has lived long in subterrene waters, and lost its eyes because it had no use for them. Like the fish of the Mammoth Cave, in Kentucky, these artesian fish have the eye-socket and a blind eye in it. The wells that produce these fish and shell-fish are mostly shallow, not more than one hundred and fifty feet deep. If put into water fresh from wells two hundred and fifty or three hundred feet deep, they soon die, as do superterrene fish; either, it is supposed, because the water is too warm, or because it has not enough air in it. The deeper the well, the warmer the water.

Many of the wells have gone dry—"been drained by other wells," as people say; but yet how can one well "drain" another, the mouths of both being on a level with each other? The wells whose mouths are at a lower level may take water from those farther up the valley; but the theory that the water deserts one well, to flow out of another of equal or higher elevation, is not sound. There is very little difference of elevation, perhaps ten feet, between San José and Alviso; and the wells near the latter place throw their water about five feet higher above the surface than do those of the former. One cause of the failure of the wells may be the filling up of the pipes. From many of them great quantities of sand, gravel, and stones half a foot in diameter, have been thrown up; and if a large stone should happen to lodge crosswise in the pipe, other smaller stones and gravel might soon stop it up entirely, or break the force of the current so that the water could not rise to the top. In many cases the pipe has not been driven down to the foundation; and the water, whirling round at the bottom of the pipe, has torn away the earth and made an excavation, thus preparing the way for a caving in of the ground, and filling up of the well.

It is the general opinion in Santa Clara valley that the artesian wells have drained away the surface-water, and the soil is much drier than it was before the wells were bored. In 1849, Dr. Bascom found water west of Santa Clara by digging three feet; and since then he has been going deeper every year, until now his surface-well is fifty feet deep. In Pellier's garden, at San José, the surface-water was six feet below the surface in 1849; now it is fourteen. Ten years ago, there was a constant stream of water along the Alameda, between Santa Clara and San José; but that ditch has been entirely dry for several years. A multitude of such observations are mentioned; yet there is no conclusive proof that the artesian wells have taken away the surface-water. It seems that the soil began to get dry before these wells were bored. The artesian wells cannot draw the water from the soil immediately around them, for they throw their waters above the earth; it may be, however, that their supplies are derived from the soil in the upper part of the valley—supplies which, if the wells were not there, would not be drained away into subterranean channels, but would go to moisten the whole valley. It is to be observed that, at the very time when the soil of the Santa Clara valley was becoming so dry, a similar disappearance of the surface-water was noticed far beyond the influence of the artesian wells —Honey Lake, on the plateau of the Sierra Nevada, and Lake Elizabeth, in the Great Basin, both disappearing about the same time, in 1859; and several other little lakes and ponds in other parts of the country following their example, soon after.

There are artesian wells at various places in the state besides Santa Clara valley, but they offer nothing new in a geological point of view.

§ 48. *Paleontology.*—It is a general rule, that the animals of former geological eras, in any given district, appear to have been the gigantic ancestors of those of the present time. Thus the kangaroo and emu of Australia, found in no other part of the world, were preceded by gigantic kangaroos and emus, whose fossil remains are found in New Holland only. So, too, South America, in antediluvian times, had gigantic sloths and tapirs, akin to the animals now found within her limits. Each continent has a fauna of its own, to which its antediluvian animals were nearly akin. Every continent has several zoölogical districts; and the ancient and modern fauna of these districts are sometimes as clearly related to each other, and as distinctly separate from those of other parts of the continent, as are the fauna of different continents from each other. But the antediluvian animals of California possessed no peculiar relationship to the animals now indigenous in the state: the former fauna was totally distinct from that of the present age; the fossil bones found are not numerous, and no large and valuable skeletons have been brought to light—only fragments here and there. Of quadrupeds, we have the remains of a mastodon, an elephant, and a new species of horse. Of birds and reptiles, nothing noteworthy has been found. We have no entire fossil fishes, but a few teeth. Dr. W. O. Ayres found near Pit River the teeth of a shark, of the genus Lamna—a genus now extinct on this coast. There are numerous beds of marine shells, the most remarkable being on the shores of San Pablo Bay, on the sides of Mount Diablo, and on the slopes of the Sierra Nevada. In the bluffs of the coast, near the Lake House, are shells identical with those now found alive in the vicinity. All our fossils are of the tertiary period, save a few ammonites of the secondary era, found in the northern part of the state.

§ 49. *Relics of Early Humanity.*—In May, 1859, an Indian arrow-head was found, eighty feet below the surface of the earth, at Buckeye Hill, Nevada county. About the same time, another arrow-head was found three feet deep, in undisturbed alluvium, near Freeman's Crossing, in the same county.

In April, 1859, the skeleton of a man was found sixteen feet deep, at Tehaehepe, in Los Angeles county.

In October, 1855, two stone mortars, such as were used by the Indians for grinding acorns and grass-seeds, were found near Diamond Springs, El Dorado county, at a depth of one hundred feet below the surface.

In October, 1854, the skeletons of two men were found at Rattlesnake Bar, fourteen feet below the surface, and under ancient strata, which had apparently not been disturbed from the time of their deposition.

These are a few only of the fossil evidences that California has been inhabited by men many thousands of years.

§ 50. *Mineral Springs.*—Mineral springs are very numerous in California. The greatest number are found in the coast valleys, from latitude 40° southward to 32°. Nearly every little vale has one or more; many of them warm or hot. The most common temperatures range from 60° to 120°. Some of these springs yield a large quantity of water, and are in romantic sites, destined to become places of fashionable resort when our population grows dense. There are so many of these springs in the state that there is not room here to mention them all.

In San Bernardino valley there are a number of warm springs. Their temperatures are thus reported: 108°, 128°, 130°, 166°, 169°, and 172°. The heat of the springs at Aguas Calientes, in San Diego county, is thus given: 58°, 74°, 130°, 136°, and 140°.

Near Warner's ranch, in San Diego, is a spring with a temperature of 135°, rising from a cleft in the granite rock.

§ 51. *Cortes Shoal.*—About one hundred miles west of San Diego is Cortes Shoal, twenty miles long and three miles wide, with a depth of only fifteen feet in one place. This shoal is evidently the summit of a submarine ridge of mountains, parallel with the other ridges of the coast. The shoal was discovered in December, 1852, by Captain Cropper, of the steamship Cortes, who asserted that there was evidently a submarine volcano in operation there. The water was in violent commotion, and at intervals was thrown up into the air in columns; there was an escape of steam, and he suddenly found the depth of water change from forty-five to nine fathoms. He saw also light and smoke, and at one time the place looked as though it were a ship on fire. The general opinion is, that he saw only the waves breaking upon the Bishop Rocks, as the rocks at the shallowest place are called; but some persons adhere to his opinion of a submarine volcano.

NOTE.—The chief writers upon the Geology of California are W. P. Blake, J. S. Newberry, and Jules Marcon, in the United States Pacific Railroad Survey reports, and Dr. J. B. Trask's reports to the state legislature, and Jules Marcon's book on the Geology of North America. For the chemical fineness of the gold in the various mining districts, I am indebted to Henry Van Valkenburg.

CHAPTER IV.

SCENERY.

§ 52. *Introductory.*—California has much beautiful scenery. The atmosphere is remarkably clear, giving the eye a wide range. The mountainous character of the state not only prevents monotony and secures a rich variety of landscapes, but gives them extent and grandeur. The large rivers, the high snow-peaks and ridges, wide bays, forests of the largest and most graceful evergreens, parks of majestic oaks, natural meadows covered in the spring with brilliant grasses and flowers, are all magnificent in their kind. The valleys are mostly bare of timber, with here and there a grove of oaks, and lines of trees and bushes along the water courses. The coast valleys are very beautiful, and in the course of ten or fifteen years, when ornamented with thorough cultivation, will be as pretty as any places in the world. Sonoma, Napa, Amador, San Ramon, and Suñol valleys may be made as beautiful as any part of the world.

§ 53. *Coast Valleys.*—Napa valley, which is now the most beautiful of these valleys, because most thickly settled and most thoroughly cultivated, is thirty miles long, five miles wide at its mouth, gradually growing narrower toward the head. Napa River, a small stream, runs through the whole length of the valley, which is of level land, bounded on both sides by steep mountains, about two thousand feet high. These mountains, brown near the foreground and blue in the distance, oak groves, brilliant laurel and madroña, fields of wheat and barley, ploughed fields, good fences, elegant farm-houses, and numerous gardens and orchards, go to make up the landscape.

The valley should be seen from the mountain-top, whence it appears spread out as level as a floor. The fields, differing in color according to the season and their condition of cultivation, lie like a great checker-board, over which are scattered numerous farm-houses, and irregular streaks of timber marking the position of the river and its tributaries. The oak-trees form a most important part of the scene. They are wide in proportion to their height, thick in the trunk, heavy in the main boughs, many of which have a horizontal or downward course. The top of the tree has the semicircular shape, and the smaller branches have the pendant grace seen in the Eastern states only in the elm. The large upper boughs of the Californian white-oak have at their extremities some branches or twigs that hang perpendicularly down from three to twenty feet, and many of the trees for this reason look in the summer as though they were covered with vines. Add to these peculiarities the abundant gray Spanish moss, hanging like long and venerable beards from all the twigs and boughs, and the dark druidical mistletoe, and we have one of the most important and characteristic features of the Californian landscape.

Suñol valley, a little dale about three miles in diameter, nearly circular in shape, and shut in on all sides by mountains, is destined to become famous at some future day for its beauty. Now it is in a state of nature, but art will give it new charms.

The places in the state most visited on account of their natural scenery are the Yosemite valley, the big-tree groves, and the Geysers.

§ 54. *Yosemite Valley.*—Yosemite valley is a dell of matchless cliffs and cascades, with more scenes of grandeur and beauty than can be found within an equal space in any other part of the world. Shut in closely by walls of rock almost perpendicular, from two thousand to four thousand five hundred feet high, it has within a radius of five miles five cascades, one of which is two thousand feet high, another nine hundred and forty, another seven hundred, another six hundred, and another three hundred and fifty, and their waters flow through a natural meadow ornamented by beautiful trees and brilliant verdure.

The valley is a chasm in the Sierra Nevada, four thousand feet above the level of the sea, and distant about one hundred and twenty miles in a direct line from San Francisco, and in a nearly due eastward direction. It is watered by the main branch of the Merced River, which above and below makes its way through the mountains in deep and dark gorges, the bottom of which is rarely seen by the sunlight. The valley is ten miles long and nearly three wide in the middle, from which it decreases each way. It is bounded on all sides by walls of yellowish granite, from two thousand to four thousand feet high, in some places perpendicular, and everywhere precipitous. It is only at the ends of the valley that it is possible for travellers to get in or out of it, and even there the entrance and exit are difficult for horses and impassable for wagons.

The general course of the valley is east and west. The main entrance is at the western end, where a steep path leads down a descent of two thousand five hundred feet. The view from the ridge overlooking the valley is splendid. The chasm is seen winding away amidst the cliffs; a cascade is in sight, and numerous mountain-peaks rise in various directions. At the bottom of the dell are seen the meandering river, the green grass, and lofty trees diminished to the appearance of shrubs. The waterfall seen on the right several miles distant, is a mere white streak on the face of the rock, and does not appear grand in the least, but it is nine hundred and forty feet high, and becomes imposing as the traveller approaches it. The body of water is about seventy feet wide on the first of June. The fall is called the Cascade of the Rainbow, from the beautiful colors which always, in sunlight, adorn the mist floating about it.

Nearly opposite this cascade, on the northern side of the valley, and about three-quarters of a mile distant, but apparently much nearer when the tourist looks up at it, is the Capitan (or Captain), a rock which projects into the valley and rises up perpendicularly from the level green-sward three thousand and ninety feet. Continuing our course up the valley, we come soon to another high peak on the same side of the valley, known as the Signal Rock, two thousand nine hundred and twenty-eight feet high. Four miles above the Rainbow cascade we come to the great falls of the Yosemite, where the stream of that name, eighty feet wide, leaps down two thousand and sixty-three feet in three falls, of which the first is one thousand three hundred feet high, the next two hundred and fifty, and the third four hundred and fifty. About three hundred feet from the top of the upper fall there is a projecting ledge on which the stream breaks when the water is low, but up to the middle of June, while the current is large and swift with melted snow, the great body of the water leaps clear of the ledge, and pitches sheer down into the hill of rocks below. The Yosemite fall, sometimes called by the Indian name of "Cholook," is, in so far as height is concerned, the greatest cataract in the world; but it does not impress the observer like Niagara. The body of water, never large, is almost lost in spray before reaching the bottom; and in the late summer, the stream dries up entirely. Niagara is sublime, overwhelming the soul with the idea of power; Yosemite is beautiful and romantic—that is all. The tremendous precipices here, as throughout the valley, are greater and more impressive than the cascades, which have not enough water to confound. Besides, the falls cannot be approached from those points whence they might be seen to the greatest advantage; and looking from a distance, the Yosemite somewhat resembles a great sheet of white satin hanging over the cliff. But inferior as this one cascade is to Niagara, the valley, taking all its scenery together, is far superior in variety and romantic beauty, and equal in grandeur. A day or two at Niagara is enough; while a lover of nature may stay at Yosemite for months and continually find new delights in the study of the scenery. I have given the total height of the three falls of the Yosemite, all of which are very near together, at two thousand and sixty-three feet, which is the figure given by the official surveyor of that county; but others have estimated the height at two thousand three hundred and two thousand five hundred feet.

Across from the Yosemite Falls, on the southern side of the valley, is the Pyramid Rock, so named from the shape which it bears when seen from some points of view. It is three thousand two hundred feet high. Three miles further up and at the head of the valley is "Mirror or Tocoya Lake," a beautiful body of water covering about eight acres. The northern side of this lake washes the foot of the North Dome, a huge mountain of rock crowned with a dome-like knob, three thousand six hundred and thirty feet high; and near the southern edge of the lake is the perpendicular face of the South Dome, a still higher mountain, which rises up four thousand four hundred and eighty-one feet, towering above all the peaks in the vicinity. This peak is a sublime sight, with its perpendicular wall, which, as you look up at it, seems as if it would keep going up forever.

Winding back now along the southern side of the valley, we soon come to the southern fork of the Merced River, which rushes down through a gorge. We ascend this gorge on foot, climbing with great labor over rocks and through the brushwood, and at the distance of a mile and a half come to the Vernal or Canopah Falls, where the stream, about one hundred feet wide, falls three hundred and fifty feet into a basin surmounted by large evergreen trees. This cascade possesses one great advantage over all the others of the Yosemite valley, and that is, it can be approached from above, where we look down upon it from the top of the granite cliff, leaning over a natural parapet of rock, as convenient as though made expressly for the accommodation of picturesque tourists.

About half a mile above the Vernal Fall the river takes a another leap, called the Nevada or Awanee Falls, but it costs a mile and a half of roundabout clambering to get to it. The fall is seven hundred feet high, half of which the water shoots plumb down through the air, and strikes the projecting rock, breaking into spray.

About two miles west of Nevada Falls is the cascade of Tusayac, about six hundred feet high, but it is very difficult of access.

A few hundred yards above Lake Tocoya is Lake Tesahae, which has an area of about six acres, and is forty feet deep.

No description can convey a clear idea of the great variety of scenery in the valley. There are a thousand nooks and corners and woody dells, full of enchanting picturesqueness. The rocky cliffs take all manner of queer forms, resembling pyramids, castles, and domes, chimneys and spires. In one place there is a narrow cleft one hundred feet deep in one of the rocks, as though some giant had commenced to split off part of the mountain and had left his work unfinished.

The river, as it meanders through the valley, is a great addition to its beauty; and its waters, as well as those of the lakes, are clear as crystal in the summer, though turbid in the spring. Mountain trout are found in all these streams.

The climate of the valley is cool. The numerous cascades agitate the air, and near the fall there are often gusty winds.

There is much difference between the vegetation and temperature of the two sides of the valley; the northern side, where the sunshine is felt throughout the day, being much warmer than the shadows of the southern cliffs. Shrubs and flowers are in the full glory of foliage, and flower along the northern wall in May and June, while the same species are still bare or budding a mile or two to the southward; but the more delicate annual shrubs are usually more healthy on the southern than on the northern side of the stream, because those in the warmer spots are stimulated to come out so early as to be badly nipped by the frosts, which prevail here all through the spring and into the summer. The valley is almost inaccessible, on account of snow, before the middle of May, and the best time for a visit is in June. In the late summer and fall the quantity of water in these streams decreases greatly, and the Yosemite cascade becomes a mere trickling brooklet.

There are a couple of houses for the accommodation of travellers, but the fashionable way with those who visit the valley is to go in parties on horseback, provided with pack animals, carrying tent, bedding, provisions, and cooking utensils along. Ladies dress in the Bloomer style. Wagons do not come within forty miles of the valley. There are some quail and hare, but no larger game. The dell was inhabited by a warlike tribe of red men eight years ago, but they undertook to fight with the whites and have all been cut off, and scarcely a sign of their existence remains, save here and there the dim vestige of a trail.

The valley was first entered by white men in 1848, if rumor be true, and afterward in 1850 and 1852, but its wonders attracted no notice from the press, and were unknown to the public until 1854, and did not attract many visitors until 1856.

§ 55. *Mammoth Tree Groves.*—The next great natural wonder of California is the big-tree grove in Mariposa county. It is a grove of four hundred and twenty-seven mammoth trees, the largest of which are thirty feet in diameter and three hundred feet in height. This is the largest species of tree in the world, and this is the largest grove of them. The grove is about twenty miles from the Yosemite valley, and thirty miles southeast of the town of Mariposa, and about four thousand five hundred feet high on the western slope of the Sierra Nevada. When the traveller enters the grove he sees on all sides of him numerous giants of the forest, varying from twenty to thirty-four feet in diameter, and from two hundred and seventy-five to three hundred and twenty-five feet in height. Sublime sight! Each tree fills him with wonder as he looks at it. A glance at one of these immense trunks conveys a new idea of the magnificence of nature; "glorious as the universe on creation's morn" is this grove. The Titans and the gods fought with such tree-trunks as these for clubs, when the attempt was made to carry heaven by storm, as recorded in Grecian mythology. The trees are so high that you must look twice before you can see their tops, and then you must keep on looking before you can comprehend their height. The best way to see them is to lie down and look up, and remember that the spire of the New York Trinity Church, which is the highest artificial structure in the United States, towering far above all the rest of the American metropolis, though two hundred and eighty-four feet high, would be entirely lost to distant view if set down among these trees.

The grove covers a space half a mile wide and three-quarters of a mile long. Classifying its trees according to their size, we find that there is one tree thirty-four feet in diameter; two trees of thirty-three feet; thirteen between twenty-five and thirty-three; thirty-six between twenty and twenty-five; eighty-two between fifteen and twenty; making a total of one hundred and thirty-four trees between fifteen and thirty-four feet in diameter; and then there are two hundred and ninety-three between one and fifteen feet through.

One very large tree has fallen, and a considerable portion of it has been burned; but appearances indicate that it was nearly forty feet in diameter, and four hundred feet high.

The Mammoth Tree is a cone-bearing evergreen, belonging to the botanical genus named *Cupressus* (cypress) by Linnæus. After the time of that naturalist, his genus of the *Cupressus* was divided; so that the Mammoth Tree would have come under the head of the *Taxodium*, which, about the year 1850, was again divided by Endlicher, the German botanist, and the redwood-tree was declared to belong to a new genus, called *Sequoia*.

In 1853, the mammoth trees first came to the notice of the public. The botanists in San Francisco, engaged in the turmoil of business, looked at the specimens, but had not time to examine them, and supposed them to be of the same species with the redwood, to which the mammoth tree certainly does bear a very close resemblance. Thinking the tree, however, to be very remarkable on account of its great size, they sent some of its cones, leaves, and wood, to botanists in New York, but they were unfortunately lost on the way. A few months later, an English collector sent some specimens to Professor Lindley, who not only found the tree to be of a new species, but determined to make a new genus of it, and he affixed to it the name *Wellingtonia gigantea.* When the news of the selection of this name arrived in California, a foolish and pretentious fellow, who meddled with matters of science of which he knew nothing, wrote a ranting article against Lindley, for trying to confer the honor of the great tree of America upon a Briton like Wellington, and declaring that the only proper title for the tree would be *Washingtonia gigantea.* If there had been any bad taste in conferring the name of a Tory and a man of blood upon such a magnificent tree, still the rules of botanical nomenclature are well established, and the matter of the name is left entirely to the discretion of the man who first gives a technical description of the plant and determines its genus. American botanists, therefore, never recognized the name *Washingtonia*, because Lindley's name was of undoubted priority; and to acknowledge the priority, and yet recognize the *Washingtonia*, would be equivalent to proving their own stupidity. And yet English botanists have, in scientific records, accused American botanists and "Americans" of making an agitation to establish the name as *Washingtonia.* These facts are part of the history of botany, and facts of interest relating to the big trees.

The general opinion among botanists is, that Lindley was wrong in declaring the mammoth tree to be of a new genus: it is a *Sequoia*, related in the closest manner to the redwood. When the redwood and the mammoth tree come to be held as of a distinct genera, then nearly every difference heretofore considered merely specific may be made the basis for establishing new genera. Dr. Seeman called the mammoth tree the *Sequoia gigantea*, and it bears that name with botanists generally.

The *Sequoias* are found only in California; the *Sequoia*

gigantea only on the western slope of the Sierra Nevada, between latitudes 34° and 41°. The tree has the great peculiarity that it bears two kinds of leaves: those on the young trees, and on the lower branches of larger ones, are about five-eighths of an inch long and an eighth wide, and are set in pairs opposite each other, on little stems; the other kinds of leaves, growing on the branches which have borne flowers, are triangular, about an eighth of an inch long, and they lie close down to the stem. The cones are not much larger than a hen's egg, whereas the cones of many smaller conifers of the coast are larger than pine-apples. The seeds of the *Sequoia gigantea* are not more than a quarter of an inch long, a sixth wide, and almost as thin as writing-paper. The bark is reddish-brown in color, of a coarse, dry, stringy, elastic substance, and very thick—on the largest trees not less than eighteen inches. The wood is soft, elastic, straight-grained, free-splitting, light when dry, and red in color. It bears a close resemblance to red cedar, but the grain is not quite so even. The wood is very durable.

The mammoth tree grows in a deep, fertile soil, and is always surrounded by a dense growth of other evergreens, such as various species of pine, fir, spruce, and Californian cedar. The scenery in these forests is beautiful. The trees grow very close together; and the trunks, usually from a foot to two feet in diameter, rise in perfect perpendicularity, and with little or no diminution of size, more than a hundred feet without a limb: and while all is perfect stillness and rest and shadow on the ground, the traveller, looking to where the sunbeams are perceptible here and there on the thick foliage, can see the flexible tops swinging from side to side in the roaring mountain-breeze. The soil, being never visited by the sun, is always moist, and produces a luxuriant and beautiful little undergrowth of mosses, flowers, and berries. When in such forests, I have at times compared myself to a merman, who, while at the bottom of the ocean, amid a large growth of queer sea-weed, and surrounded by beautiful shells and the treasures of a thousand wrecks, should look from his abode of peace, and see the surface of the water, far above him, raging in a terrific storm.

Many young trees of the *Sequoia gigantea*, produced from the seed, are growing in gardens in California, in the Eastern states, and in Europe.

The mammoth tree is found only in a few small groves, of which six or seven are known, though probably there are many in unexplored parts of the Sierra Nevada. Three of these groves are in Mariposa county, one in Calaveras, one in Tuolumne, and one in Tulare.

The three Mariposa groves are within two miles of each other. The second one in size contains eighty-six trees; the third thirty-five. The Tuolumne grove contains ten trees, one or two of which are said to be thirty-five feet in diameter.

The Calaveras mammoth grove was the first discovered, and attracts the greatest number of visitors. There are in this grove ten trees thirty feet in diameter, and eighty-two between fifteen and thirty, making ninety-two over fifteen feet through. One of the trees, which is down, must have been four hundred and fifty feet high and forty feet in diameter. The "Horseback ride," one of the notabilities of the place, is a hollow trunk, which a man can ride upright through on horseback, seventy-five feet.

In 1854, one of the largest trees, ninety-two feet in circumference and three hundred feet high, was cut down. Five men worked twenty-two days in cutting through it with large augers. On the stump, which has been smoothed off, there have been dancing-parties and theatrical performances; and for a time a newspaper, called the *Big Tree Bulletin*, was printed there.

At the same time that this tree was cut down, another was stripped of its bark for a distance of one hundred and sixteen feet from the ground. This tree continued green and flourishing two and a half years after being thus denuded, and did not begin to show signs of dying until a very hard frost came in the winter of 1856–'57. Although seven years have passed since its bark was stripped off, some of its branches are yet green.

A section of bark and part of the wood of the felled tree are now in the English Crystal Palace. The rings of this tree were counted; and its age was variously estimated, according to the different methods of counting, at from nineteen hundred to three thousand years. Probably its age was about two thousand years. It sprouted while Rome was in her glory. It is older than any kingdom, language, or creed, of Europe or America. It was a large tree before the foundation of the Christian Church, and was fifteen hundred years old before the period of modern civilization began. Twenty centuries look down upon the tourist from the tops of the larger trees; and some of the little ones will still flourish for a thousand years from now, when all our present kingdoms and republics shall have disappeared, and our political and social systems shall have been swept away as full of evil, and replaced by other and better systems, under which men will live in civilized society without each being forced to rob his brother by means more or less legal and respectable.

In many of the trees in all the groves, hollows are burned at the foot, and some of them have been burned so as to stand on three legs. One of these, in the Calaveras grove, called "Uncle Tom's Cabin," has an open space under it of more than a dozen feet square. The largest trees seem to end abruptly at the top, having been broken off by the snow, which often falls to a great depth so high up on the Sierra Nevada. The trees, in some places, grow very near together; in others, they are comparatively far apart; and occasionally two or three will be seen which are united at the ground, although they may have been twenty or thirty feet apart when they sprouted.

It is said that the big-tree grove of Tulare county is eight miles long, and contains larger trees than either Calaveras or Mariposa, the largest measuring one hundred and twenty-three feet in circumference twelve feet above the ground. We have, however, no detailed description of this grove.

§ 56. *Geysers.*—The Geysers, in the northern part of Sonoma county, are among the wonders of the state. They are in a deep and steep ravine, amid a district filled with the marks of violent volcanic action. Down the western slope of the mountains which separate Clear Lake from the basin of Russian River, runs a stream called the Pluton River; and near this are the Geysers, a multitude of springs, boiling with heat, and emitting large quantities of steam, with a hissing, roaring, and sputtering noise. Near them are many tepid and cold springs, which add to the wonderful character of the place. Hot and cold springs, quiet and boiling springs, are found within a few feet of each other. And then the waters differ as much in taste, odor, and color, as in temperature and action. One is almost as fetid at times as rotten eggs; another has black water, resembling ink; a third is called the "Eye-water Spring," and its waters are reputed to be excellent for curing sore eyes and cutaneous diseases; and the waters of others are strongly purgative. The ground in the ravine is in places deeply covered with the minerals deposited by the springs: among these, sulphur, sulphate of magnesia (Epsom salts), sulphate of aluminum (alum), and various salts of iron, predominate.

The chief feature of the Geysers is called "The Steampipe," an orifice about eight inches in diameter, in the hill-side, from which rises a large volume of steam to a height varying from fifty to two hundred feet. The steam roars continuously, sometimes bursting out in puffs louder than that made by the general-escape pipe. It deposits flowers of sulphur on the objects which come within its range.

"The Devil's Punch-Bowl," called also "The Witches' Cauldron," is in a large hole, six feet across, in the hill-side. The liquid in the bowl is black and thick, and is always in commotion with the heat, and the vapor from it deposits black flowers of sulphur on the rocks around.

The sides of the cañon are bare, and smoking with heat. The Geysers are a favorite place of resort for pleasure-seekers in the state. They are seventeen hundred feet above the level of the sea.

§ 57. *Mud-Volcanoes.*—In the Colorado Desert, about latitude 33° 25′, and longitude 115° 45′, are some remarkable mud-volcanoes. They are in that part of the desert below the level of the sea; and if the water of the ocean were turned in upon that low land, they would be lost to sight. As it is now, they are very rarely visited, because they are in a region so desolate, that an excursion to them is accompanied by serious hardships. The volcanoes cover a space a quarter of a mile long and an eighth of a mile wide; this area is of soft mud, through which hot water and steam are constantly escaping. The noise can be heard at a distance of ten miles, and the steam is visible at a greater distance. The quantity of water thrown up is small; that of the steam great. The vapor rises steadily in some places, with a hissing noise; in other places it bursts out with the noise and action of an explosion, throwing the mud a hundred feet into the air, with a loud report.

There are places where the mud is in constant movement, and rises in great bubbles, and bursts as if boiling with intense heat; while in other places regular cones, apparently hardened into permanency, and with shapes varying from low hillocks to sharp points, have been formed. There are boiling springs which throw up their water twenty or thirty feet; and there are large basins, one hundred feet across, and five or six feet below the general surface, in which a bluish paste is continually boiling. Some of the springs are surrounded by incrustations and arborescent concretions of carbonate of lime; others are encircled by deposits of sulphur. The air blown from the salses is fetid with sulphur. It is very dangerous to approach the springs and cauldrons, because the whole earth is soft in the vicinity of them, and frequently the crust is broken and thrown up with great force, to establish new springs, steam-vents, and mud-cauldrons; and the boiling slime or water thrown up on these occasions would suffice to kill a man in a few seconds.

§ 58. *Santa Cruz Ruins.*—Fifteen miles northeastward from the town of Santa Cruz are "The Ruins," as they are called—forty and odd perpendicular cylinders of sandstone, from a foot to two feet in diameter, with holes from six to fourteen inches wide running through them. These cylinders were discovered in 1855, in a bed of sand, on the side of a sandstone mountain, and were at first supposed to be the remains of some work of human hands: whence their name of "The Ruins." Much curiosity was excited by their discovery, and a number of men were employed to dig away the sand, so as to expose the foundation on which the cylinders stood. The excavation was carried down in one place to the depth of forty feet, and the base of the column was found to rest on the bed-rock sandstone. The surface of the rock was sloping and rough, and there was nothing to indicate the work of man. It is now supposed that the cylinders were deposited by mineral springs, although it is believed that no similar columns have been formed elsewhere, the elevations made by mineral springs being, with this exception alone, shaped like hillocks or cones—never like cylinders. The theory of deposition by springs may be the best mode of explaining their existence, but it is not satisfactory: the cylinders rise perpendicularly, or nearly so, and are very little thicker at the base than at the top; some of them preserve the same thickness from bottom to top. The material of the shafts differs from that of the bed-rock by being coarser and darker. And besides, the texture appears in places to have a spiral form, as though it had been made of a thick paste, rolled up spirally into a cylinder, and then hardened into a solid; leaving, however, a plain trace of the manner in which it was made. And some pieces, which have been broken off, suggest such a mode of formation.

§ 59. *Mirage.*—Among the most remarkable scenes witnessed in California are the illusions of the mirage in the deserts of the Colorado and the Great Basin. "All the phenomena of mirage," says Professor W. P. Blake, "are exhibited on a grand scale upon the Colorado Desert. Mountain-ranges, so far distant as to be below the horizon, are made to rise into view in distorted and changing outlines. Inverted images of smaller objects, and apparent lakes of clear water, are often seen, and invite the traveller to turn aside for refreshment. The first exhibition of mirage that was seen [by Blake's party] was from the margin of the plain at Carriso Creek, looking toward the Gila, about ninety miles distant. It was early in the morning, and the eastern sky had that golden hue which precedes the rising sun. Tall blue columns, and the spires of churches, and overhanging precipices, seemed to stand upon the verge of the plain. Their outlines were changing gradually, and, as the sun rose higher, they were slowly dissipated. After reaching Fort Yuma, and witnessing the strangely precipitous and pinnacled outline of the mountains beyond, it was at once apparent that the mirage consisted of their distorted images. When we were upon the northern part of the desert, the peak of Signal Mountain was often distorted and raised above the horizon. The points of distant ranges also seemed at times to be elevated above the surface, precisely as the headlands of a coast sometimes appear to rise above the water at sea.

"Many of the phenomena called mirage are not due to refraction, but are believed to be the result of reflection from the sand, or smooth surface of clay, or the polished pebbles. The smooth clay forms an excellent reflector for all the rays which are incident at a slight angle, and is most frequently the cause of the appearance of water. The beautiful surface of the pebbly plain may be regarded as a combination of myriads of reflectors; for each pebble is so highly polished, that it reflects light almost like a mirror. The reflection from such a brilliant surface, when seen at a favorable angle, looks like a sheet of water, the similarity being heightened by the motion of the stratum of heated air in contact with the surface."

The phenomena of mirage are frequently witnessed in the Sacramento Basin, and also in the coast valleys, on warm, dry days.

§ 60. *Caves.*—There are a number of caves in California. Of these the most noted are the Alabaster Cave, seven miles from Auburn, in Placer county; the Bower Cave, twelve miles from Coulterville, in Mariposa county; the Cave of Skulls, in Calaveras county; and the Santa Cruz Cave, two miles from the town of Santa Cruz. The Alabaster Cave has two chambers: one about one hundred feet long by twenty-five wide; the other two hundred feet long by one hundred wide. It contains a large number of brilliant stalactites and stalagmites. The Bower Cave has a chamber one hundred feet long by ninety wide; it is reached by an entrance seventy feet long, and in one place only four feet wide. The Santa Cruz Cave has no beauty to render it attractive. The Cave of Skulls is remarkable for having contained, when first discovered, a number of human skulls and bones, all covered with layers of carbonate or sulphate of lime, from the thickness of a leaf to an inch. These bones are now in the cabinet of the Smithsonian Institute. At Cave City, and seven miles from Murphy's, in Calaveras county, is a cave in which a Know-Nothing lodge was accustomed to meet in 1855. In the bluff bank of the Middle Fork of the Cosumnes River, eighty feet above the stream, is a cavern, called Limestone Cave, with many intricate passages and some fine stalactites.

§ 61. *Waterfalls.*—Besides the cascades of the Yosemite valley, there are a number of others in the state. There is a cataract, about five hundred feet high, on Fall River, which empties into the Middle Fork of Feather River; one of three hundred and eighty feet, where the South Fork of the American River slides down over a convex rock, looking like a streak of snow when seen from a distance; one of sixty feet, in the San Antonio River, in Calaveras county; another of seventy-five, on the same stream, which falls fourteen hundred feet within a mile; and one of three hundred feet, called the "Riffle-box Falls," in Deer Creek, Nevada county.

California has five natural bridges. The largest of these is on a small creek emptying into the Hay Fork of the Trinity

River, where a ledge of rock three hundred feet wide crosses the valley. Under this rock runs the creek, through an arch twenty feet high by eighty feet across. The rock above the arch is one hundred and fifty feet deep. On Lost River, in Siskiyou county, there are two natural bridges, about thirty feet apart. The rock is a conglomerate sandstone, and each is from ten to fifteen feet wide, and the distance across the stream is about eighty feet. One of these bridges is used regularly by travellers. On Coyote Creek, in Tuolumne county, ten miles northward from Sonora, are two natural bridges, half a mile apart. The upper bridge is two hundred and eighty-five feet long with the course of the water, and thirty-six feet high, with the rock thirty feet deep over the water. The lower bridge is similar in size and height to the other.

§ 62. *Solfataras.*—In the northeastern part of Plumas county are many hot springs—perhaps numbering one thousand—covering an area of ten acres. They roar and hiss so as to be heard at a distance of a mile, and their steam can be seen from a greater distance. The whole place smells strongly of sulphur, which mineral, as well as alum and various earthy salts, abound in the soil about the springs.

In four or five places in California the earth is constantly hot, and sulphureous gases and vapors are always escaping. There is such a solfatara about fifteen miles eastward from Santa Barbara; another near Owen's Lake; another near the Geysers, in Sonoma; and another near the hot springs in Plumas county. It was rumored in 1858 that there was an active volcano in Plumas county, near Lassen's Peak, but there is no satisfactory proof of its existence, though there is a portion of country in that vicinity of which very little is known.

§ 63. *Mount Shasta.*—One of the best opportunities for romantic adventure in the state is in the ascent of Mount Shasta. Several parties have gone to its summit—no trifling undertaking. The ascent is very difficult; the sides of the peak are steep and rugged. The distance from the southern foot of the mountain to the summit is estimated at fifteen miles. Four miles from the summit there is a bench of trap-rock; seven miles farther is a second bench, of red cement or lava; three miles farther is a third bench, of black lava and obsidian. Near the second bench there are several lakes in cavities which were once probably craters. One of them certainly was once the vent of volcanic action. On the extreme summit of the mountain are a number of basaltic columns, looking like chimneys. The scenery is very grand; for, as the peak is fourteen thousand four hundred feet high, and towers far above all the mountains around it, the view has no limit in any direction save a very remote horizon. The Klamath, Trinity, Scott, Rogue, Pit, and Sacramento valleys are all visible, besides Lassen's Peak, the Downieville Buttes, the Marysville Buttes, the Three Sisters in Oregon, and so on. About a hundred yards west of the summit there are a dozen steaming-hot sulphur springs, and the earth about them is so hot as to be unpleasant. The air is so rare at the summit of Mount Shasta, that some persons ascending it have been troubled while there with dizziness, headache, spitting of blood, and difficulty of breathing.

CHAPTER V.

BOTANY.

§ 64. *Peculiar Fauna and Flora.*—California has a botany and zoölogy of her own. Her indigenous plants and animals are peculiar to her soil. Her plants, her quadrupeds, her birds, and her fishes, are different from those of other countries. The Californian vulture is, next to the condor of South America, the largest bird that flies; and he might easily migrate to other parts of the continent, but he makes his home only in this state, and is certainly never seen east of the Rocky Mountains. The grizzly bear might travel almost as well, but he is found only in California and Oregon. The Californian deer is different from that of Virginia in horns, teeth, feet, color, and size. The bird known as the roadrunner or *paisano* might fly to all parts of the continent, but is found only west of the Sierra Nevada. There is *a* blue-jay here, but it differs from the bird known to the New-Englanders as *the* blue-jay. The robin of New England differs from the robin of Old England, and the Californian robin differs from both. The sturgeon of the San Francisco market are not the same with those eaten in New York; and one species found in California is not found in a state so near as Oregon. Our trees are like, and yet are unlike, those of the Atlantic states and Europe. We have oak and pine, spruce, sycamore, and horse-chestnut trees, and yet any observant man sees at a glance that they differ in many important particulars from the trees known by those names elsewhere. California, with a little of the country adjacent, is a distinct botanical district. Her vegetation was first produced on her own soil, and has not been derived from or communicated to any other district by the course of nature.

§ 65. *Distribution of Plants.*—Most of the Sacramento and San Joaquin valleys, the Colorado Desert, the eastern slopes of the Coast Mountains, and the Coast Range south of latitude 35°, are treeless; the Sierra Nevada and the western slopes of the Coast Range north of 35°, have fine forests; and in the foot-hills of the Sierra Nevada, and in the coast valleys, there are beautiful open groves of oak-trees. The timber of the Sierra is mainly spruce, pine, and fir; that of the coast, north of 37°, redwood; and spruce and pine south of that latitude.

§ 66. *Superiority of Conifers.*—The botany of California is remarkable for containing a number of the largest and most beautiful coniferous trees in the world, growing to a height of three hundred feet and a thickness of eight and ten feet in the trunk, and some of them still larger. Among these gigantic glories of the vegetable kingdom are the mammoth tree, the redwood, the sugar-pine, the red fir, the yellow fir, and the arbor-vitæ, or *Thuja gigantea.* Other large conifers contribute to the magnificence of our forests. We have the laurel, the madroña, the evergreen-oak, and the nut-pine (*Pinus sabiniana*), evergreen trees with a growth resembling that of deciduous trees. Our deciduous trees are few, and of little value to the mechanic.

The mammoth tree (*Sequoia gigantea*) was described in the preceding chapter.

§ 67. *Redwood.*—The redwood (*Sequoia sempervirens*) is the second in size and the first in commercial value of all the trees in California, though not much superior to the sugar-pine in either respects. It grows only within thirty miles of the ocean from Monterey to Crescent City, and is never found out of the state. It bears a remarkable resemblance in color and texture of wood and bark, and color, form and distribution of foliage to the mammoth tree, to which it is not much inferior in size. A redwood-tree called "Fremont's tree," in Santa Cruz county, is two hundred and seventy-five feet high, and nineteen feet in diameter six feet above the ground; and others equally large are found in the northwestern part of the state. Trees two hundred and fifty feet high and eight feet through are not rare. The wood is very straight-grained, free-splitting, durable, soft, and light. So freely does it split, that boards twenty feet long, eight inches wide, and an inch thick, are sometimes made from it with the frow. No wood in the world splits so beautifully and regularly. There is no better wood for the general use of the farmer, and it is the chief building material of the coast. No timber is more durable either above or below ground. The color is a rich dark-red, which, when varnished, makes a fine appearance in furniture. The tree grows in dense forests, which contain an immense amount of timber. Thus, on the plain southeast of Crescent City, there are hundreds of acres of land of which every fifteen feet square, on an average, supports a tree three feet through and two hundred and twenty-five feet high—a statement that may appear incredible to those who have seen only the forests east of the Mississippi River. These trees will often furnish twenty saw-logs, each ten feet long, and every acre will supply material to make one million feet of sawn lumber, which, at the low rate of fifteen dollars per one thousand feet, is worth fifteen thousand dollars. The redwood stump, after the tree has been cut down, throws out a number of shoots, one or two of which choke down the weaker ones and become large trees. A redwood forest is almost inexterminable.

§ 68. *Pines.*—The sugar-pine (*Pinus lambertiana*) is the most magnificent tree of all the pine kind, and indeed it has no superior in the vegetable creation, save the mammoth and the redwood, the confessed monarchs of the plant kingdom. It is closely related to the white pine (*Pinus strobus*) of the Eastern states; "though," as Dr. Newberry says, "like all the conifers on the Pacific coast, it exhibits a symmetry and perfection of figure, a healthfulness and vigor of growth not attained by the trees of any other part of the world." The mature tree sometimes reaches a height of three hundred feet and a diameter of twenty, but it rarely exceeds two hundred and ten. The young trees of the sugar-pine give early promise of the majesty to which they subsequently attain. They are unmistakably young giants; even when having a trunk a foot in diameter, their remote and regularly-whirled branches, like the stem covered with a smooth, grayish-green bark, showing that, although so large, the plant is still "in the milk," and has only begun its life of many centuries. The sugar-pine conspicuously exhibits one of the most general and striking characteristics of the conifers—the great development of the trunk at the expense of the branches. Nearly the whole growth is thrown into the trunk, which generally stands without a flaw or flexure, a perpendicular cone, all its transverse sections accurately circular, sparsely set with branches, which, in their insignificance, seem like the festoons of ivy wreathing about the columns of some ancient ruin. The leaves are three inches long, dark bluish-green in color, and they grow in groups of five. The foliage is not dense. The cones are large, sometimes eighteen inches long by four thick. The wood is similar to that of the white pine—white, soft, homogeneous, straight-grained, clear, and free-splitting. It furnishes the best lumber in the state for the "inside work" of houses, and is the chief building material used in the Sierra Nevada, where it grows. The tree derives its name from a sweet resin which exudes from the duramen or hard wood of the tree. This resin is sugar-like in appearance, granulation, and taste, and could not be distinguished from the manna of the drug-stores except by a slight terebinthine flavor. The pine sugar is cathartic. It is found in small quantities only, though it is said one hundred and fifty pounds of it were collected by a man who devoted himself for a few weeks to the business of gathering it.

The Western yellow-pine (*Pinus ponderosa*) is a noble tree, next in size among the pines of California to the sugar-pine. It sometimes reaches a diameter of seven feet. Its leaves grow in threes at the ends of the branches, giving the foliage a peculiarly tufted appearance. The color of the leaves is a dark yellowish-green. The bark is of a light yellowish-brown or cork color, and is divided into large, smooth plates from four to eight inches wide and from twelve to twenty inches long, whereby the tree may be recognized at a distance. The tree is found near the snow-line in the Sierra Nevada, and east of the summit, and northward to Washington Territory.

The nut-pine (*Pinus sabiniana*) is remarkable as a conifer for its spreading top, and for its large cones full of edible seeds. It branches out somewhat after the manner of a maple; rarely more than sixty feet high, though often with a trunk four feet through—a thickness of trunk that with most other conifers would give more than double the height. About half way from the ground to the top, the trunk divides into a number of branches, which grow upward. The nut-pine is found in the lower part of the Sierra Nevada, and in the coast mountains near the head of the Sacramento valley. The seeds are larger than the common white bean, and are very palatable, with a slight terebinthine taste. The leaves are from four to ten inches long, and grow in threes. The foliage of the tree when seen from a distance, resembles that of the willow, both in color and distribution. In places where the nut-pine is found, the woodpeckers select them as storehouses for their winter food, cutting holes in their bark and putting an acorn in each. The Indians formerly relied upon the nuts for a considerable portion of their food. They climbed the tree by catching hold of the rough, strong bark with their hands, then putting their feet against the tree, without touching it with their body or knees, they walked up till they reached the limbs.

The twisted pine (*Pinus contorta*) is found in the northern part of the state. The leaves are yellowish green in color, about two inches long; and they grow in pairs. The tree does not exceed sixty feet in height.

Coulter's pine (*Pinus coulterii*) grows in the Santa Lucia mountains. It reaches a height of one hundred feet, and has a trunk three feet through. Its branches are large and spreading, the leaves a foot long and pale sea-green in color; the cones seventeen inches long, seven inches through, and like a sugar-loaf in shape.

§ 69. *Firs.*—The red fir, or Douglas spruce (*Abies douglasii*), is a tree of very large size, growing to be three hundred feet high and ten feet thick in the trunk. It is, as Dr. Newberry says, "one of the grandest of the group of giants which combine to form the forests of the West." The wood is strong, but coarse and uneven in grain; the layers of each year's growth being soft on one side and very hard on the other. The timber is much used for rough work in houses, and for ship-building. The tree grows in dense forests on the Sierra Nevada and Cascade Mountains, from 35° to 49°, and near the coast north of 39°.

The yellow-fir or Williamson's spruce (*Abies williamsonii*) bears a close resemblance to the red fir, and the two trees are usually found in company with each other.

The black fir (*Abies menziesii*) is smaller and of little value.

The *Abies bracheata* (Santa Lucia fir) grows in the Santa Lucia mountains. The height is about one hundred feet, the shape a perfect cone, the lowest branches resting on the ground. The tree produces a resin used by the Catholic priests for incense.

The Western balsam-fir (*Picea grandis*), or white fir, attains a height of one hundred and fifty feet, and a diameter of seven feet in the trunk. The bark on the trunks of the young trees contains numerous cysts full of the resinous fluid called the balsam of fir.

§ 70. *Cedars.*—The Western juniper, or cedar (*Juniperus occidentalis*), bears a strong resemblance to the juniper (*Juniperus virginianus*) of the Eastern states. Its wood, however, is white in color. It grows to be about thirty feet high. The wood of a juniper-tree found near the quicksilver mines of New Idria is so hard and fine in texture, that it would probably be valuable to engravers.

The Californian white cedar (*Libocedrus decurrens*) grows one hundred feet high, and seven feet thick in the trunk. It is found from Mount Shasta to the Tejon Pass. The trunk is usually angular. Many of the trees are affected with a dry-rot which destroys their value as timber.

The fragrant-cedar (*Cupressus fragrans*) is found along the northern coast of the state. It is a large tree, and produces a white, clear lumber, valuable for furniture and the inside work of houses. The wood has a strong, lasting, and not unpleasant odor, half way between turpentine and ottar of roses.

Lawson's cedar (*Cupressus lawsoniana*) is a tree of little value.

The arbor-vitæ, also called cedar (*Thuja gigantea*), is a most symmetrical and graceful conifer, growing to be nearly three hundred feet high.

§ 71. *Yew and Nutmeg.*—The Western yew is an upright tree, from fifty to seventy-five high, with thin and light foliage, the leaves being about an inch long. Its growth is straighter, its branches fewer, and its foliage thinner, more feathery, and lighter in color, than the European yew. It grows on the Sierra Nevada from 34° northward to British Columbia.

The coast cypress (*Cupressus macro-carpus*) is found only on Cedar Point, at Monterey, and there are not more than one hundred trees of it there. The foliage is very dense.

The Californian nutmeg (*Torreya californica*) is a graceful and beautiful evergreen found in the Coast Mountains near the bay of San Francisco. It grows from fifty to seventy-five feet high, and resembles the Western yew in foliage and general form. The fruit is like a nutmeg in size and shape, but it has a disagreeable terebinthine taste, and is never used as a condiment.

§ 72. *Laurel.*—The Californian laurel, or bay (*Oreodaphne californica*), is one of the most common and beautiful trees of the coast valleys. It is an evergreen, which grows to a height of fifty feet, with a trunk sometimes thirty inches in diameter.

The leaves are dark green, lustrous, four inches long, one inch wide, sharp at both ends, with smooth edges. The foliage is dense. The wood is grayish in color, very hard, durable, and difficult to split. Both leaves and wood have an aromatic odor, which is stronger in the former; and becomes still stronger when the leaves are bruised. The odor resembles that of bay-rum. It gives the headache to some sensitive persons.

§ 73. *Madroña.*—The madroña (*Arbutus menziesii*) is one of the most striking trees of the Californian forest. It is an evergreen, with an open growth, somewhat like that of a maple, bright-green and lustrous leaves, and a bright-red bark. Its height is sometimes fifty feet; its diameter in the trunk two feet. The leaves are oval in shape, three inches long, pea-green underneath, and dark and shining above. The bark is smooth, and it peels off at regular seasons; the new bark is a pea-green, which changes to a bright red. The wood is very hard, and is used to some extent in the arts, especially for making the wooden stirrups commonly used in the state. The tree bears a bright-red berry in clusters, of which the birds are fond.

§ 74. *Manzanita.*—The manzanita (*Arctostaphylos glauca*), another prominent feature in the Californian forest, is a dense, clump-like shrub, which grows as high as twelve feet, and nearly as broad as it is high. The trunk divides near the ground into several or many branches, and these terminate in a great multitude of twigs, so that the shrub is a dense mass of branches and branchlets, all of which are very crooked. The wood is dense, hard, and dark-red in color. The bark is red and smooth, occasionally peeling off and exposing a new light-green bark, which soon turns red. The leaves are regularly oval in form, about an inch and a half long, thick and shining, and pea-green in color; they set vertically upon their stems. The manzanita bears a pinkish-white blossom in clusters, and these are replaced by round red berries about half an inch in diameter; they have a pleasant, acidulous taste, and are often eaten by the Indians and grizzly bears, but there is too little meat on them to pay white men for the trouble of gathering them. The shrub grows in the coast valleys, and in the Sierra Nevada, up near to the limit of perpetual snow. The name means "little apple," *manzana* being the Spanish for apple.

§ 75. *Ceanothus.*—The ceanothus, sometimes called the Californian lilac, of which there are many species, is a beautiful evergreen shrub, growing about ten feet high, with clusters of lilac-like flowers, of various shades of blue, violet, and red, according to the species. The tree produces a multitude of little twigs, and a dense foliage, and may be trimmed into almost any shape.

§ 76. *Oaks.*—The Californian white oak (*Quercus hindsii*), or long-acorned oak, is a very large tree, and the characteristic oak of California. It resembles the white oak of the Atlantic slope in the color of its bark and the shape of its leaves; but its growth is very different. It seldom reaches a greater height than sixty feet, and is often wider than high. Sometimes it measures one hundred and twenty-five feet from side to side. The trunk, which occasionally grows to be eight feet through, throws out large horizontal boughs within ten feet of the ground, and above that point the trunk is soon lost among the large branches. The tree furnishes no straight timber, and the wood is so soft and brittle as to be of no use in the arts; whereas the white oak of the Mississippi valley is a most valuable tree, with a trunk so tall and straight, that sills and beams of it sixty feet long are common, and with a wood so tough, that it supplies all the axles and plough-beams of the country. The Californian white oak is not even fit for fence-rails. The tree, however, is very beautiful and majestic, and the open groves of it in the valleys and foot-hills form, as Dr. Newberry says, "the most important element in those scenes of quiet beauty which so often excite the admiration of the traveller in California." The tree bears much resemblance in form and size to the oak of England, the groves of it appear-

ing like the English parks. At the ends of the large boughs are branches which hang down like vines—giving the tree, when seen from a distance, something of the appearance of an elm. The acorns are large, sometimes two and a half inches long. They once formed the chief article of food of the Californian Indians.

The evergreen oak (*Quercus agrifolia*) is a low, spreading tree, much like an apple-tree in size and shape. The foliage, however, is darker and denser. The acorns are small, thin, and sharp-pointed. The wood is hard, crooked in grain, and valuable for knees in ship-building.

The Californian chestnut oak (*Quercus densiflora*) is a low, handsome evergreen tree, with a leaf very much like that of the chestnut. The bark is very rich in tannin, and is extensively used for the tanning of hides. The tree is rare north of latitude 39°, and is most abundant in the mountains about Santa Cruz.

The Western chinquapin (*Castanea chrysophylla*), or golden-leaved chestnut, is an evergreen shrub that grows in the Sierra Nevada. At the height of three feet it bears an edible and palatable fruit, something like the beechnut in shape, but larger. The flowers and ripe fruit are often found on the same bush. The leaves are dark-green above, and covered with a yellowish powder beneath. The Western chinquapin grows to be a tree thirty feet high in some parts of Oregon.

The fulvous oak (*Quercus fulvescens*) is a deciduous tree, that grows about thirty feet high, with leaves somewhat like those of the Western chinquapin. The acorn, when young, is concealed in the cup, the two together resembling a little wheel; but the acorn, when mature, is an inch and a half long, and projects considerably beyond the cup. The wood is tougher than that of most of the oaks of California.

Kellogg's oak (*Quercus kelloggii*) is a large deciduous tree, found only in California. Its leaves are deeply sinuate, with three principal lobes on each side, terminating in several acute points. It bears fruit only in alternate years. or at least most

abundantly every other year. An idea prevails that the acorns give to swine a disease of the kidneys.

The huckleberry-leafed oak (*Quercus vaccinifolia*) is a shrub, from four to six feet high, which grows on the mountains in the northern part of the state. Its leaves, in size and form, resemble the huckleberry; the acorn is of the size and shape of a small hazel-nut.

§ 77. *Buckeye.*—The Californian horse-chestnut, or buckeye (*Æsculus californica*), is a shrub, or low, spreading tree, abundant in the Sacramento, San Joaquin, and coast valleys. It likes to grow about rocky ledges, in ravines, and on the banks of streams. Sometimes it throws up a dozen stems, which grow to a thickness of three or four inches each; but usually it has one trunk, six or eight inches through. The tree rarely exceeds fifteen feet in height, and it has a hemispherical shape, very dense foliage, rising from the ground in a globular form. It continues to put forth large clusters of fragrant blossoms from early spring till late summer. The leaves are among the first to open of the deciduous trees of the state. Five leaves grow together on one stem. The fruit has a close resemblance to that of the buckeye-tree of the Mississippi valley, but is larger and more abundant. It is a staple article of food with those few Californian Indians who still depend upon wild fruits and game for their subsistence.

§ 78. *Sycamore.*—The Mexican sycamore (*Platanus racemosa*) exhibits a striking resemblance to the Western sycamore of the Atlantic slope. It has the same straggling, irregular growth; the same smooth, white, scaly bark; the same large, yellowish leaf: but instead of having only one ball on a stem, like the Atlantic sycamore, it has several, the stem running through one or two, and terminating in the last one.

§ 79. *Pitahaya.*—The pitahaya (*Cereus giganteus*), a gigantic cactus, is one of the most prominent features of the botany of the deserts in the southern part of California. It grows to a height of fifty feet, with a trunk thirty inches in diameter. Sometimes the trunk has no boughs, but usually it throws out from two to six, which are about half the thickness of the trunk; they run out horizontally for a foot or two, and then turn upward and rise parallel with the trunk. There are no twigs or leaves, but flowers and fruit grow on the tops of the trunk and branches. The whole plant resembles a huge candelabrum. The flowers are three inches long, as wide, with stiff, curling, and cream-colored petals. The fruit is as large as a hen's egg, and the meat is a red pulp, full of little seeds. The taste is insipid; but when the fruit is dried, according to the Indian custom, it acquires a flavor somewhat like that of a fig.

§ 80. *Yucca.*—The yucca, or bayonet-tree, is a kind of palm, —an endogenous tree that lives in the southern deserts. It sometimes grows to be thirty-five feet in height, with a trunk two feet through; but usually it is about ten feet high, with a trunk eight inches in diameter. It has no twigs or branches, but sometimes it divides into two trunks. The foliage, consisting of leaves eighteen inches long, and shaped like the blade of a bayonet, hangs down from the tops of the trunks.

§ 81. *Mezquit.*—The mezquit (*Algarobia glandulosa*) is a low tree of the Colorado Desert. It sometimes reaches a height of twenty feet, with a trunk fifteen inches in diameter. The lower branches are very near the ground, and the whole tree has a very regular, semispherical form. The leaves are like those of the black locust, and the foliage thin. The tree bears numerous pods, from three to five inches long, full of sweet, nourishing beans, about the size of the common white bean. The mezquit-bean is often eaten by men, and horses and mules are very fond of it.

The curly mezquit (*Strombocarpus pubescens*) is a similar shrub, and bears a crooked bean, called the "screw-bean." It also grows only on the desert.

§ 82. *Miscellaneous Trees and Shrubs.*—A few walnut-trees grow along the Sacramento River, and it is said that some chestnuts have been found in Mendocino county, but they are unknown in the greater part of the state. We have no indigenous beech, elm, hickory, locust, acacia, or sassafras. Our wild cherry and wild plum are bushes, but their fruits resemble the wild plums and cherries of the East. We have willows and cottonwood, which differ little in appearance from those of the Mississippi valley. There are wild grapes, blackberries, gooseberries, huckleberries, raspberries, salmon-berries, and strawberries. A truffle, or a root resembling it, is found in the valleys of the coast and the Sierra Nevada. The grizzly bear considers it a delicacy, and frequently digs it up. A shrub called the "joint-fir" (a species of *Ephedra*), sometimes used for making tea, is found in Calaveras and Tuolumne counties. In the valleys of the Coast Mountains is found the *yerba buena* (Spanish for "good herb"), a creeping vine, bearing some resemblance in its leaf and vine to the wild strawberry. It has a strong perfume, half-way between peppermint and camphor. The *yerba de la vibora* (Spanish for "rattlesnake-herb," known to botanists as the *Daucus pusillus*) is a carrot-like vegetable, the leaves of which are said to be a specific for the bite of the rattlesnake.

§ 83. *Poison Oak.*—The poison oak, or poison ivy (*Rhus toxicodendron*), grows abundantly in the Sacramento Basin, and along the coast. It thrives best on a moist soil, and in the shade. In a thicket with other bushes it sends up many thin stalks eight or ten feet high, with large, luxuriant leaves at the top; in the shade, the leaves are green. In the open, dry ground, exposed to the sun, and without support from other bushes, the poison oak is a low, poverty-stricken little shrub, with a few red leaves. If it can attach itself to an oak-tree, it becomes a parasitic vine, and attains a thickness, though very rarely, of four inches in the trunk, and climbs to a height of forty feet. The touch of the leaf is poisonous, and causes a very irritating eruption of the skin. It rapidly communicates by the touch from one part of the body to another, causing severe inflammations and swellings. The most delicate parts of the body are most affected by the poison. The eyes are sometimes closed up entirely by the swelling round them; and many cases are recorded of faces so swollen, that they could not be recognized by intimate friends. Some persons are not affected by the tc uch of the Rhus; but instances have occurred wherein persons supposing themselves, after long experience, to be free from danger, have at last been poisoned: and when the poison has once taken hold, the system is always very easily affected from that time forward. Even passing to the leeward of the bush on a windy day, or going through the smoke of a fire in which it is burning, will bring the poison to the surface again.

§ 84. *Amole.*—The amole (*Chlorogalum pomeridianum*), or soap-plant, has an onion-like, bulbous root, which, when rubbed in water, makes a lather like soap, and is good for removing dirt. It was extensively used for washing, by the Indians and Spanish Californians, previous to the American conquest. The amole has a stalk four or five feet high, from which branches about eighteen inches long spring out. The branches are covered with buds, which open in the night, beginning at the root of the boughs, about four inches of a branch opening at a time. The next night, the buds of another four inches open, and so on.

§ 85. *Nutritious Herbage.*—Of indigenous nutritious grasses there are a number in the state. The wild oat, though not a grass, may be mentioned under this head. It resembles the cultivated oat so nearly, that there has been some doubt whether they are not identical, but the opinion among botanists is that they are a distinct species. The wild oat, in the year 1835, was found only south of the bay of San Francisco; but about that time, when the white men crossed frequently from the southern to the northern side of the bay, the oat was sown in a natural way by horses and cattle, and it spread rapidly over the Sacramento valley and the coast region. It grew very luxuriantly, and in some places surpassed in the height, size, and abundance of stalks, any field of cultivated oats which I have ever seen. It is said that in some localities the oat-stalks were so high, that men sitting erect on horseback could not see each other at a distance of ten feet. The soil and climate were evidently very favorable to it. During the last six or eight years, the wild oats have been eaten down so closely by cattle, that in many places they have been killed out. They are propagated from year to year, not by the roots, but by the seeds, many of which fall into cracks in the earth, where they lie in safety until the rains come, when the ground closes up and the grain sprouts. The earth cracks in the summer, in many parts of the state; and in places where the wild oats grow, the position of the cracks of one year may be traced the next season by the position of the stalks of the grain.

The wild oat grows on hill and plain, and furnishes a large part of the wild pasture of the state. It is wholesome, nutritious, and palatable for cattle. Much of it is cut for hay. The amount of grain which it furnishes is small in proportion to the quantity of straw, and it is never threshed.

After the wild oats, in importance to the herdsman, comes the "burr-clover," so named from a spherical burr, about a quarter of an inch in diameter, which it bears in clusters of three. This burr-clover is found in all the settled parts of the state. Cattle do not like it when green; but after it dries, the burrs fall upon the ground, and are picked up by the cattle, while the stranger is astonished at seeing them eating and keeping fat on what appears to him to be bare earth. On examining the surface of the ground, he will find that it is covered with the dry stalks and burrs of the burr-clover. The bloom consists of three very small yellow flowers. It is said that the stalks of this clover take root whenever the joints touch the ground.

The alfilerilla (*Erodium cicutarium*) is another indigenous nutritious herb of much importance to the herdsman. It is succulent, sweet, hardy, bearing clusters of spikes or pins an inch and a half long. These spikes have given it the name of pin-grass; and the resemblance of its leaves to the geranium has suggested the name of "wild geranium," by which title it is also known to some persons. It has a large root, which it sends deep into the ground, thus enabling it to resist the drought, while above the surface it puts forth a dense mass of stalks and leaves, spreading out sometimes several feet in every direction. Cattle prefer it to every other indigenous herb of the state. The seeds seem to abound throughout the soil, for wherever the earth is ploughed up for the first time, there the alfilerilla appears, though it may never have been seen there before. It is common in gardens, cultivated fields, and fallow lands.

The white Californian clover has a large yellowish-white bloom, from an inch to an inch and a half in diameter, a beautiful plant, well suited as an ornament for yards and gardens. It grows very large, and two feet high in moist, favorable situations; while in dry places it will also mature its seed without rising more than two or three inches above the ground. It is very sweet, and it is often eaten by the Indians, who like it both raw and boiled. Cattle are also extremely fond of it.

Another species of clover has a round bloom about a third of an inch in diameter, composed of violet-tinged flowers.

Another clover has a bloom from a sixth to a quarter of an inch in diameter, the flowers of which are subdued green, tipped with pink at the end.

The *Melilotes officinalis*, another herb, commonly called a clover, though not strictly entitled to that name, likes a very moist soil, and then grows luxuriantly, crowding out nearly every thing else. Its bloom consists of a drooping head about an inch long and a sixth of an inch thick, hung with little yellow flowers. Cattle are not fond of this herb in any shape, but they like it better in hay than when green.

Of nutritious grasses there are a number, but they do not form a sod. The drought of summer and fall seems to kill the roots.

Of wild flowers there are a great variety and abundance in California, and they have their different seasons for blooming; and in cañons where the soil is always moist, flowers may be seen in every month of the year. In the spring-time the hill-sides are frequently covered with them, and their red, blue, or yellow petals hide every thing else. Each month has its flowers: in March the grass of a valley may be hidden under red, in April under blue, and in May under yellow blossoms. There is such a variety that within an hour I have counted twenty species on a spot not more than twenty feet square. This was on dry, sandy soil, in Sonoma valley, in the month of May. None of the flowers are large, brilliant in color, or rich in sweet, strong perfume.

The tulé is a reed which covers all the large tracts of swamp lands in the state. It has no leaf, but a plain, round stalk, varying from half an inch to an inch and a half at the butt, and tapering gradually to a point. It is usually not more than eight or ten feet high, but at the Tulare Lake it grows to fifteen or twenty feet.

The grass and herbage begin to grow and clothe the landscape in green after the first heavy rains of the rainy season. These rains may come in December, January, or February; and until they do come, the earth, in the districts not covered with timber, is brown. The grass continues green until June, when it begins to dry up and turn yellow and brown, which colors then predominate in the landscape until the rains come again. The death of the grass, except at high elevations, is caused not by the cold but by the drought; and in those months when the prairies of Indiana and Illinois are covered with snow, the valleys of California are dressed in the brilliant green of young grass.

The mistletoe grows abundantly on the oak-trees of California. The Spanish moss, which hangs in long lace-like gray beards from the branches, also serves to give beauty to the groves in the valleys.

NOTE—Most of my information about the botany of the state has been derived from the reports of Dr. J. S. Newberry, in the United States Pacific Railroad Survey, and from the conversation of Dr. A. Kellogg, Dr. H. Behr, and Mr. H. G. Bloomer, of San Francisco.

CHAPTER VI.

ZOOLOGY.

§ 86. *General List.*—Among the indigenous animals of California are the grizzly bear; the black bear; the panther, the wild-cat; the gray wolf; the coyote; three foxes; the badger; the raccoon; the opossum; the mountain-cat; the weasel; two skunks; one porcupine; three squirrels; two spermophiles; two ground-squirrels; three rats; three jumping-rats; one jumping-mouse; nine mice; one mole; the elk; one deer; one antelope; the mountain-sheep; three hares; two rabbits; the seal; the sea-otter; the sea-lion; the beaver; two vultures; the golden eagle; the bald eagle; the fishhawk; eighteen other hawks; nine owls; the road-runner; twelve woodpeckers; four humming-birds; eleven flycatchers; one hundred and nine singers; one pigeon; two doves; three grouse; three quails; one sandhill crane; forty-one waders; sixty-six swimmers, including two swans and five geese; about two dozen snakes, including the rattlesnake; half a dozen salmon; two codfish; and one mackerel.

§ 87. *Bears.*—The grizzly bear (*Ursus horribilis*) is the largest and most formidable of the quadrupeds of California. He grows to be four feet high and seven feet long, with a weight, when very large and fat, of two thousand pounds, being the largest of the carnivorous animals, and much heavier than the lion or tiger ever get to be. The grizzly bear, however, as ordinarily seen, does not exceed eight hundred or nine hundred pounds in weight. In color the body is a light grayish-brown, dark brown about the ears and along the ridge of the back, and nearly black on the legs. The hair is long, coarse, and wiry, and stiff on the top of the neck and between the shoulders. The "grizzly," as he is usually called, is more common in California than any other kind of bear, and was at one time exceedingly numerous for so large an animal; but he offered so much meat for the hunters, and did so much damage to the farmers, that he has been industriously hunted, and his numbers have been greatly reduced. He ranges throughout the state, but prefers to make his home in the chaparral or bushes, whereas the black bear likes the heavy timber. The grizzly is very tenacious of life, and he is seldom immediately killed by a single bullet. His thick, wiry hair, tough skin, heavy coats of fat when in good condition, and large bones, go far to protect his vital organs; but he often seems to preserve all his strength and activity for an hour or more after having been shot through the lungs and liver with large rifle-balls. He is one of the most dangerous animals to attack. There is much probability that when shot he will not be killed outright. When merely wounded he is ferocious; his weight and strength are so great that he bears down all opposition before him; and he is very quick, his speed in running being nearly equal to that of the horse. In attacking a man, he usually rises on his hind-legs, strikes his enemy with one of his powerful fore-paws, and then commences to bite him. If the man lies still, with his face down, the bear will usually content himself with biting him for a while about the arms and legs, and will then go off a few steps and watch him. If the man lies still, the bear will believe him dead, and will soon get tired and go away. But let the man move, and the bear is upon him again; let him fight, and he will be in imminent danger of being torn to pieces. About half a dozen men, on an average, are killed yearly in California by grizzly bears, and as many more are cruelly mutilated.

Fortunately, the grizzly bear is not disposed to attack man, and never makes the first assault unless driven by hunger or maternal anxiety. The dam will attack any man who comes near her cubs, and on this account it is dangerous to go in the early summer afoot through chaparral where bears make their home. Usually a grizzly will get out of the way when he sees or hears a man, and sometimes, but rarely, will run when wounded. It is said that grizzlies in seasons of scarcity, used to break into the huts of the Indians and eat them. No instance of this kind, however, has been reported for some years past.

The greater portion of the food of the grizzly is vegetable, such as grass, clover, berries, acorns, and roots. The manzanita, service, salmon, and whortleberries, are all favorites with him. The roots which he eats are of many different species, and it was from him that we learned the existence of a Californian truffle, very similar to the European tuber of the same name. The grizzly is very fond of fresh pork, at least after he knows its taste, and if swine come within his reach, he soon learns the taste. The farmers in those districts where the bears are abundant, shut up their hogs every night in corrols or pens, surrounded by very strong and high fences, which the bears frequently tear down. After having killed a hog, if any part of the carcass is left, the grizzly will return the next night and feast upon the remains, and go until it becomes putrid. He prefers, however, the fresh pork if it can be had. Not unfrequently the grizzly discovers the carcasses of deer, elk, and antelope, killed by hunters, who have gone off for horses to carry their game home. In such case, the hunter usually finds little left for him when he gets back. They do not like climbing, and rarely attempt to ascend trees. The grizzly, though he often moves about and feeds in the day, prefers the night, and almost invariably selects it as the time for approaching houses, as he often does, in search of food. The cub is one of the most playful, good-humored, and amusing of animals. He will tumble somersets, sit up on his haunches and box, and in some of his pranks will show a humor and intelligence scarcely inferior to that of very young children. The grizzly may easily be tamed, and it becomes very fond of its master. Adams, the Californian mountaineer and bear-hunter, trained several grizzlies so that they accompanied him in his hunting excursions, defended him against wild animals, and carried burdens for him. The meat of the young grizzly resembles pork in texture and taste, exceeding it in juiciness and greasiness; but the meat of the old he-bear is extremely strong, and to delicate stomachs it is nauseating.

The black bear (*Ursus americanus*) is found in the timbered portions of California, but is not abundant. It is more often seen near the coast north of Bodega than in any other portion of the state. Dr. Newberry, speaking of the food of the black bear on this coast, says: "The subsistence of the black bears in the northern portion of California is evidently, for the most part, vegetable. The manzanita, wild plum, and wild cherry, which fruit profusely, and are very low, assist in making up his bill of fare. Rarely, too, we saw trees of yellow-pine bearing marks of bears' teeth, where they had torn off the outer bark to get at the succulent inner layer, which is capable of sustaining life, and to which the Indians very generally have recourse when pressed with hunger." It is believed that neither the grizzly nor the black bear hybernates in California.

§ 88. *Panther and Wild-Cat.*—The panther of California, supposed by Dr. Newberry to be the *Felis concolor*—the same with the panther found on the Atlantic slope of the continent—has a body larger than that of the common sheep, and a tail more than half the length of the body. Its color is dirty-white on the belly, and elsewhere a brownish-yellow, mottled with dark tips on all the hairs. The panther is a cowardly animal, and, except when driven by some extraordinary motive, never attacks man. A friend of mine, who was out hunting, dressed in a buff coat, was creeping through some brush to get near a deer, when he felt a heavy animal strike his back. He sprang up very suddenly, and saw a panther, which had jumped down upon him from a tree, probably mistaking him for a calf or a deer. The brute seemed very much astonished and frightened at seeing a man there, and immediately fled at full speed. The panther is nocturnal in his habits, and always prefers the night as a time for attacking colts, which are a favorite prey with him. He is found in all parts of the state where there is timber, but he never stops long in any place, unless he can find bushes to hide in.

The American wild-cat (*Lynx rufus*) is common in California, particularly in the vicinity of the bays of San Francisco and San Pablo, where he often catches fish and water-fowl as well as land-animals. His color is a light brown, with dim, dark spots on the sides, and longitudinal lines along the middle of the back.

§ 89. *Wolves and Foxes.*—The gray wolf (*Canis occidentalis*) is found in all the inhabited parts of California, but is not abundant.

The coyote is very common in the state, and occupies the same place here with that occupied in the Mississippi valley by the prairie-wolf. Dr. Newberry thinks the two belong to the same species (*Canis latrans*), but I am inclined to believe that they are specifically different. The color of the coyote has more of a reddish tinge, he howls more, does not bark so much, and is more cunning. His food consists chiefly of rabbits, grouse, small birds, mice, lizards, and frogs; and in time of scarcity he will eat carrion, grasshoppers, and bugs. He is very fond of poultry, pigs, and lambs, and will destroy almost as many of them as would a fox. He is one of the worst enemies and most troublesome pests of the farmer. His method of catching chickens is to hide near the hen-roost about daylight, and, as the hens come down, he pounces upon them from his hiding-place; and his motions are often so quick, that the victim has not even time to squall before she dies. In the spring and autumn, when wild geese and ducks are abundant, many coyotes make their homes in the tules, where they catch the birds which have been wounded by the hunters.

The coyote loves nothing better than a young pig. When he sees an old sow with her young ones, he will hide, and wait a long time, in hopes that a little one will come within his reach; but if there be no hiding-place, he goes up boldly. The sow will at once face the assailant, and start to attack him. He allows her to come up within a few feet of him, and then moves off slowly; and she, like a fool, thinking she will catch him, continues the chase. While running, he keeps his head turned to one side, partly to watch her, and partly to watch the pigs; and when he has seduced her far enough away, he suddenly makes a dash at the pigs, and, getting one of them, runs off with it, leaving the agonized and furious sow far behind. If the coyote does not succeed in getting a pig at the first attempt—that is, if he does not lead the sow far enough away—he tries it again and again, till he succeeds, the sow being so stupid as to follow him, after having repeated opportunities to see his purpose.

The coyotes frequently go in packs, and sometimes will undertake to attack a cow. On such occasions, they have a concerted plan of operations: they surround their intended victim, and while those in front rush at her as a feint, those behind attempt to cut her hamstrings. As their teeth are very sharp, they often succeed. The cow's hamstrings once cut, she falls, and is completely at their mercy; and they quickly pick her bones.

The coyote is a great thief, and will steal the pillow from under a sleeping man's head; for it happens in California that bags of provisions are often used as pillows. When the coyote is hungry, he will gnaw any thing that is greasy, and for that reason he frequently cuts off the hemp and raw-hide ropes with which horses are tied out at night; but he never bites into hair-ropes, which for that reason were formerly used exclusively for staking out horses.

The coyote is nocturnal in his habits, and is very fond of howling or yelping. He begins with a shrill, quick bark, and follows up with a succession of yelps, ending in a long-drawn, quavering, melancholy howl. When one begins, all others within hearing take up the cry. Ten years ago, the traveller in the Sacramento valley rarely passed a night without hearing their music. They are not so numerous now, but still they are frequently seen in the most densely-settled parts of the country.

The red fox (*Vulpes fulvus macrourus*) is found north of latitude 37°; the gray fox (*Vulpes virginianus*) in all the timbered parts of the state. The coast fox (*Vulpes littoralis*) is found only on the island of San Miguel, off the coast of Santa Barbara. In its color it bears a great resemblance to the gray fox, but it is not more than half as large, is less cunning, and is slower in its motions. Its tail is only one-third the length of its body. The specimens observed were very bold and stupid, allowing themselves to be caught, over and over again, in the same manner.

The desert fox (*Vulpes macrourus*), which is found in the central deserts of the continent, crosses the Sierra Nevada, and is often killed in Calaveras and Tuolumne counties.

§ 90. *Badger, etc.*—The American badger (*Taxidea americana*) is abundant in the plateau of the Sierra Nevada, and is occasionally found in other parts of the state. It is very shy, and is rarely seen by the traveller.

The black-footed raccoon (*Procyon hernandozii*) is found in the timbered portions of the Pacific slope of our continent from Santa Barbara to British Columbia. It is longer than the Atlantic raccoon (*Procyon lotor*), but resembles it very closely in its mental character and capacity, habits and appearance. The raccoon is fond of grapes, and when he enters a vineyard selects those of the finest flavor.

An opossum (*Didelphys californica*) is found in the wooded portions of the state, but is not abundant.

The yellow-haired porcupine (*Erethizon epixanthus*), a native of California, is the largest of its genus. The spines are a couple of inches long, yellowish in color, with brown tips. On the lower part of the sides the spines are replaced by long, stiff bristles.

The mountain-cat, or striped bassaris (*Bassaris astuta*), is abundant along the western base of the Sierra Nevada, between latitudes 36° and 39°. The body is about the size of that of the domestic cat, but the nose is very long and sharp, and the tail very long and large. The color of the animal is dark gray, with rings of black on the tail. The miners call it the "mountain-cat," and frequently tame it. It is a favorite pet with them, becomes very playful and familiar, and is far more affectionate than the common cat, which it might replace, for it is very good at catching mice.

The pine-marten (*Mustela americana*) is found in California, but is rare.

The yellow-cheeked weasel (*Putorius xanthogenys*) is found along the coast, in the vicinity of the bay of San Francisco.

The common mink (*Putorius vison*) has a skin as valuable as that of the beaver; the fur is of a dark, brownish, chestnut color, with a white spot on the end of the chin.

California has two skunks (*Mephitis occidentalis* and *Mephitis bicolor*), very common animals. The *Mephitis bicolor*, or little striped skunk, is chiefly found south of latitude 39°; the other in the northern and central parts of the state. The colors of both are black and white.

§ 91. *The Squirrel Family.*—The Californian gray squirrel (*Sciurus fossor*), the most beautiful and one of the largest of the squirrel genus, inhabits all the pine-forests of the state. Its color on the back is a finely-grizzled bluish gray, and white beneath. At the base of the ear is a little woolly tuft, of a chestnut color. The sides of the feet are covered with hair in the winter, but are bare in the summer; the body is more slender and delicate in shape than that of the Atlantic gray squirrel. It sometimes grows to be twelve inches long in the head and body, and fifteen inches long in the tail, making the entire length twenty-seven inches. Dr. Newberry says: "The Californian gray squirrel is eminently a tree-squirrel, scarcely descending to the ground but for food and water, and it subsists almost exclusively on the seeds of the largest and loftiest pine known (*Pinus lambertiana*), the 'sugar-pine' of the Western coast. The cones of this magnificent tree are from twelve to sixteen inches in length, and contain each one hundred or more seeds of the size and shape of the small white bean of commerce. These cones would be unmanageable by the squirrel in the tree, and he has the habit, so common in the family, of dropping them to the ground, where he can dissect them at leisure. This he usually does early in the morning, climbing to the extremities of the topmost branches, where the cones hang, and cutting off a sufficient number to supply his wants for the day. He then descends, and, commencing at the base of the cone, tears off the scales in rapid succession, and skilfully possesses himself of the seeds which they conceal. He is compelled, however, to supply other wants than his own, for the smaller pine-squirrel (*Sciurus douglasii*) and the ground-squirrel (*Tamias townsendii*) appropriate a large share of his booty. When oak-trees are near, and acorns are ripe, he has recourse to them for subsistence; as often as opportunity offers, robbing the woodpeckers of their stores, in which also he has the active co-operation of his more diminutive congeners. From the fact that he feeds upon the ground, it has been supposed that he was less active and less fitted for climbing than most tree-squirrels. This, I think, is not true. He is exceedingly quick and graceful in his movements; and if less frequently seen to spring from tree to tree than the black and gray squirrels of the eastern states, it is because he inhabits coniferous trees, which are remarkable for the insignificance of their branches compared with the size of the trunk, the limbs never stretching out and interlocking, as those of the oak and maple and other trees, in which our common species live."

The Californian pine-squirrel (*Sciurus douglasii*) inhabits the pine and redwood forests of the state. He is gray above and red beneath, with a black stripe separating the two colors. He lives in a burrow or hollow log, but climbs well, and obtains his food chiefly from the pine-cones, which he cuts off in numbers at a time, and tears to pieces at his leisure, after they have fallen to the ground. He lays up a store of the seed in his burrow, for his winter supply. He is quick in his motions, graceful in his attitudes, and shy in his habits.

The Missouri striped ground-squirrel has five dark-brown stripes on the back, separated by four gray stripes; the sides are reddish-brown, the belly grayish-white, and the tail rusty-black above and rusty-brown beneath. The animal is four or five inches long. It is found in the northern parts of the state. It eats acorns and the seeds of the pine, manzanita, and ceanothus, in the thickets of which last-named bush it prefers to hide its stores.

The *Spermophile* has two species in California, which resemble each other so closely, that they are usually supposed to be the same; they are popularly known as the Californian ground-squirrels, the little pests which are so destructive to the grain-crops. Their bodies are ten or eleven inches long in the largest specimens; the tail is eight inches long, and bushy; the ears large; the cheeks pouched, and herein consists the chief difference between them and squirrels; the color above black, yellowish-brown, and brown, in indistinct mottlings, hoary-yellowish on the sides of the head and neck, and pale yellowish-brown on the under side of the body and legs. They dwell in burrows, and usually live in communities in the open, fertile valleys, preferring to make their burrows under the shade of an oak-tree. Sometimes, however, single spermophiles will be found living in a solitary manner, remote from their fellows. Their burrows, like those of the prairie-dog, are often used by the rattlesnake and the little owl. Dr. Newberry says: "They are very timid, starting at every noise, and on every intrusion into their privacy dropping from the trees, or hurrying in from their wanderings, and scudding to their holes with all possible celerity; arriving at the entrance, however, they stop to reconnoitre, standing erect, as squirrels rarely and spermophiles habitually do, and looking about to satisfy themselves of the nature and designs of the intruder. Should this second view justify their flight, or a motion or step forward still further alarm them, with a peculiar movement, like that of a diving duck, they plunge into their burrows, not to venture out till all cause of fear is past. Should you in the mean time have seated yourself with your back against a tree, and have remained for a time as immovable as the trunk against which you lean, you will soon see sundry little heads protruding from the burrows, with as many pairs of eyes and ears skilled to detect the least sign of danger from their equally-feared enemies, the coyote, the Californian vulture, the red-shouldered and red-tailed hawk, and man himself. If, however, your silence and quietness persuade them that you are none of these, they will swarm forth from their holes, and at first timidly, but, gaining confidence, more fearlessly, engage in all the sports and antics for which the *sciuridæ* are noted, and in which none excel the species under consideration. It is a pretty sight, and one to which I have often treated myself, to sit down quietly under these old oaks, and watch the squirrels running about over the grass and trees, gambolling and playing together. As far as the eye could reach through the vista, the sprightly movements of these innocent animals could be discerned."

The two species are called Beechey's spermophile (*Spermophilus beecheyi*) and Douglas's spermophile (*Spermophilus douglasii*). The size, habits, and general appearance of the two species are the same, but they differ in the color of a stripe along the spine from the base of the head to the middle of the back: in Beechey's spermophile it is yellowish-hoary, in Douglas's it is dark-brown. The former species is found very abundantly south of the straits of Carquinez; the latter north of it, and fewer in number.

Beechey's spermophiles are among the most formidable enemies of the farmer in those districts where they make their homes. They increase very rapidly in the vicinity of farms, and do great damage in grain-fields and gardens; they eat grain and garden vegetables in all stages of their growth; they peel young fruit-trees and vines; they are, in short, dangerous to nearly every thing that is cultivated. They are very industrious, and lay up large stores for the winter, spending several hours every pleasant summer's day in gathering food.

They go considerable distances to fields; and the traveller, whose approach scares them, sees them in hundreds running across the road before him, with their tails erect, hurrying from the field to hide themselves in their burrows. Many a large wheat-field, which would have yielded forty bushels to the acre if there had been no spermophiles to trouble it, is so despoiled by them, that the crop will not pay for harvesting. They are particularly abundant in the Santa Clara, Amador, and Pajaro valleys; and their number is an important consideration in the estimate of the price of land. They will not live in moist land, nor very near the ocean, where the fogs prevail. They are poisoned with strychnine and phosphorus, drowned by irrigation, and shut out by tight board-fences. In wet winters many of them are drowned; after a dry winter they are always numerous. Away from cultivated fields they depend for food chiefly upon grass-seeds, grass-roots, and acorns.

The Californian gopher (*Thomomys bullivorus*) is, next to Beechey's spermophile, the most abundant and most troublesome rodent of the state. When full grown, it has a body six or eight inches long, with a tail of two inches. The back and sides are of a chestnut-brown color, paler on the under parts of the body and legs; the tail and feet are grayish-white; the ears are very short. In the cheeks are large pouches, covered with fur inside, white to their margin, which is dark-brown.

The gopher inhabits the fertile valleys of the coast from latitude 34° to 39°. He spends nearly all his time under ground, and does most of his mischief there, gnawing off the roots of fruit-trees and garden vegetables, eating newly-sown grain and seeds, and nibbling at flowers and sweet bulbs. He is not a climber, nor is he very agile: if he gets into a trench eight inches wide and a foot deep, with perpendicular sides, he will run a long distance in it rather than clamber out; and one of the best methods of catching him is to make such a trench round a field, and place square tin boxes, fifteen inches deep, eight inches square, and open at the top, in the bottom of the trench, at regular intervals of about fifty yards. The gopher frequently travels at night, and when he comes to this trench he falls in. He then runs along in it, looking for a convenient place to get out, and, coming to a tin box, falls into that. He can neither jump out nor gnaw through; so he remains a prisoner till he starves to death, or the farmer comes along and kills him.

The Colorado gopher (*Thomomys fulvus*) is found in that portion of the state south of latitude 34°, but is not abundant. It is smaller than the Californian gopher, and has more of a reddish tinge in its colors. Its habits and appearance otherwise are very similar to those of its northern congener.

The broad-headed gopher (*Thomomys laticeps*), found in the vicinity of Humboldt Bay, is about five inches long. Its color on the back, sides, and belly, is yellowish-brown, with a reddish tinge between the fore legs.

§ 92. *The Rat Family.*—California has a number of indigenous jumping-rats, jumping-mice, and other rats and mice, too many and not sufficiently singular, or interesting to the general reader, to deserve a complete description here. I shall content myself, therefore, with giving a simple list of them, with the districts they inhabit, their entire length from the point of the nose to the end of the tail, and the main color of the back:

Philip's jerboa, or jumping-rat (*Dipodomys philipi*); Sacramento Basin and the southern valleys; twelve inches; yellowish-brown.

The dun jumping-rat (*Dipodomys agilis*); coast valleys south of San Francisco; twelve inches; lead-color.

The Colorado jumping-rat (*Perognathus penicillatus*); Colorado Desert; nine inches; tawny.

The King's River jumping-rat (*Perognathus parvus*); King's River valley; four inches.

Hudson's jumping-mouse (*Jaculus hudsonii*); valley of Canoe Creek; eight inches; yellowish-brown.

Black rat (*Mus rattus*); coast from Humboldt Bay to San

Diego; fourteen inches. There is much doubt whether it is indigenous.

The long-tailed mouse (*Reithrodon longicauda*); coast near San Francisco; five inches; dark-brown.

Gambel's mouse (*Hesperomys gambelii*); from Tomales Bay to Kern River; five inches; glossy-brown.

Boyle's mouse (*Hesperomys boylii*); valley of the American River; eight inches; glossy-brown.

Californian mouse (*Hesperomys californicus*); Santa Clara valley; six inches; sooty-brown.

Desert-mouse (*Hesperomys eremieus*); Colorado Desert; five inches; grayish-yellow.

The bush rat (*Neotoma mexicana*); near San Diego and in the Colorado desert; thirteen inches; yellowish brown.

The *Neotoma fuscipes*, a rat; coast valleys, from 38° to 40°; fifteen inches; reddish above.

The *Arvicola montana*, a mouse; near Petaluma, Monterey, and Lost River; six inches; yellowish brown.

The long-faced mouse (*Arvicola longirostris*); Pit River valley; six inches; yellowish brown.

The Californian ground-mouse (*Arvicola edax*); coast valleys south of San Francisco; six inches; yellowish brown.

The *Arvicola californica*, a mouse much like the species last named.

The Oregon mouse (*Arvicola oregona*); near Tanales Bay; four inches; yellowish brown.

The Oregon mole (*Scalops townsendii*) is found near the bay of San Francisco, and perhaps in other parts of the state. It is six or seven inches long, nearly black in color, with faint-purplish or sooty-black reflections in the hair.

§ 93. *The Deer Family.*—The American elk (*Cervus canadensis*) is found in California as well as in many other parts of the continent. The animal is nearly as large as a horse, and has some resemblance to it in general shape, though smaller, and slimmer in the head, neck, and legs. Its length from the nose to the tail is seven feet; its height five feet; its greatest weight one thousand pounds. The color is a chestnut brown, dark on the head, neck, and legs, lighter and yellowish on the back and sides. The horns are very large, sometimes more than four feet long, three feet across from tip to tip, measuring three inches in diameter above the burr, and weighing, with the skull, exclusive of the lower jaw, forty pounds. The horns of the old bucks have from seven to nine, perhaps more, prongs, all growing forward, the main stem running upward and backward. The elk were very abundant in California previous to 1849, and they were frequently seen in large herds; but within the last ten years they have become rare, and before the close of another decade they will be extinct in our state. A few are found in the San Joaquin valley, but the best place for hunting them is in Mendocino county. Several hundred carcasses find their way every year to the San Francisco market. The young fat elk furnishes a very juicy and sweet venison.

The white-tailed Virginian deer, once common in the states east of the Mississippi, is not found in California, but in its place we have the black-tailed deer (*Cervus columbianus*), which is a little larger and has brighter colors, but does not furnish as good venison, the meat lacking the juiciness and savory taste of the venison in the Mississippi valley. The average weight of the buck is about one hundred and twenty-pounds, and of the doe one hundred pounds, but bucks have been found to weigh two hundred and seventy-five pounds. The summer-coat of the black-tailed deer is composed of rather long and coarse hair, of a tawny brown, approaching chestnut on the back. In September this hair begins to come off, exposing what the hunters call the "blue coat," which is at first fine and silky, and of a bluish-gray color, afterward becoming chestnut brown, inclining to gray on the sides, and to black along the back. Occasionally deer purely white are found. The horn, when long, is about two feet long, and forks near mid-length, and each prong forks again, making four points, to which a little spur, issuing from near the base of the horn, may be added, making five in all. This is the general form of the horn; sometimes, however, old bucks are found with but two points. The deer likes the hills and the timber; the prong-horned antelope (*Antilocapra americana*) loves the valley and the open land. Before the Americans took California, the Sacramento and San Joaquin valleys abounded with herds of antelope; but now they are rare in the northern part of the state, and not abundant in the southern part. The traveller in the Tulare valley and in the Great Basin near the Coast Mountains, sees herds of them every day. Thousands are killed yearly for the market. In size the antelope is not quite so large as the Californian deer, which it resembles closely in form and general appearance. They are distinguished at a distance by their motion: the antelope canters, while the deer runs; the antelope go in herds, and more in a line following the lead of an old buck, like sheep, to which they are related, while deer more frequently are alone, and if in a herd they are more independent, and move each in the way that suits him best. In color, the back, upper part of the sides and outside of the thighs and forelegs are yellowish brown; the under parts, lower part of the sides, and the buttocks as seen from behind, are white. The hair is very coarse, thick, spongy, tubular, slightly crimped, or waved, and like short lengths of coarse threads cut off bluntly. The horns are very irregular in size and form, but usually they are about eight inches long, rise almost perpendicularly, have a short, blunt prong in front, several inches from the base, and make a short backward crook at the top. The female has horns as well as the male. The hoof is heart-shaped, and its print upon the ground may be readily distin-from the long, narrow track of the deer. The antelope is about two feet and a half high, and four feet long from the nose to the end of the tail.

The mountain sheep (*Ovis montana*) is found on the Sierra Nevada, from the Tejon Pass to the Oregon line, but it is a rare and very shy animal, and is seldom killed. Its length is about five feet, and its weight sometimes three hundred and fifty pounds, considerably greater than that of the deer or domesticated sheep. The color is white beneath, grayish brown elsewhere. The horns of the ram are very large, sometimes five inches through at the base and three feet long. The horns, after starting upward, turn backward, then downward, and so round with a circular or spiral shape, the tip inclining outward. Mountaineers assert that these horns are used by the sheep in getting down from the high cliffs which he is fond of frequenting. Instead of clambering down toilsomely over the rugged and broken rocks, he makes an easy job of it by leaping headlong, confidently down, over precipices fifty, yes, one hundred feet high, and alights head first on his horns, which are strong enough to be unbroken by the shock, and elastic enough to throw him ten or fifteen feet into the air—and the next time he alights on his feet all right.

§ 94. *The Hare Family.*—The Californian hare, or "jackass rabbit," as it is commonly called (*Lepus californicus*), is one of the largest of its class, growing sometimes to be two feet long from the nose to the end of the tail. Its ears are very large, and have suggested the vulgar name. It was once abundant in all the valleys from the Klamath to the Colorado; it is more rare now. The color beneath is a pale cinnamon; above it is mixed black and light cinnamon, the longest hairs being of a light smoky-ash color for about half the length, then dark sooty-brown, then pale cinnamon-red, and finally black at the tip.

The prairie hare (*Lepus campestris*) also, one of the largest hares, inhabits the plateau of the Sierra Nevada, Pit River valley, and the country about the Klamath lakes. It is all white in winter; in summer yellowish gray, with brownish tinges above and white beneath. The length, from the tip of the nose to the root of the tail, is from seventeen to twenty-three inches; and the tail and ear each measure about four inches.

Audubon's hare (*Lepus audubonii*) inhabits the coast valleys from Petaluma to San Diego. It is fifteen inches long, with a tail measuring to the end of the hairs on it three inches. The color is mixed yellowish-brown and black above, white beneath, thighs and rump grayish.

Trowbridge's hare (*Lepus trowbridgii*) is found along the coast southward from 39°. The length is from eleven to fifteen inches; the tail, with hair and all, less than an inch. The back is yellowish brown mixed with dark brown, paler on the sides, and ash-colored beneath.

The sage rabbit (*Lepus artemisia*) is found in all the open parts of California north of the Straits of Carquinez. It is from eleven to sixteen inches in length; in color, brown above and white beneath, with a yellowish tinge, the under part of the neck a yellowish brown. The fur on all parts of the body is lead-colored at the base.

§ 95. *Aquatic Mammals.*—The American beavers (*Castor canadensis*) were once very abundant in all the large streams of California, and it was chiefly for their sake that the first American trappers entered the country some thirty-five or forty years ago. They are still found in nearly all parts of the state, and even numerous, it may be said, in some of the sloughs near the junction of the Sacramento and San Joaquin Rivers. They rarely build dams in California, but live in burrows in the banks. When they dive they slap the water with their tails, making a noise that can be heard at a considerable distance in a still night. Their skins, which once commanded very high prices, have lost much of their value since the adoption of silk for making hats.

The common mink (*Patorius vison*) is found in California, but is not abundant. The general color of the animal is dark brownish-chestnut, with a white spot on the end of the chin. The skin of the mink is as valuable as that of the beaver.

The Californian otter (*Lutra californica*) is found all along this coast, and was formerly abundant on all the large streams. It is carnivorous, living entirely on fish and shell-fish. It prefers large streams and lakes for its home, while the plant-eating beaver prefers small streams. The Californian otter is sometimes five feet long from the point of the nose to the tip of the tail. When in the water, its hair is at times beautifully iridescent.

The sea-otter (*Enhydra marina*) is larger than the Californian otter, and is also carnivorous. It generally makes its home near islands, and roams about in the water within ten or twenty miles of land. The sea-otter was at one time very abundant along the coast of California, and it was one of the attractions which induced the Russian Fur Company to establish a post at Fort Ross, in latitude 38° 30′, where a number of Aleutian Indians were employed from 1812 to 1840 in the otter fishery. They would start out in their little single canoes, made water-proof with a covering of fish-bladders, so that there was no danger of their sinking if the sea should sweep over them, and thus they would go out fifty miles to sea and travel up and down the coast, usually coming home well laden with sea-otter skins worth sixty or eighty dollars each. The sea-otter is still abundant on the southern coast, and there are men in Santa Barbara county who make it a business to hunt them.

"The otter," says Mr. W. A. Wallace, "is very harmless, and always seeks to escape from human observation. When attacked they make no resistance, but endeavor to escape by sinking in the sea. If closely pursued and there is no escape, they scold and grin like an angry cat. If they escape the enemy, as soon as they are safe they turn and deride him with various diverting tricks, such as standing on end in the water, jumping over the waves, holding the paws over the eyes, as if to shade them from the sun while looking at the enemy—then lying flat upon the back and stroking the belly. In their escape they carry their sucklings in their mouths, and drive before them those not fully grown. They were formerly taken, by the Russians and Indians, by means of nets, clubs, and spears. The young are said to be delicate eating, the flesh resembling lamb. The flesh of the old ones is insipid and tough.

"The otter is never seen upon land. He is purely an aquatic animal. When he swims he turns upon his back and propels himself with great rapidity. The fore-paws are rounded like a cat's, but the claws of the older ones are generally worn off. The hind-legs, or propellers, are broad and flat, like paddles, and are used very dexterously. The seal much resembles the otter, seen at a distance, but he swims upon his belly, and the hunter seldom mistakes the one for the other. The otter sleeps in the water, lying upon his back, and anchors himself from the motions of winds and waves by drawing a string of kelp across his breast, just below his fore-legs. When discovered in this position, they are often approached very near by the hunters. They are very buoyant in the water, but when the chase has been long continued, and the blood of the otter becomes heated by the exercise, on being shot the body sinks rapidly to the bottom, and never rises. More than half the otters shot are lost in this way.

"Once a day the otter comes near the shore for food. He eats every thing that grows in salt water, and is particularly fond of abelones (*haliotus*), mussels, and sea-eggs. At high water the abelone loosens its shell from the rock, to receive the nourishment which the overflowing waters bring to it, and it is then easily taken from the rock and removed from its shell. The otter is well acquainted with all the peculiarities of this fish, and this opportunity to capture it for food."

The common seal, a species of phoca, is abundant along the coast.

The sea-lions, of the *Otaria* genus, frequent the coast from May to November, making their homes during the winter in some other clime, but where is not known. They delight to collect on clear summer days on rocks near the water's edge, and bask in the sun. They may often be seen on the rocks near the Golden Gate, and heard too, for they keep up a kind of barking or growling in chorus, which grows louder as they see any one approaching. They do not wait, however, to let a man come near, but pitch off into the sea before he is within two hundred yards of them. Their color varies from light yellowish-brown to dark brown.

§ 96. *Vultures.*—The Californian vulture (*Cathartes californianus*), sometimes improperly called "condor," the largest bird on the continent, and next to the condor the largest flying bird in the world, inhabits all parts of the state, though it is not abundant in any place. It is as prominent and peculiar a feature of the birds of California as the grizzly bear among the quadrupeds. It is very shy, and is rarely killed. The total length of the Californian vulture is about four feet, and its width from tip to tip of the outstretched wings, ten feet or more. Its color is brownish black, with a white stripe across the wings. The head and neck are bare, and red and yellow in color. The bill is yellowish white, and the iris carmine. Dr. Newberry says: "A portion of every day's experience in our march through the Sacramento valley, was a pleasure in watching the graceful evolutions of this splendid bird. Its flight is easy and effortless, almost beyond that of any other bird. As I sometimes recall the characteristic scenery of California, those interminable stretches of waving grain, with here and there, between the rounded hills, orchard-like clumps of oak, a scene so solitary and yet so home-like, over these oat-covered plains and slopes, golden yellow in the sunshine, always floats the shadow of the vulture."

Dr. Heermann, of the United States Pacific Railroad Survey, wrote thus: "Whilst unsuccessfully hunting in the Tejon valley, we have often passed several hours without a single one of this species being in sight, but on bringing down any large game, ere the body had grown cold these birds midst be seen rising above the horizon and slowly sweeping toward us, intent upon their share of the prey. Nor in the absence of the hunter will his game be exempt from their ravenous appetite, though it be carefully hidden and covered by shrubbery and heavy branches; as I have known these marauders to drag forth from its concealment and devour a deer within an hour. Any article of clothing thrown over a carcass will shield it from a vulture, though not from the grizzly bear, who little respects such flimsy protection. My coat, used on one occasion to cover a deer, was found on our return torn by bruin to shreds, and the game destroyed. The Californian vulture joins to his rapacity an immense muscular power, as a sample of which it will suffice to state that I have known four of them, jointly, to drag off, over a space of two hundred yards, the body of a young grizzly bear weighing upward of one hundred pounds."

The turkey-buzzard, or turkey-vulture (*Cathartes aura*), specifically the same with the bird known by that name in the Atlantic states, is found in all parts of California. From the tip of the bill to the end of the tail it is about thirty inches long, and six feet from tip to tip of the outstretched wings. The head and neck are bare, covered with a bright-red wrinkled skin. The plumage commences below that, with a circular ruff of projecting feathers. The color of the plumage is black, with a purplish lustre, many of the feathers having a pale border. The bill is yellowish in color.

§ 97. *The Eagle Family.*—The golden eagle (*Aquila canadensis*) inhabits California, and indeed all parts of North America. Its length is thirty or forty inches; its color on the head and neck is yellowish brown, white at the base of the of the tail, and brown, varying to purplish brown, and black elsewhere.

The bald eagle (*Haliætus leucocephalos*) was abundant in California ten years ago, and is still often seen along the Sacramento, San Joaquin, and Klamath Rivers. It frequents rapids for the purpose of catching fish, which seem to furnish the larger part of its food. It is from thirty to forty inches long, white on the head and at the base of the tail, and brownish black on the breast, wings, and back.

The fish-hawk (*Pandion carolinensis*) is found along all our large rivers. It is from twenty to twenty-five inches long. The head and under parts are white, with pale yellowish-brown spots on the breast. The back, wings, and tail are dark brown.

The goshawk (*Astur atricapillus*) is of the same size with the fish-hawk, and in color is dark—a bluish-slate above, and mottled-white and light ashy-brown beneath.

There are seventeen other hawks in the state, most of them small and rare.

§ 98. *Owls.*—California has nine species of owls, namely: the barn, great-horned, screech, long-eared, short-eared, great gray, saw-whet, burrowing, and pigmy owls. All of them are found extensively on the continent, beyond the limits of our state, and all save the last two are common east of the Mississippi.

The burrowing owl (*Athene cunicularia*) is ten inches long, ashy-brown above and whitish-brown beneath, variegated by spots and bands of white and dark-brown. Dr. Newberry says: "The burrowing owl is found in many parts of California, where it shares the burrows of Beechey's and Douglas's spermophiles. We usually saw them standing at the entrance of their burrows. They often allowed us to approach within shot, and, before taking flight, twisting their heads about, bowed with many ludicrous gestures, thus apparently aiding their imperfect sight, and getting a better view of the intruder. When shot at and not killed, or when otherwise alarmed, they fly with an irregular, jerking motion, dropping down much like a woodcock at some other hole."

The pigmy owl (*Glaucidium gnoma*) is seven inches long, and inhabits the wooded districts. It flies about actively in the daytime, and appears to subsist chiefly on sparrows, which it catches in daylight. The general color is brownish-olive above and brownish-white beneath.

§ 99. *Road-runner.*—The paisano, or road-runner (*Geococcyx californianus*), is one of the most remarkable birds of the state. It lives almost entirely upon the ground, very rarely flies, and frequents the highways, along which it will run from any one approaching. Its speed is nearly equal to that of a common horse, and it often furnishes an exciting chase to the solitary rider. It is abundant in the valleys and low hills, and makes its home among the bushes. The bird is akin to the cuckoo, and its generic name signifies "ground-cuckoo." Its length is from twenty to twenty-three inches, of which twelve are taken up by the tail. The general color is olive-green above and white beneath; the central tail-feathers are olive-brown, the others dark-green—all edged and (except the central two) tipped with white. Dr. Heerman says: "I have not witnessed the following feat, but am assured by many old Californians that this bird, on perceiving the rattlesnake coiled up asleep, basking in the sun, will collect the cactus and hedge him around with a circle, out of which the reptile, unable to escape, and enraged by the prickly points opposing him on every side, strikes himself, and dies from the effects of his self-inoculated venom."

§ 100. *Woodpeckers.*—There are eleven species of woodpecker in the state; of which two, the Californian (*Melanerpes formicivorus*) and Lewis's (*Melanerpes torquatus*), are worthy of special mention.

The Californian woodpecker is called by the Spanish Californians the *carpintero*, or carpenter, because he is in the habit of boring holes with his beak in the bark of the nut-pine, redwood, Californian white oak, and Western yellow pine, and then storing acorns in them for his winter use. The holes are just large and deep enough to hold each an acorn, which is hammered in so that there is no danger of its falling out. The acorns on the northern side of the trees, where they are protected from the rains, which come from the southward, often keep good for years. The bark of the nut-pine is preferred, probably being softer and more regular in grain than other bark. The holes are bored to within two or three feet of the ground, and to a height of fifty feet—sometimes, but rarely, in the limbs as well as the trunk. From thirty to fifty holes are often found in a square foot. In seasons when or places where acorns are rare, the woodpecker will put away hazel-nuts in the same manner. The squirrels often plunder the stores, and then the birds attack the thieves, darting down upon them and pecking them with their beaks. When the squirrel sees the property-owner coming, he hurries to a hole, or gets under a limb, where the woodpecker cannot conveniently strike him. Sometimes Indians and even white men are glad to avail themselves of the woodpecker's stores as a protection against starvation.

The length of the bird is nine inches; the anterior part of the body above and the tail are black; the belly, rump, a patch on the forehead, and a collar on the neck, white; and the crown, and a short occipital crest, red. Dr. Newberry says: "This beautiful bird, the rival and representative of the red-headed woodpecker [of the Atlantic slope of the continent], is an inseparable element of the scenery of the Sacramento valley. While we were encamped under the wide-spreading oaks of that region, I had a very good opportunity to study their habits, as they would come into the trees in the shade of which I was lying. They are not shy, and frequently came round in considerable numbers. Their manners are the very counterpart of the Eastern 'red-head,' and their rattling cry is not unlike his. Like the 'red-head,' I have seen two or three of them amuse themselves by playing 'hide and seek' around some trunk or branch; and like the 'red-head,' too, they delight to sit on the end of a dry limb, and fly off in circles for the insects which come near them."

Lewis's woodpecker is in color dark glossy green above and gray beneath, with dark-crimson patches on the sides of the head and belly. The feathers on the under part are bristle-like. It prefers an elevated home, and is found ten and twelve thousand feet above the sea.

§ 101. *Humming-Birds.*—There are four humming-birds in California, all different from those found in the Atlantic states. The white-throated swift, a bird resembling the swallow, but smaller, is common in the Colorado Basin. We have a whip-poor-will different from the one known in the Eastern states. Two night-hawks are found in our state, one of them appearing on this slope of the continent only in the vicinity of the Colorado, and on the other slope not extending far beyond the

Rio Grande. The belted king-fisher (*Ceryle alcyon*) is at home in California as well as in all other parts of the continent.

§ 102. *Fly-catcher.*—The family of fly-catchers (*Colopteridæ*), which connects the non-melodious with the true singing birds, is represented in California by eleven species, most of which are not seen in the Atlantic states. They are small birds, from five to nine inches in length, and their colors are usually dull. Most of them have their upper mandible bent down abruptly at the tip; and they always have twelve feathers in the tail. One of the most common and the best-known of the fly-catchers is the bird called the "pewee."

§ 103. *Singers.*—The zoölogical sub-order called *Oscines*, or singers, has one hundred and nine species in our state, including two mocking-birds, three thrushes, two blue-birds, three robins, three larks, five black-birds, eleven finches, six wrens, six swallows, six warblers, one martin, one bunting, six titmouses, one snow-bird, two grosbeaks, one cow-bird, one oriole, one crow, three ravens, three jays, one water-ouzel, two magpies, and so on. Some of these birds are not called "singers" in common language, but they all belong to the *Oscines* sub-order, which is marked by a peculiar muscular apparatus for singing, composed of five pairs of muscles in the throat. Though there are many species of *Oscines* in the state, yet the birds are not so numerous, so melodious, nor are they heard so often, as the feathered songsters in the Eastern states. The traveller may proceed for days in the Sacramento Basin, during the summer season, without hearing more than a few chirps. Our singing-birds have been multiplying very rapidly of late, because of the settlement and cultivation of the land, whereby their supply of wholesome and palatable food is much increased, and their enemies the hawks are driven away. Most of our swallows, one mocking-bird, one black-bird, and one raven, found in California, are also seen east of the Mississippi; but all our jays, robins, blue-birds, and magpies, and our oriole, are of species not found in the Atlantic states. The majority of the *Oscines* indigenous on this coast are unknown in the older states. Our mocking-birds are never domesticated, and are not to be compared to the mocking-bird of Virginia.

§ 104. *Scratchers.*—The ornithological order of *Rasores*, or scratchers, is represented in California by eleven species, namely: one pigeon, two doves, three grouse, two quails, one partridge, and one sand-hill crane. The pigeon, partridge, grouse, quails, and one of the doves, are specifically different from the birds known by the same name east of the Mississippi. The wild-turkey is not found in our state.

The most abundant and prominent of our scratchers, the Californian quail (*Lophortyx californicus*), is found in all the valleys of California and Oregon. Its breast and upper parts are lead-colored, with an olive-brown gloss on the back and wings; the chin and throat are black, with a white line running backward from the eye; the forehead is brownish-yellow; the belly is pale-buff, with an orange-brown round spot in the middle, changing to white at the sides; the feathers on the back and sides have a central streak of white, and those on the top and sides of the neck have black edgings. The head bears a crest numbering from three to six feathers, usually five, about an inch and a half long. The shafts are bare, very slender, and, though all are in a straight line on the longitudinal medial line of the head, they are so near together as to look like but one shaft, more especially as the fine, fur-like bushes at their tops all combine to form a compact little plume. These feathers are usually erect, the plume leaning forward when the bird is trying to look its best in the presence of company; but when running about in the grass, and not thinking of its appearance, the crest is lowered, falling forward over the bill.

The Californian quail has two notes—the song and the call. The song of the Atlantic quail is in two notes—the well-known whistle, sounding like "Bob-White." The song of the Californian quail has but one note, beginning like the "Bob" and ending like the "White" of its Eastern relative. The calls of the Atlantic and Pacific quails are nearly alike, and may be represented by the syllables "hi-re-he."—"As a game-bird," says Dr. Newberry, "the Californian quail is inferior to the Eastern one, though perhaps of equal excellence for the table. It does not lie as well to the dog, and does not afford a good sport. It also takes a tree more readily than the Atlantic quail. Like its Eastern relative, the cock-bird is very fond of sitting on some stump or log projecting above the grass and weeds which conceal his mate and nest or brood, and especially in the early morning, uttering his peculiar cry."

The plumed quail (*Oreortyx pictus*), likewise called the "mountain quail," while the *Lophortyx californicus* is often styled the "valley quail," is peculiar to this coast, and is one of the most beautiful features of its ornithology. It is a partridge ten inches long, very plump in shape, handsome in color, majestic in its bearing, and graceful in motion. Its head is surmounted by a crest of two straight feathers, three and a half inches long, which hang backward, one immediately over the other. The breast and neck are lead-colored, the upper parts generally olive-brown; the throat, and head beneath the eyes, orange-chestnut; the abdomen white. There are numerous variegations of white, black, and minor shades, on the plumage, all contributing to heighten its beauty.

The mountain partridge lives in the hills and mountains, from the Tejon Pass to the Columbia River. Its song suggests the sound represented by the word "whoit," whistled fuller and louder than the song of the Californian quail. It roosts upon the ground; and if bushes be near, in which to hide, it will rather run than fly from its enemies. It seldom flies more than two hundred yards at a time. The cock is equally attentive with the hen to the young brood, which usually varies from eight to twelve in number. The families seem to be much attached to each other, and if they are scattered, they are very uneasy until all are collected again. In such cases, the hunter can entice them to come to him by imitating the call of either old or young. They are easily domesticated

—more readily than their brethren of the valley. The mountain partridge hates the quail, and when brought into its presence always attacks it; the smaller bird makes no resistance.

Gambel's quail (*Lorphortyx gambelli*) is a bird differing from the Californian quail only in having duller colors, and is perhaps specifically the same, the difference in color being a mere accident of climate. Occasionally, white quails, very similar in form and size to the *Lophortyx californicus*, are found near Humboldt Bay.

The sage-cock, or cock of the plains (*Centrocercus urophasianus*), the largest of the American grouse, often weighing five or six pounds, inhabits the dry plains in the vicinity of Pit River. It is sometimes twenty-nine inches long, and forty-two inches across from tip to tip of outstretched wings. Its color above is variegated with black, brown, brownish-yellow, and whitish-yellow; its breast is white, its belly black. The male has bare, flame-colored patches of skin on the neck, which are ordinarily hidden by the feathers, but which are plainly visible when he struts about before the hen, with his neck puffed out like a pouter pigeon's.

The sharp-tailed grouse (*Pediocœtes phasianellus*) is also found in the northeastern corner of the state. It is eighteen inches long, light brownish-yellow above, varied with black, and white beneath, the feathers on the breast and sides having brown marks shaped like a V. The tail is long and sharp, the central feathers and the others growing gradually shorter as they approach the sides; there are eighteen feathers in the tail.

The dusky grouse (*Tetrao obscurus*) inhabits the coniferous forests of the Sierra Nevada, in the northeastern part of the state. The cock, according to common report, is the handsomest of all the American grouse. It is twenty inches long, dark-brown above, mottled with lead-color, and lead-color beneath. There are twenty feathers in the tail, which is broadly tipped with a light slate-color.

The band-tailed pigeon (*Columba fasciata*), the only wild

pigeon found on the Pacific coast, bears a strong resemblance in form, size, and color, to its congener in the Atlantic states, and has similar habits, but is not numerous. Small flocks migrate through the state every spring and autumn, and some of them spend the summer here.

The white-winged dove (*Melophelia leucoptera*) has been seen in the southern part of the state, but is very rare. It has white spots on its wings, whence its common and technical names are derived.

The common dove (*Zenaidura carolinensis*) of the Atlantic slope is found on the Pacific slope as well.

The sand-hill crane (*Grus canadensis*) are found from the meridian of Cincinnati to the Pacific, and are not rare in California. They spend the winters in our valleys, and in the spring migrate to the Klamath Lakes and farther north, where they spend their summers and breed. Subsisting upon vegetable food exclusively, they are themselves good to eat, and are frequently seen in the San Francisco market.

§ 105. *Waders.*—The order of waders (*Grallatores*) is represented in California by forty-one species of birds, namely: one crane, two herons, two bitterns, one fly-up-the-creek, one ibis, six plovers, one oyster-catcher, two turnstones, one avoiset, three phalaropes, one stilt, one willet, one godwit, one curlew, five snipes, five sand-pipers, one sanderling, three rails, and one coot. The oyster-catcher, one turnstone, one plover, and one heron, are the only species in the list not found east of the Mississippi, and none of them have such value or peculiarities as would give interest to a particular description of them.

§ 106. *Swimmers.*—California has sixty-six species of the order of swimmers (*Natatores*): of these there are two swans, six geese, twenty-two ducks, four albatross, two petrels, seven gulls, four terns, three pelicans, three cormorants, four guillemots, one loon, and various miscellaneous species. One swan, all the albatrosses, five gulls, the two petrels, the loon, and one guillemot, are found only on this coast.

The trumpeter-swan (*Cygnus buccinator*) is a very large bird, measuring five feet from the point of the bill to the end of the tail, and six feet across from tip to tip of the outstretched wings. The plumage is snowy-white in color, its legs and bill black. The name of "trumpeter" is given to it because of its clarion-like scream, which is heard as it flies. It frequents the lakes in the northern and northeastern parts of the state, and is sometimes seen in the rivers. It is a shy bird, and is rarely killed.

The American swan, found also on the Atlantic slope of the continent, is similar in appearance and size to the trumpeter, but lacks its loud voice, and is otherwise distinguishable from it chiefly by having an orange-colored spot on its bill in front of the eye, whereas the bill of the *Cygnus buccinator* is entirely black.

Wild geese are very abundant in California during the spring and fall, when they pass through on their migrations. Among them are the Canada goose (*Bernicla canadensis*), the snow-goose (*Anser hyperboreus*), the white-footed goose, or "speckled belly" (*Anser erythropus*), Hutchins's goose (*Bernicla hutchinsii*), and the black brandt (*Bernicla nigricans*). Hutchins's goose is more abundant than any of the others. Some of them, while in the state, get all their food in the tules; others in the spring resort to the fields of young grain, where they pasture. Dr. Newberry says: "I was much interested in noticing the perfect harmony of intercourse which seemed to exist among the smaller species. They intermingled freely while feeding, and when alarmed arose without separation; and I have seen a triangle flying steadily high over my head, composed of individuals of three species, each plainly distinguishable by its plumage, but each holding its place in the geometrical figure, as though it was composed of entirely homogeneous material; perhaps unequal members of the darker species, with three, four, or more pure snow-white geese flying together somewhere in the converging lines."

Among the ducks of California are the mallard and canvas-back. The meat of the latter has not so fine a flavor as in the Eastern states, probably because it does not here find the wild celery upon which it feeds along the streams of the middle states.

Many of the geese and ducks pass the winter in California, where they find an abundance of food in the grain-fields and tules.

The murre, or foolish guillemot (*Uria ringvin*), is similar to the gulls, seventeen inches long, dark-brown above and white beneath, with transverse stripes of ashy-brown on its sides. Its throat is brown in summer and white in winter. It frequents the islands along the coast, and lays its eggs there on the bare ground or rocks. These eggs are wonderfully irregular in form, size, and color, but are generally about three and a half inches long, sea-green in color, with dark-brown spots of angular shapes on them. Quantities of these eggs are obtained every year at the Farallones, and are sold in the San Francisco market at about half the price of hen's eggs per dozen, or, if taken by weight, at one-fourth. Their taste, however, is rank, and they are not used by those who can afford to buy the hen's eggs.

Dr. Heermann says: "At one o'clock every day during the egg season, Sundays and Thursdays excepted (this is to give the birds some little respite), the egg-hunters meet on the south side of the island. The roll is called, to see that all are present, that each one may have an equal chance in gathering the spoil. The signal is given, every man starting off at a full run for the most productive egging-grounds. The gulls (*Larus occidentalis*, Western gull), understanding, apparently, what is about to occur, are on the alert, hovering overhead, and awaiting only the advance of the party. The men rush eagerly into the rookeries; the affrighted murres have scarcely risen from their nests, before the gull, with remarkable instinct, not to say almost reason, flying but a few paces ahead of the hunter, alights on the ground, tapping such eggs as the short time will allow, before the egger comes up with him. The broken eggs are passed by the men, who remove only those which are sound. The gull, then returning to the field of its exploits, procures a plentiful supply of its favorite food."

§ 107. *Fishes.*—The fishes of the coast and rivers of California are all different from those of the Atlantic side of the continent, with the exception, perhaps, of one species of the halibut. The cod and shad, two of the most important fishes of the sea of the Eastern shore, and the lobster among crustaceans, are here wanting, as also the cat-fish kind in the rivers. Otherwise our waters are probably as rich in game for the fisherman as those of any country.

§ 108. *Salmon.*—The most important fish of California is the quinnat salmon (*Salmo quinnat*), a species found from Point Conception to the Columbia River. Its color above is olivaceous brown, changing to salmon-color beneath. The largest one ever caught weighed sixty-two pounds; the common size is from ten to thirty pounds. The salmon are born in the rivers, but go down to the sea, where they spend part of every year. They commence to enter the bay of San Francisco in November, and continue to come in for three or four months. They ascend the Sacramento and San Joaquin Rivers and some of their smaller tributaries, deposit their spawn, and in June go out to sea again. They come in lean and go out lean, but in the late winter and early spring they are fat. There are two common popular errors, that the salmon do not eat after leaving the sea, and that they never get back alive. The former error is owing to the fact that no large articles of food are found in its stomach, and the latter to the fact that when going out all are lean, and that many are found dead along the banks of salmon-streams. But the salmon find their chief food in minute animalculæ, and not in fish, for catching which they seem to be so well fitted, with their large mouths and sharp teeth. It is well known that the salmon bite like trout, and furnish excellent sport in clear water to the skilful fisherman with the fly. They dislike the mud with which the streams emptying into San Francisco Bay are filled by the miners, and therefore do not go far from the sea or ascend the small tributaries; but elsewhere they ascend every little brook, up to points where there is scarcely enough water for them to swim; and in these expeditions they are so much exhausted and bruised that they soon die; but the number thus killed is as nothing compared with those which go out to sea again. The female salmon having found a suitable place, uses her nose to dig a trench in the sand about six feet long, a foot wide, and three inches deep, and having deposited her spawn in it, throws a little sand over it with her tail, and departs, leaving her eggs to be hatched and the offspring to be fed as best they can. In the month of May the young salmon are found on their way to the sea, from three to six inches long. It is supposed that the salmon always return to the river in which they were born; so that the salmon born in the Klamath River never enter San Francisco Bay, nor do those born in the Sacramento and San Joaquin Rivers ever enter Humboldt Bay. Although the season in which salmon are abundant in the rivers extends from November to June, yet some of them are found in the streams of California at all seasons, and they can be had fresh in the San Francisco market every day in the year.

The quinnat is the chief salmon of all the streams and bays of California, but Gairdner's salmon (*Fario gairdneri*) is found in the Klamath River, and the stellatus salmon in Humboldt Bay and its tributaries. Gairdner's salmon has a silvery-gray back, silvery sides, and a yellowish-white belly. The body has numerous indistinct, blackish spots. The stellatus salmon is light-olive in the back, yellowish-white on the belly, and rarely exceeds two or three pounds in weight.

§ 109. *Halibut.*—There are two species of halibut on the coast of California, the Californian (*Hippoglossus californicus*) and the common (*Hippoglossus vulgaris*). There is some doubt whether the latter species is properly named; if it be, then we have one species of fish found on the Atlantic coast. The Californian halibut is a slender fish, weighing at the largest twenty-five pounds, in color grayish-brown above and white below. The halibut prefer a colder climate, and are not sufficiently abundant in this latitude to sustain a special fishery; but a few are in our market throughout the year. They live in deep water, and in places where the bottom is rocky. They eat little fish and shell-fish, and bite readily at the hook. Their meat is very delicate.

§ 110. *Turbot.*—The turbot (*Pleuronychthys rugosus*) is the only large flat-fish, except the halibut, found along our shore. It sometimes grows to weigh twenty pounds, but the common size is from three to ten pounds. The turbot inhabits deep waters and rocky bottoms, eats fish, and bites readily at the hook. It is one of the best fish in our market.

§ 111. *Sole.*—We have four species of small flat-fish, commonly called soles (*Psettichthys sordidus*, *Psettichthys melanostictus*, *Parophrys vetulus*, and *Platessa bilineata*). They are so much alike, that they are not distinguished from one another by fishermen generally. The *Platessa bilineata* is the largest, sometimes weighing two pounds; the others rarely exceed one pound. They frequent the shallow waters of the bay of San Francisco, and are caught abundantly in nets at all seasons of the year. The flat-fishes do not bury themselves in the mud here through the winter, as they do in the North Atlantic. The soles feed on crustacea, little fishes, and marine animalculæ.

§ 112. *Mackerel.*—The mackerel (*Scomber diego*) is found north of Point Conception. It is good, but not more than half as large as the Atlantic mackerel. The Californian mackerel rarely exceeds ten inches in length. It lies near the surface of the water at sea, and is not fond of entering bays or going very near the shore. Like its Eastern congener, it bites readily at any white rag or shining white substance jerked through the water.

§ 113. *Rock-Fish.*—The rock-fish furnish the main supply of fish in the San Francisco market. All belong to the genus *Sebastes*, of which there are eight species, the most important being the red (*rosaceus*), black (*melanops*), and wharf rock-fish (*auriculatus*). The red rock-fish grows to weigh twenty pounds; the other species rarely exceed four or five. The wharf rock-fish is the only one caught in the bay; the others live out at sea, in deep water and on rocky bottoms: they eat crabs and shell-fish, and bite freely at hooks. They are always in market, and their meat is excellent at all seasons.

§ 114. *Sturgeon.*—We have three species of sturgeon, of which the only important one is the Californian sturgeon (*Acipenser brachyrinthus*), which sometimes grows to be seven feet long and to weigh two hundred pounds. The sturgeon is a sea-fish, which enters fresh water to spawn, but it is caught in the bay of San Francisco and tributaries at all seasons of the year; whereas in the Eastern states there are seasons for sturgeon in the market, as there are for beans and peas.

The sturgeon eats the slimy matter, both animal and vegetable, at the bottom of the sea. It never bites, its mouth being circular in form, and fitted only for sucking. It has a habit of shooting up from the bottom and springing out of water, and then falling flat upon its belly, making a loud splash—very different from the porpoise, which also darts out of the water, but always strikes head first, making little noise. Some ichthyologists suppose that the object of the sturgeon in thus falling on the water is to free itself from parasites; others that it is merely a kind of play. The spawning-season is not known precisely, but it is probably from December to May. The meat of the sturgeon is coarse, and in the market is worth only about one-fourth or one-sixth of that of the better table fishes; but the sturgeon-fishery is profitable, because of the abundance and large size of the fish.

§ 115. *Jew-Fish.*—The Jew-fish (*Stereolepis gigas*), one of the largest scale-fishes—weighing sometimes five hundred pounds—is abundant south of Point Conception, and rarely straggles as far north as San Francisco Bay. Only two have been caught near the Golden Gate, and one of them filled the city with wonder. It is a bottom-fish, living in deep and shoal water, and frequenting lagoons and kelp. It often comes to the surface, and, according to report, goes to sleep there. It bites readily at the hook, and may be taken with harpoons. The meat is very good.

§ 116. *Sun-Fish*—The sun-fish (*Orthagoriscus analis*) is found occasionally south of Point Conception, where it is seen floating on the surface, in accordance with the habits of the genus everywhere. It weighs from one to a hundred pounds. Its form suggests the idea that the body has been cut off near the broadest part, and the tail sewed on.

§ 117. *Green-Fish.*—The green-fish (*Opplomona pantherina*), generally called cod in the San Francisco market, but having no relationship to the true cod, is abundant along the coast. It grows to be about two feet in length. The meat is coarse, and green in color; and the fish has little commercial value.

§ 118. *Sea-Bass.*—The sea-bass (*Johnius nobilis*) is a plain, oval fish, bluish-gray in color above, silvery below, weighing from fifteen to forty pounds. It is closely related to the weak-fish of the New York market. The meat is white and delicate, and always commands a high price in the market. It is a surface-fish, and sometimes enters the bays, but is not abundant anywhere. It is caught from March to November.

§ 119. *Sheepshead.*—The Californian sheepshead (*Labrus pulcher*) is a black fish, with a broad, bright-red band surrounding the body, and weighs from one to twelve pounds. It has white, broad, projecting teeth, like those of a sheep. It has no relationship to the Atlantic sheepshead, but is a congener of the black-fish of the New York market. The meat has a very fine flavor when fresh, but loses its delicacy after being dead a day or two. It is found south of Point Conception, on rocky and kelpy bottoms, from April to October. Its food is chiefly shell-fish.

§ 120. *Smelts.*—We have four species of fish called smelts (*Atherinopsis californiensis*, *Atherinopsis affinis*, *Osmerus preciosus*, and *Osmerus similis*). The *Atherinopses* are not true smelts, but belong to the same genus with the sanderlings of the Atlantic, which last are thrown away, or used only as bait; whereas our *Atherinopses* are valuable fishes. The *Atherinopsis californiensis* forms the great bulk of the smelts in our market. It is the largest of the Pacific smelts, sometimes reaching a length of fifteen inches, and a pound in weight. The *Osmerus* species are small. All of them have bright silver bands along their sides. The smelts are more abundant here than on the Eastern coast, and are the best of our small fishes. They are caught at all seasons of the year; in the bays with nets—never at sea, or with hooks.

§ 121. *Anchovies.*—There are two anchovies (*Engraulis mordax* and *Engraulis nanus*) on the coast of California. They are so nearly alike, that they are undistinguishable except by ichthyologists. Both are small, from four to six inches long, very delicate in flavor, but very bony. They are fully equal to the European anchovy for the table. They feed on minute animalculæ, go in shoals, and are caught with nets in the bays at all seasons of the year.

§ 122. *Sardine and Herring.*—The sardine (*Meletta cerulea*) is abundant from Humboldt Bay to San Diego. It grows to a length of eight or nine inches, and is therefore much larger than the Mediterranean sardine, to which it is fully equal in flavor. It is found along the coast from April to October, and is caught in the bays with nets.

The herring (*Clupea mirabilis*) is not so abundant as the Atlantic species, nor so large, but is equal in flavor. It comes in the spring, and goes in the autumn.

§ 123. *Viviparous Fishes.*—The viviparous or embiotocoid fishes of this coast are a peculiar feature of its ichthyology. They constitute, perhaps, the most remarkable natural group of fishes in the world, and their discovery caused a marked sensation among zoölogists. Other viviparous fishes had been previously known, but their young are brought forth in a very immature condition; whereas the little embiotocoid fishes are born with a fulness of development similar to that of warm-blooded animals, and the moment after they leave the mother they are seen swimming and taking care of themselves. There are seventeen or eighteen species belonging to the several genera, among which the *embiotoca* and *holconotis* are prominent. All are marine fishes save one, which is found in fresh water. They weigh from half a pound to three pouuds, and most of them are grayish brown above and silvery beneath. They are abundant in the market at all seasons of the year, and are called "perch" by the fishermen, though they have no relationship to the true perch. The meat is not good. The young are born from April to August.

§ 124. *Fresh-Water Fishes.*—Among the fresh-water fishes the most important is the brook-trout (*Salar iridea*), which is found in all the mountain-streams of the state, and offers fine sport for fly-fishing. It not unfrequently grows to weigh two pounds, and, if report is to be believed, sometimes reaches ten and twelve pounds. In appearance and flavor it is similar to the trout of other countries.

A fish called the salmon-trout (*Ptychocheilus grandis*), but not related to the salmon, the trout, or the salmon-trout, is found in all the large rivers and lakes of California. It grows to weigh thirty pounds. Its teeth are not in the mouth, but in the throat, where it crushes such shell-fish as it feeds upon. It bites voraciously, and is caught with the hook and with nets. The meat is poor, bony, and insipid. It is brought to the market in winter. The small ones are called pikes.

A chub (*Tygoma crassicauda*), and two suckers (*Catostomus labiatus* and *Catostomus occidentalis*), never weighing more than three pounds, are also found in our rivers. They are not valuable.

§ 125. *Shell-Fish and Crustaceans.*—We have five species of shell-fish valuable for the table: one oyster, two muscles, one cockle, and a soft-shelled clam. The oysters are small, not finely-flavored, nor abundant. We have no lobster, but a prawn (*Palinuris*), very similar to the lobster in size, color, flavor, habits, and general appearance, except that it lacks the large claws. Crabs are abundant. The abelone or aulone (*Haliotis*) is found as far north as Point Reyes, and abounds south of Point Conception. It is a mollusk with one shell, from five to seven inches across; the shells are beautifully iridescent, and are much used in the arts for buttons, knife-handles, inlaying, &c. Many vessels are engaged in fishing for them. The abelones stick to the rocks and to each other, collecting in some places in masses two feet thick; the fishermen break them off from the rocks with a spade. When the abelones do not suspect danger, they loosen their hold and raise their shells from the rock, and then the fisherman may easily thrust his spade down along the surface of the stone; but if he alarms the abelone beforehand, he finds the shells fastened down to the rock with great power, and all the strength of a man is scarcely sufficient to pry one of them off. The meat of the abelone is eaten by the Chinese, who dig them; the dried meat resembles horn in its color and hardness, and in shape looks as though it might be the hoof of a calf.

The shrimp (*Crangon franciscorum*) is found in the bays of California, and was very abundant a few years ago, but lately is getting scarce, at least in San Francisco Bay.

§ 126. *Reptiles.*—The snakes of California are not large, numerous, nor remarkable. Only one of them, the rattlesnake, is poisonous.

The scorpion is found in the warmer portions of the state, but is not abundant.

Tarantulas are common in Calaveras, Mariposa, Fresno, and Tulare counties. They belong to the same genus with the spiders, but the body grows to be three inches long and an inch wide, and the entire length from end to end of outstretched legs is five inches. The body and legs are covered with silky, brown hair. The tarantula eats little insects of various kinds, but, unlike most other spiders, has no net. It lives in a hole in the ground not much larger than itself when pressed into the smallest compass, and the hole is covered by a little door on a hinge, which closes by its own weight or by a spring. In the top of the door are several little holes, into which the tarantula can insert its claws when it wishes to enter; and so quick are its motions when terrified, that it often disappears suddenly under the eyes of men pursuing it, and they have great difficulty in finding its hiding-place. The door fits tightly, and is larger on the outside, so that it never sticks fast.

The bite of the tarantula is poisonous, but not fatal—or at least has never, so far as I know, proved fatal in California. It rarely bites men, and generally flees when it discovers their approach. The tarantulas have dangerous enemies in several species of wasps, the females of which kill them by thrusting eggs into their bodies. When the larvæ of the wasp are hatched, they make food of the carcass. So soon as the tarantula dies, the wasp drags it to her hole, usually the deserted burrow of a spermophile, where she may collect twenty or thirty dead tarantulas in one season. There are three different species of these wasps; one kind is blue, another yellow. Sometimes the wasp darts down repeatedly upon the tarantula, and does not touch him except with her egg-planter, depositing an egg at every thrust. On other occasions the two grapple, and the wasp continues to insert her eggs until the tarantula dies. The editor of a newspaper of Mariposa thus describes the killing of a tarantula: "Some of our readers may have heard of the tenacity with which the venomous tarantula is pursued by an inveterate enemy, in the form of a huge wasp—invariably resulting in the defeat and death of the former. We were an eye-witness to one of these conflicts last week, while on a ramble among the adjacent hills. This is the season when the poisonous tarantula leaves his well-fashioned abode to perambulate the dusty roads and smooth paths so often trod by the industrious miners, and about their haunts a dozen or so may be seen any day, of this hideous enlargement of the spider-race, within a circuit of a few yards, leisurely wending their way along the roads and by-ways. Often have we marked, with attentive curiosity, his awkward gait while lifting his long, unwieldy legs above the short blades of grass, and wondered for what uses and purposes this ugly little monster was placed upon this beautiful globe. While attentively watching the motions of one of these insects during our walk, we were much surprised to see the object of our attraction suddenly stop short in his wanderings and raise itself up to its full height, as though watching the coming of some unwelcome visitor. We at first supposed that it had just espied us, and was expecting danger at our hands; but upon our retreating a few steps, he quickly crouched behind a tuft of dried grass, and remaining very quiet, seemed to make himself as small as possible. A slight buzzing was heard in the air, and in a moment a wasp passed near, hovering on the wing over his trembling victim, the much-dreaded tarantula. Like some bird of prey, the wasp remained thus poised a moment, and then, quick as thought, darted down upon the enemy, and stung him many times with great rapidity. The tarantula, smarting under the pain, began a retreat, with all the speed of which he was capable; but the wasp hung over him with wonderful tenacity, and again and again struck him with his venomous sting. Gradually the flight of the tarantula became slower and more irregular, and at length, under the repeated thrusts of his conqueror, he died, biting the grass with his terrible fangs."

Locusts and grasshoppers are abundant in the valleys; musquitoes in the tules, and along the streams in the Sacramento Basin; and fleas everywhere.

§ 127. *Honey-Dew Aphis.*—Among the noteworthy insects of the state is one which secretes a sweet liquid called "honey-dew," and deposits it on trees. It is transparent, thick like honey, and sweet, sometimes with a bitter after-taste, but more frequently having a flavor like parched corn. The leaves and twigs are covered with it, the deposit usually being nearly even, occasionally in spots or drops. The honey-dew is more frequently found on oak-trees than on any other tree or bush; and oftener in dry seasons, and remote from the coast, than in wet weather or within the reach of the sea-fogs. A kind of molasses may be made by breaking off the twigs covered with the secretion, and boiling them in water.

Honey-dew is found in most countries where the soil is barren or the climate dry, and may be the same with the manna of the Hebrews. It is known, too, that various insects secrete sweet liquids; and some of the *Aphis* genus are kept as milch-cows by the ants, which stroke them down or tickle them with their antennæ, when they want some of the sweet milk, and the captive *Aphis* obligingly squeezes out the secretion through her sides, which is industriously gathered by the milk-ants.

NOTE.—Nearly all the information about the quadrupeds and birds of California, heretofore printed, may be found in the papers of Dr. J. S. Newberry and Professor S. F. Baird, in the United States Pacific Railroad Survey Reports. Most of my information about the fishes and fisheries, and much even of the language, is derived from the conversation of Dr. W. O. Ayres, of San Francisco; and I hope that, as he is the most competent man, he will some day treat the subject in a special work.

CHAPTER VII.

AGRICULTURE.

§ 128. *General Remarks.*—Of the 160,000 square miles in the area of California, about 60,000 may be tillable; of which 16,000 are in the coast valleys, 30,000 in the low lands of the Sacramento Basin, 12,000 in the Sierra Nevada, and 2,000 in the Klamath Basin: while the 25,000 square miles of the Great Basin, the 15,000 of the Colorado Desert within the limits of this state, 30,000 of the Sierra Nevada, 26,000 of the Coast Mountains, and 6,000 of the Klamath Basin, may be put down as unfit for the plough. The 60,000 square miles of tillable land contain nearly 40,000,000 acres, but only 1,000,000 are cultivated in the state: of the remaining 39,000,000, one-fourth have a soil very thin, or not fertile because of the presence of alkaline substances; one-half are too remote from market, even where the soil is good; and a considerable portion is tied up in lawsuits, so that the ownership is doubtful, and the claimants dare not improve it for fear of losing the improvements. Only a small portion of the state is, therefore, fit for the plough. Not more than one acre in ten could now be tilled profitably, and I suppose that not more than one acre in four will be tilled during this century.

As compared with the great agricultural states of the Mississippi valley, in so far as relates to the proportion of rich land fit for the plough, California is at a great disadvantage, and is probably inferior in this respect to every state on the Atlantic slope of the continent. In Illinois and Indiana, nearly every foot of land has a rich soil and a level position. Again. California is at a great disadvantage as compared with her sister-states east of the Rocky Mountains, in the proportion of land fitted for growing a variety of crops. In the northern part of the Mississippi valley, nearly every acre will produce all the main articles of cultivation—fruits, maize, potatoes, and garden vegetables, as well as wheat and oats. In this state, however, of the 40,000,000 tillable acres, at least 30,000,000 are so dry, that they cannot, because of the want of moisture, and the impossibility of irrigation, be made to produce any crop save small grain; and of the remaining 10,000,000 acres, three-fourths will not yield fruits, maize, potatoes, pumpkins, or garden vegetables, without irrigation.

These are undoubtedly very serious drawbacks to the agriculture of the state, but we have great advantages in many other points. The climate in the valleys, for instance, is so warm and the sky so clear through the winter, that vegetable life upon moist ground is almost as active in January as in July; and our trees and shrubs have nearly twice as much time to grow and mature as in the free states of the East, where frost reigns from October to May. It is a well-known fact that California has produced larger specimens of garden vegetables, more thrifty growth and rapid development of fruit-trees, and larger crops of small grain to the acre, than any state in the Union, and many persons have supposed our soil to be richer. This supposition is erroneous, as I am satisfied; the superiority of the Californian productions is owing to the more favorable climate. I am not aware that any comparison of our soils has been made by chemical analysis with those of Illinois, Missouri, Indiana, and Ohio; but the probability is, that the latter are more fertile. The loam is deeper; the vegetation has been greater, and it has enriched the soil by the accumulation of its decomposed remains through thousands of years; whereas in the valleys of California, the vegetation is comparatively scanty, and the air is too dry to permit a decomposition of wood or grass to enrich the soil. The bottom-lands of the Sacramento and San Joaquin are far inferior in depth, blackness, and fertility of loam, to the valleys of the Miami, Wabash, and Illinois Rivers.

§ 129. *Agricultural Districts.*—Let us now consider the different districts in the state suitable for agriculture. These districts, as I have said before, compose only a small part of California, and have strongly-marked boundaries. They are nearly all valley-land, shut in by mountains. The Great Utah Basin has very little tillable land in the state; there are small patches of fertile soil, but too slight to deserve special mention. The Colorado Basin is in about the same condition. It is possible that a considerable tract of land will be rendered fit for tillage by turning the Colorado into the low part of the desert; but this is a remote contingency, and we have no accurate information about the character of the soil which it is proposed to irrigate in this manner. In a few little valleys, however, just at the eastern foot of the Coast Range, the soil is fertile, and the climate so warm, that fruits ripen six weeks earlier than on the western side.

The largest tracts of tillable land in the Klamath Basin are the Scott and Shasta valleys, each about thirty miles long and four wide. They are elevated from three to four thousand feet above the sea; the winters are severe, and frosts common in spring and autumn, and not rare in summer. Most of the soil is a gravelly clay, with a rich, sandy loam, along the immediate borders of the streams. Wheat, oats, apples, and potatoes, do well; but maize, peaches, melons, tomatoes, and sweet-potatoes, require a warmer climate. There is some level land in the eastern part of the Klamath Basin, near the Klamath Lakes, but the soil is barren, and the vegetation like that of a desert. Del Norte county, which may be said to belong to the Klamath Basin, has 44,117 acres of land, of which 15,240 are covered with redwood, 7,277 with spruce, 19,204 are prairie, 2,400 are sand-ridges, and 4,712 are in lagoons. Most of the redwood land is level and fertile, but the timber is dense almost beyond example, and could not be cleared profitably, because all the redwood stumps throw out sprouts, which grow up to trees. Much of the prairie-land is fertile and suitable for cultivation, but it is remote from the market. Klamath and Trinity counties are almost destitute of valley-land. Both, however, have numerous small spots of rich soil in their mountains, and both have a mining population who must be fed, and have the means and disposition to pay well for the necessaries and delicacies of life. The farms must be small, but the farmers are protected by the rugged mountains from the competition of those in the large valleys. Hay-Fork valley, one of the best little tracts of tillable land in Trinity county, is three thousand six hundred feet above the sea-level.

The largest tract of level land in the plateau of the Sierra Nevada is the valley of Suisun River and Honey Lake, about sixty miles long and ten wide. The elevation is about four thousand five hundred feet above the sea; much of the soil is sandy and alkaline, with no indigenous vegetation save the worthless wild sage. Portions of the soil, however, are fertile, and there are a number of farms under cultivation. Honey Lake valley has the advantage of proximity to Washoe, and good roads for communication. There are 115,000 acres in the valley, of which 20,000 are swampy, at least in wet seasons. Eagle Lake valley is about one-third the size of Honey Lake valley, and of similàr soil.

All along the western slope of the Sierra Nevada there are little spots of fertile soil, well suited for cultivation; but with the exception of a few, they are too small to deserve special mention. The soil is usually a red clay or a black loam. The largest tillable spots on the western descent of the Sierra Nevada are in Plumas county. Sierra valley is forty-five miles long by six wide, and is drained by a tributary of Feather River. Much of the land is barren sand; much is tule-swamp, and only a small portion is fit for tillage. No water flows from the valley after July. The Big Meadows contain 100,000 acres; Beckworth's valley, 90,000; Indian valley, 20,000; Mountain Meadows, Red Clover valley, and Mohawk valley,

10,000 acres each; and American valley, 5,000. In all of them the soil is very sandy, but not barren. They are from three to five thousand feet above the sea; all shut in by high mountains; and all containing a considerable portion of swampy land.

The low land of the Sacramento Basin comprises 20,000 square miles—about 3,000,000 acres. The Sacramento valley has several benches. The lowest bench is about twenty feet above the low-water mark of the river, and has a soil of sandy loam, richer immediately along the stream than farther off. The next bench, very irregular in height and width, has a soil of red, gravelly clay, which extends back to the mountains. In some places this clay becomes very soft in wet seasons—so soft, that weak cattle may mire down in it and be unable to extricate themselves. I knew a case where a team of weak oxen, exhausted by hard driving and scanty food, sank down in a wet gravel-ridge, so that only their heads and a little of their necks and shoulders appeared above-ground; and passing the place some months later, when the ground had become as hard as clay and gravel ever are, I saw the six bare skulls of the oxen resting with their chins on the earth where they had sunk down, and behind them were the projecting spines of the back-bone, with the yokes still on the necks of each pair of oxen. This gravel is seldom cultivated at present, but in many places it will produce good crops of barley. It forms at least one-half of the Sacramento valley. Very little of it can be irrigated; and the general belief is, that no corn, potatoes, garden vegetables, fruit, or grapes, can be grown on it without irrigation. The sandy loam produces large crops of wheat, barley, and oats, without irrigation. Fruit-trees and grape-vines thrive without it after they grow to be three or four years old, but in most places require it till they have taken a good start. Garden vegetables cannot be grown without irrigation, unless planted very early, and of such kinds as ripen before July. In the level valley there are no springs, nor are there any artesian wells; so the only method of getting water is by pumping it up from common wells, for which purpose windmills are extensively used, and thus most of the water used for irrigation is obtained. A large part of the valley, especially of that near the rivers, is subject to overflow; and about once in five years a flood comes, sweeping away houses, fences, and cattle, destroying gardens, and covering the earth with a thick clay, which, instead of enriching the soil, is far poorer than the ancient deposits of sandy loam, made before the miners had commenced to tear down the mountains for their golden treasures. In these times of flood, so much of the valley is covered with water, that it looks like a great lake, and the pilots of the river steamers know the channel only by the rows of trees along the banks, for the banks themselves are completely hidden from sight. The flood rarely comes earlier than January or later than March.

The bottom-lands along the Feather River are considered richer than those near the banks of the Sacramento. Tributary to Sacramento valley on the western side is Cache Creek valley, about twenty miles long by five wide; and connected with it is Clear Lake valley, a basin nearly circular in shape, and twenty miles across, surrounded by mountains. The lake is about one thousand feet above the sea. South of Cache Creek, and also tributary to the Sacramento valley, is Putah Creek, which drains Berreyesa valley, twenty miles long by two wide. These little valleys have very rich land; and being shut in by near mountains, the soil is much moister than out in the open plain. The nearer to the coast and the farther north, the greater the moisture as a general rule; and it may almost be said that the value of the land depends upon the moisture.

In the northwestern corner of the Sacramento Basin, along the banks of Cottonwood Creek, there are some beautiful, moist, and fertile little vales. The Sacramento valley has very few trees, save along the banks of the streams and stream-beds, where oaks, sycamores, laurels, willows, buckeyes, birch, and wild grape, are the principal growth, marking in summer the places where the water runs in winter. The only trees growing away from the watercourses are oaks, which are usually found in groves, and almost invariably without undergrowth.

In the Sacramento valley there are about two hundred square miles, or one hundred and twenty-eight thousand acres, of tule-land, most of it above high tide, and covered by water only in times of flood. Very little of it has been drained or cultivated, and therefore we do not know its value. All the tule-land is covered five or six feet deep with water in times of flood.

The northern part of the San Joaquin valley is much like the southern part of the Sacramento valley. When the San Joaquin River approaches within fifty miles of Suisun Bay, it divides into three channels, which are separated from one another by islands of low tule-land. In times of high flood, the river spreads out and covers a space fifteen or twenty miles wide. There is less gravel and clay but more sand in the San Joaquin valley than in the Sacramento valley; the soil is drier, and contains more of alkaline substances, and the vegetation is more scanty. From Pachecos Pass across to Firebaugh's Ferry, a distance of about fifty miles, there is not a tree, and in the autumn the country looks like a desert. At Fresno City the soil is nearly a pure sand, and the river at low water is not more than six or eight feet below the surface of the plain. From the bend of the San Joaquin River, southward, a district sixty miles wide by one hundred and fifty long, most of the soil is a barren sand, in many places covered with an alkaline efflorescence.

The country about Kern River is very desolate, and between that river and the Tejon Pass is a desert plain, covered with a scanty and useless vegetation. East of Tulare Lake, however, there is some rich soil, particularly in the "Four-Creek country," where the Cahuilla River, issuing from the mountains, divides into half a dozen streams, which spread out over a space twelve miles wide, and then unite again, filling a large tract of fertile land with abundant moisture. Tulare Lake is fifty-seven feet above Fresno City, and might be drained, and its land converted into cultivated fields, but whether with a profit is a question now unanswerable. This country is so remote from the main centres of population, that probably the chief occupation for residents here must be the breeding of sheep and cattle.

Turning our attention now to the valleys in the Coast Mountains, we find that in all the low land between latitudes 39° 30′ and 40° 40′, the soil is a rich and moist loam, very favorable for pasture and for maize, but not so well suited for small grain. In some valleys the season is six weeks later than it is across the mountain in the Sacramento valley. Russian River valley produces more maize than all the remainder of the state. Lying between the bend of Russian River and the head of Petaluma valley is Santa Rosa plain, which has a soil of rich sandy loam, excellent for grain, and probably favorable for the grape. Near the mouth of Russian River is the plain of Bodega, the best place in the state for potatoes. The soil is a light, sandy loam, which is kept moist by the sea-fogs. Petaluma valley is the chief dairy district of the state. The soil is a rich, moist loam. Sonoma valley has a soil of red gravelly clay near the mountains, and a warm, sandy loam near the creek. This is the chief grape district in the northern half of the state. Much of the soil is too thin to produce good crops of wheat. The grape is grown without irrigation, the distance from the ocean (about twenty-five miles) not being so great as entirely to cut off the fogs. The bed-rock is in some places trap, in others sandstone, and in others magnesian limestone. The latter is supposed to be particularly favorable to the growth of the grape.

Next to Sonoma valley is Napa, which has a deep, clayey soil, the strongest in the state, and therefore the best for wheat. In proportion to its size, it produces more wheat than any other part of the state. The upper part of the valley has a great deal of gravel, and may be good for grapes. A larger proportion of the land is cultivated in Napa valley than in any other part of the state, and the cultivation is more thorough. Suisun valley has a rich, sandy loam, good both for barley and wheat. Vaca valley, a small vale near Suisun, has a very warm, fertile soil, and is shut in by the hills from the wind. It will be an excellent place for fruit, and every thing will ripen early there.

South of the straits of Carquinez is Diablo valley, which has an excellent soil for wheat and barley. So also has San Ramon valley, but the fruit has been badly nipped by frost during the last three or four years. Amador valley has a soil of rich sand at the sides and strong loam in the centre, all of it moist and fertile. Livermore valley, which may be considered as the eastern half of Amador, is a bed of gravel, of little value for tillage. Suñol valley has a rich, sandy loam. The Alameda plain, on the eastern side of San Francisco Bay, from San Pablo to San José, is one of the richest agricultural districts in the state; the soil is fertile—in some places clay, in others sand. The soil on the western side of the bay is similar in character, but there is not so much of it. The lower part of Santa Clara valley has a fertile, black, sandless loam, changing to sand and then to gravel, which last is abundant toward the head of the valley, where very little of the land is tilled. The principal fruit district in the state is in the vicinity of San José. The plain east of Monterey Bay, in Santa Cruz county, has a fertile soil, and a climate peculiarly favorable to beans; excellent crops of wheat and barley are also grown here. The soil of Pajaro valley is one of the richest and strongest in the state, and its crops of wheat and potatoes are unsurpassed. The Salinas has a rich, sandy loam in the lower part of its valley and near the river, but the sides and head of the vale contain much gravel; the climate and soil are very dry, and only a small portion of the land is cultivated. The Cuyama, Santa Inez, and Santa Clara River valleys, are sandy and dry, and have but little tillage; the last-named valley has a soil that is in places almost pure sand, too thin to secure a covering of grass in a wet winter. Most of the level land in Los Angeles, San Bernardino, and San Diego counties, is sandy and dry, and very little of it is cultivated. Irrigation is necessary for fruit, vines, and vegetables. Wheat and barley do not produce well. Los Angeles is the principal grape district in the state; the largest vineyards are planted in the bottom-lands of the Los Angeles, San Gabriel, and Santa Ana Rivers, where the soil is almost pure sand: and yet vineyards which have been in bearing twenty-five years, and have never been manured, are now as productive as ever. Allow a stream of water to run twenty-four hours through a field, and at the end of that time the bottom of the ditch will contain nothing but white sand, all the earthy particles in the soil having been dissolved and carried away. At the Monte the San Gabriel River sinks, and, after flowing two or three miles under-ground, reappears. The place where it sinks is very moist, covered with abundant vegetation, and, after Russian River valley, is the best district in the state for maize.

§ 130. *Agricultural Produce.*—California has 1,000,000 acres of land in cultivation, about six-tenths of which are used for growing grains and roots employed as food for men and domestic animals. Of these grains and roots in 1860 (the agricultural statistics for which year were more fully reported than those of 1861) the following amounts were grown, namely: 5,700,000 bushels of barley; 5,000,000 of wheat; 1,500,000 of oats; 1,500,000 of potatoes; 500,000 of maize; 65,000 of beans; 55,000 of peas; 55,000 of sweet potatoes; 50,000 of buckwheat, and 40,000 of rye, making an aggregate of 14,470,000 bushels—an average of twenty-four bushels to the acre, and, estimating the population of the state at 400,000, an average of thirty-eight bushels to the person. I purposely omit the consideration of all articles not enumerated in the list, either because we have no statistics, or because the insertion would serve rather to confuse than to instruct. We grow very few roots of the turnip kind as food for cattle.

Examining, then, the amounts of the crops above mentioned

in their proportion to the total 14,470,000 bushels, we find that arley forms 39 per cent., wheat 34 per cent., oats 10 per cent., potatoes 10 per cent., maize 3 per cent., and peas, beans, sweet potatoes, buckwheat, and rye, about one-half of one per cent. each. This proportion will be found to differ greatly from that of any other country. Compare our state with Ohio, which may fairly be considered as a representative of the free agricultural states of the Union. Ohio has 10,000,000 acres under cultivation—that is, ten times as much as California—and, according to the census of 1850, in the previous year produced 92,644,000 bushels of the above-mentioned articles, six times as much as California. The population of Ohio was then 1,980,000; so the yield was an average of forty-six bushels to the person. Coming down to particulars, we find that in 1849 Ohio produced 59,000,000 bushels (or 63 per cent.) of maize; 14,000,000 bushels (15 per cent.) of wheat; 13,000,000 bushels (14 per cent.) of oats; 5,000,000 bushels (5 per cent.) of potatoes; 638,000 bushels (one-half of 1 per cent.) of buckwheat; 425,000 bushels of rye, 354,000 of barley, 187,000 of sweet potatoes, and 60,000 of peas and beans. Comparing the proportion of the several items to the total of the crops, we find that California grows eighty times as much barley as Ohio, twice as much wheat and potatoes, twelve times as much peas and beans, and only one-twentieth as much maize. The proportion of oats, buckwheat, and rye, is about the same in the two states.

§ 131. *Rotation of Crops.*—Rotation of crops, as the phrase is understood in the Atlantic states and Europe, receives very little attention from the farmers of California, and indeed is impossible on the greater part of the land, because its dryness will not permit the growth of roots or common grasses. The soil is too dry for corn, potatoes, turnips, clover, and timothy or herd's grass. Peas and beans yield well in only a few localities. Alfalfa or lucerne will thrive, but it needs several years to get deep root and make a thick sod. Horses and cattle find food in the open plains and hills throughout the year; and the wild oat is cut for hay. The farmers generally are anxious to make as much money as possible, and as soon as possible, without regard to the future value of the land. Some of them are not permanent residents of the state, and intend to leave it so soon as they can get a certain number of dollars together; others are farming land the title of which is in dispute, and, as they feel uncertain about its ownership, they are indifferent to its exhaustion. Many of them come from the Western states, where the land had not, previous to their migration, become poor; and as rotation of crops had never been a necessity within their experience, they have never adopted it.

The soil of California is not exhausted; much of it will continue tc produce large crops of small grain without interruption and without manure for a score of years to come. Previous to 1853, the valleys were filled with cattle, and their manure had contributed for a long time—along the southern coast for half a century—to enrich the soil.

The prominence of barley in Californian farming is owing to the facts that it is an almost certain crop, produces largely, can be grown on all the cultivated land in the state, yields good volunteer crops, is excellent food for horses, can be kept for years, and always commands a market abroad.

§ 132. *Ploughing.*—Ploughing commences with the first heavy rain. The heat and drought of summer and autumn bake the ground, and render it too hard for the plough; so the farmer must wait for the rains. The sooner they come, after the first of October, the more convenient for him, and the more work he can do. The rain must be sufficient to wet the earth down six or eight inches deep; a little shower will not suffice. The soils of loam and clay are so hard, that no ordinary plough is strong enough to break through them; and ploughing would do no good, because the earth would be in large clods, which would furnish little nutriment to the grain.

§ 133. *Advantages, etc.*—The Californian farmer has a great advantage over those of the Northern Atlantic states, in the mildness of the winters. Here we have no snow or ice, and no time is lost because of cold. Neither are our frosts so severe as those east of the Mississippi. But, on the other hand, farmers in other parts of the Union have a great advantage over us in the security of title and the equal division of the land among the tillers of the soil. Here most of the titles are in dispute, and much of the land held under undoubted title is owned in large tracts by a few persons.

Barns are not used in California. The grain, after cutting, is put into a stack, or thrown into a heap, until a threshing-machine can be obtained, and the grain is then placed in the granary. Between harvest and threshing-time there is little danger of rain; and to such slight danger as there is, every farmer exposes himself. Barns in other countries are necessities; here they could not be used if we had them. Not unfrequently the grain, within two weeks after cutting, is stored in a warehouse in San Francisco; often it is left lying in sacks upon the field until it is sold—a period of months. In August and September, the square piles of white sacks in the stubble-fields are a common and prominent feature of the Californian landscape in the farming districts.

The straw, elsewhere saved, is usually burned here; and a most improvident practice it is. Not more than a dozen or two of farmers save the straw. The others burn it after the first rain, when there is no danger of the fire spreading; and the columns of smoke rising all over the country make a singular appearance.

Farms in California are usually larger than those in the Eastern states, and many of the ranches—containing from ten to fifty thousand acres—are large enough for principalities.

As our valleys are not covered with sod, so the first ploughing is nearly as easy as any of the subsequent ones; and the severe task of breaking prairie, so common in the states of the upper Mississippi valley, is unknown here.

§ 134. *Fences.*—In the matter of fences, the Californian farmer is at a disadvantage, as compared with his Eastern brethren. Throughout the United States, the system has prevailed of permitting horses, cattle, sheep, and hogs, to run at large, with no right of indemnity for any damage which they might do in cultivated fields, unless surrounded by a "lawful fence." This may be a good system for the pioneer, who tills little land, and wishes his horses and cattle to have a wide range; and it was well suited to the pastoral life of the Spanish Californians previous to the American conquest: but it is of doubtful policy as applied to the present condition of affairs, at least in the principal agricultural valleys, where all the land is under plough. For instance, in the Alameda plain, along the eastern shore of San Francisco Bay—a district fifty miles long by three wide, and containing one hundred and fifty square miles, of which one hundred and twenty-five are cultivated—there are five hundred farms, and probably two hundred miles of fencing, made at a cost of one hundred thousand dollars. This is a severe tax upon farming, and it is levied chiefly to protect the grain and fruit-trees from the depredations of horses and cattle belonging to neighbors and strangers. A farmer would rarely go to the expense of fencing his own cattle and horses out: it would be much cheaper for him to keep them in a yard or stable.

The legislature has prescribed what kind of a fence is "lawful;" and if any domesticated animal breaks through a lawful fence, its owner is liable for the damage done in the enclosure; and if the trespass be repeated by any neglect of the owner of the animal, he is liable for double damages. The farmer may take up the trespassing animal as an estray; but if he injure or kill it, he becomes responsible to the owner. The requirements of lawful fence are not the same in all parts of the state. In the counties of Butte, Amador, Tuolumne, Calaveras, San Diego, Nevada, San Bernardino, Colusi, Placer, Santa Barbara, Yuba, Shasta, Klamath, Trinity, and Siskiyou, "every enclosure" (I quote the words of the statute) "shall be deemed a lawful fence which is four and a half feet high, if made of stone; and if made of rails, five and a half feet high; if made upon the embankment of a ditch, three feet high from the bottom of the ditch, the fence shall be two feet high; said fence to be substantial and reasonably strong, and made so strong that stock cannot get their heads through it, and if made to turn small stock [sheep, goats, hogs, &c.], sufficiently tight to keep such stock out. A hedge-fence shall be considered a lawful fence, if five feet high and sufficiently close to turn stock." In other counties the requirements are so complex and lengthy, that I shall not try to describe them all. In Marin, Alameda, Sacramento, San Francisco, Stanislaus, Yuba, Santa Clara, Yolo, San Mateo, Santa Cruz, San Joaquin, San Bernardino, Sutter, Santa Barbara, San Luis Obispo, Los Angeles, Tuolumne, Tehama, Colusi, Butte, Napa, Humboldt, Merced, Trinity, Monterey, and Solano, hogs are not permitted to run at large; or at least any person finding them trespassing upon his "premises"—which means land, whether enclosed or not—may take them up, keep them at the expense of the owner, and treat them as estrays.

Board-fences are the best. They are usually made five feet high, with redwood posts set eight feet apart, and five spruce boards six inches wide and an inch thick in each panel. Such a fence, well made, costs five hundred dollars a mile. Worm and post-and-rail fences are common near the redwood districts —for instance, in Sonoma, Mendocino, Humboldt, Marin, Napa, San Mateo, Santa Clara, and Santa Cruz counties. The farmers generally make their own fences of these kinds, and the cost is of time, not money. When the work is done by the job, it costs from three to six hundred dollars a mile, according to the distance and position of the timber, and the quality of the wood; the price increasing in proportion as the trees are far off, or situated in deep cañons, and as the wood is tough and cross-grained. Ditches are common in the tule-lands. Hedges are made with willows and cactus in Los Angeles, San Bernardino, and San Diego counties. There are a few hedges of osage-orange and gorse, for ornament, in the counties about San Francisco Bay, but few or none for use. The osage-orange grows thriftily about San José, where it can be irrigated, but hedges are liable to much damage from gophers, which are fond of the roots; and if a hole is made, it is difficult to get young plants to grow, the older ones choking them down. After the third year irrigation is not necessary. In dry land, where water is not abundant for irrigation, the hedges do not grow up regularly. In the general opinion of farmers, osage-orange hedges will not pay, even in the land best suited for them: the labor of planting the seed, transplanting the sprouts, irrigating, replanting, and trimming for three years, costs more than a board-fence, which is useful from the first day, and is in no danger from gophers, whereas the hedge is useless for three years, and is in constant danger.

The willow-hedge is the most common fence in Los Angeles county, and is a prominent feature of the scenery near the towns. The fence is made with cuttings, the larger the better; the largest are three inches in diameter and eight feet long. These are planted perpendicularly three feet deep and nine inches apart, and then irrigated freely, when nearly all will grow. If larger cuttings cannot be had, small ones, half an inch thick and two feet long, are taken, and only an inch or two is left above-ground. If the cuttings are of the largest size, the fence is good in the second year; if small, four or five years may be required to make a tight fence. Twigs and poles are woven horizontally through the hedge. In the course of eight or ten years, the willows grow to be trees, with trunks five or six inches in diameter, and with dense tops from fifteen to thirty feet high. They thus not only shut out trespassing animals, but furnish a large amount of firewood, an item of no small importance in the woodless plains of the south, and throw a pleasant shade over the roads which they line. The willow-fence requires frequent irrigation, for its growth will usually depend upon the amount of water supplied to it.

The cactus was used extensively for fences at the old missions, and some fields are still enclosed with it. The plant is merely thrown upon the ground, where it takes root, no matter how dry or barren the soil, and grows up in a dense mass of thick leaves, six feet high and from five to ten feet wide. It is covered with thorns, and is feared by all large animals, but spermophiles and gophers are fond of burrowing under it, for it protects them against their enemies, and its leaves furnish them with food.

Several machines have been made to cut ditches through the tules, and throw the dirt up as an embankment on one side, but none of them have been very successful; and the spade is still considered the best instrument for making fences in the tules.

§ 135. *Barley.*—The soil and climate of California appear to be particularly favorable to the growth of barley, which forms a larger proportion of agricultural produce here than in any other part of the world. It is a hardy grain, preferring a sandy or gravelly soil, and dry weather. Three kinds are grown in California—the common, the Nepaul, and the chevalier. Only a few acres of the Nepaul have been raised, as an experiment; the chevalier is cultivated to a small extent, and chiefly for pearl-barley, of which a little is made in the country. The yield of the chevalier is from ten to twenty per cent. less than that of the common barley.

The sowing commences with the first heavy rain, which comes in some years as early as the first of November, and continues to the first of April. The ground used for small grain bakes hard during the heat and drought of summer and autumn; and ploughing is not possible until the rain comes, and rain enough to wet the earth thoroughly, at least six inches deep. The ploughs are then set to work immediately, running from four to eight inches deep. One ploughing is usually considered sufficient. The grain is sown according to convenience, soon after the ploughing, or after the lapse of weeks, and is immediately harrowed in. The amount of seed sown to the acre varies from a bushel and a half to two bushels. The sowing is usually done broadcast, but some farmers prefer the drill. Early sowing gives the best yield, if the winter rains be light; but when the rains are abundant, the late-sown fields are the best. There is always danger that small grain in California, if sown early, will get more rain than it wants. The same barley is sown early and late; our farmers do not know any thing of "winter barley" as distinct from "spring barley"—a division familiar in the Atlantic states.

The harvest precedes that of wheat; commencing in the Sacramento Basin early in June, and in the coast valleys late in the same month. The grain is all cut with reaping-machines, and is never housed, but is threshed on the field, with or without stacking. Sometimes it is bound; frequently it is gathered in a tight wagon-bed, and hauled into a pile in the centre of the field, where it remains until the threshing-machine can come. The rarity of rain from June to October renders this course pretty safe; though it has happened, on one or two occasions during the last ten years, that grain in the field has been injured by September rains. The same land is cultivated year after year in barley; and there has been very little, if any, decrease in crops during the last ten years.

When men are hired to plough and sow by the job, they charge three dollars per acre; reaping and binding cost two dollars per acre; threshing costs from one-twelfth to one-tenth of the grain, and sacks holding one hundred pounds cost fifteen cents apiece. The common yield is from thirty to thirty-five bushels per acre, and fifty per cent. more than the average barley-crop in the Eastern states. In 1856, according to the reports of the county assessors, the average yield of barley in Alameda county was 45 bushels; in Sonoma, 39; in Marin, 39; in Sacramento, 26; in Amador, 34; in Santa Cruz, 30. In 1857, according to the same authorities, the average yield in Alameda was 40 bushels; Sonoma, 25; Marin, 39; Sacramento, 24; Amador, 25; and Santa Cruz, 30. In 1859, Alameda reported an average of 29 bushels; Contra Costa, 30; Napa, 25; San Joaquin, 17; Sonoma, 40; Santa Cruz, 30; Yolo, 10; Sacramento, 25. In 1860, the assessors' reports show an average of 30 bushels for Alameda, 40 for Butte, 40 for Amador, 35 for Calaveras, 40 for Fresno, 45 for Marin, 40 for Los Angeles, 20 for Mendocino, 31 for Merced, 33 for Monterey, 28 for Nevada, 17 for Sacramento, 20 for San Joaquin, 21 for Santa Clara, 30 for Santa Cruz, 16 for Shasta, 60 for Sonoma, 60 for Yolo, and 40 for Yuba.

Many of these figures are merely guessed at by the assessors, who, however, are compelled to travel all over their respective counties, and converse with all the farmers. Their conjectures, therefore, are worthy of respectful consideration. But an average of sixty bushels per acre for a whole county looks almost too large to be believed unless supported by some special authentication more than we have. Nevertheless, crops of sixty bushels to the acre are not rare. In 1853, a field of one hundred acres in the valley of the Pajaro produced ninety thousand bushels, and one acre of it yielded one hundred and forty-nine bushels! It was grown by J. B. Hill; was mentioned as undoubtedly true by the assessor of Monterey county in his official report; and a prize was granted by an agricultural society for the crop. The field which took the prize of the State Agricultural Society, in 1859, yielded sixty-seven bushels to the acre. The field was a large one, and ten acres (a fair sample of the whole) were measured. The crop which takes that prize is not necessarily the largest crop in the state, but only the largest among those offered for competition. No doubt, many larger crops were harvested in 1857. In 1859, ninety bushels of Nepaul barley were grown to the acre by Mr. Burrell, in Santa Cruz county, but in a small field.

There is probably no part of the world where volunteer crops do so well as in California, and barley seems to produce better on the volunteer system than any other grain. Volunteer crops are those grown from the seed which falls out in harvesting; there is, therefore, no sowing or planting. Sometimes the land is ploughed and harrowed; sometimes it is left untouched. Large amounts of volunteer barley are grown every year, and sometimes the yield is excellent. One case is reported of a field in Yolo county which produced five successive volunteer crops of barley, the last and least crop amounting to thirty bushels per acre!

In 1860, the largest barley counties of the state were—Yolo, which produced 1,541,640 bushels; San Joaquin, 912,500; Alameda, 630,750; Contra Costa, 350,000; Sacramento, 300,683; Santa Clara, 300,000; Yuba, 243,761; Butte, 241,340; Santa Cruz, 212,000; and Nevada, 207,000.

§ 136. *Wheat.*—Many kinds of wheat are cultivated here, of which the main are Chili, Australian, Odessa or Old Californian, Red Mediterranean, Sonora, Oregon White, Bald, and Egyptian. The general division of wheat into "winter" and "spring," common in the wheat-growing districts of the Eastern states, is unknown. All our wheat may be set down as spring wheat. When winter wheat is brought here from abroad, it does not thrive the first year; but in the second year, having been converted into spring wheat and acclimated, it yields well. The Chili gives general satisfaction, and is more cultivated than any of the others. The Australian has a tendency to smut, but this is corrected with blue vitriol. These two form three-fourths of the crop; the other fourth is made up chiefly of Mediterranean and Sonora. The Egyptian yields largely, but has little gluten, and is fit only for coarse bread or maccaroni. All the acclimated wheat of the state is white; though imported red seed shows its color the first year, but in the second year it loses its redness.

The qualities in which the best wheat excels are glutinousness or strength, flintiness or dryness, whiteness of color, thinness of skin, cleanness, plumpness and size of berry, and weight.

The value of wheat depends, to a great extent, upon its strength. In this point lies its chief difference from potatoes, which always do and must occupy an inferior place upon our tables. Much gluten in flour renders the dough tough, makes handsome bread, with the air-bubbles in it small and uniform in size, and retains moisture, so that the bread will weigh much in proportion to the flour used; while if the amount of gluten be small, the grain of the bread will be uneven, the dough will give way in places, allowing the formation of large cavities, and less moisture will be retained. The wheat of different countries varies greatly in glutinousness, and California occupies a very high position. Our wheat is far more glutinous than that of any other North American state, and, although I have no precise information, I am inclined to believe that we have a like superiority in this respect over European countries. The consequence is, that our wheat is now in demand in New York to mix with their weak grain, so that a tolerably strong flour may be made.

But the wheat of California is not all equally glutinous; some of it is much weaker than other. The most glutinous is that grown in Santa Clara valley; the southeastern part of San Mateo county; the southern part of Alameda county; and Diablo, San Ramon, and Suisun valleys. That of Santa Rosa, Pajaro, Salinas, Petaluma, and Sonoma, is considerably inferior in glutinousness, but is better than that of the Sacramento, San Joaquin, and Napa valleys, the vicinity of Half-Moon Bay, and Alameda opposite the Golden Gate. The strongly glutinous is about one-third of the crop of the state. It is not known why the wheat in one district is more glutinous than in another. None of that grown very near the coast is strongly glutinous; so the moisture seems to be injurious. Napa wheat is inferior in glutinousness to that of Sonoma, though farther from the coast, and more free from ocean-fogs, but the soil of Napa is much moister.

In Oregon and Washington, where the climate is very moist, the wheat is as weak as at Half-Moon Bay. In the Mississippi valley, where a great amount of rain falls, the wheat is also weak; and just in the Gallego and Haxull district, if report be true, the rain-fall is less than in any wheat-district east of the Alleghanies. And yet in the Sacramento and San Joaquin valleys, which are among the driest parts of California, the wheat is very weak. This is accounted for—by those adopting the theory that glutinousness depends entirely upon the climate—by saying that those valleys are visited, while the grain is in the milk, by weather so hot that the berries are burned, and are prevented from attaining their perfect development. It would be well if this matter were thoroughly studied, for it is one of much importance to the merchant and ship-owner, as well as to the farmer, the baker, and the consumer.

The wheat grown on the clayey loam about Alviso is not so glutinous as that produced on the sandy loam about Santa Clara and the gravelly clay in other parts of the valley. It is worthy of remark that the soil of the Putah and Cache valleys, tributary to the Sacramento, differs in no noteworthy particular from the soil in Suisun, Diablo, and San Ramon, which latter yield strong while the former produce weak wheat. It has been observed that during the last three years the wheat of a large farm in San Mateo county, said to be the best cultivated in the state, has been gradually decreasing in strength. It is not known whether the change is caused by a difference in the seasons, or by a progressive exhaustion of the soil. So far as observations have been made in California, the amount of gluten is not affected by early or late sowing, thorough or careless cultivation, largeness or smallness of the yield, or cleanness of the crop.

In flintiness or dryness, Californian wheat has no superior, and no equal save in the Chilian. It may be stored in bulk, or it may be thrown into the hold of a ship within two weeks after harvest, and then sent twice through the tropics, and there is no danger that it will heat or sweat. The same may be said of its flour. No wheat or flour from the Atlantic states is near it in this respect. In August, 1860, J. B. Frisbie loaded a vessel at Vallejo with wheat taken from the harvest-field—it had never been inside of a house, but had lain upon the ground for several weeks after threshing—and that cargo of wheat, when discharged at Liverpool, was as sweet and clear from mustiness, mould, sprouting, or fermentation, as it was when harvested. The Atlantic flour, when kiln-dried and pressed, does not keep like ours as it comes from the mill, after having gone there fresh from the threshing-machine and the harvest-field.

The flour made from flinty wheat is peculiarly suited for shipment to tropical countries, where the moister flour soon ferments and sours. These are excellent markets, for they are certain, they pay well, and there is little competition. Most of the flour now exported to the West Indian islands and the Malaysian archipelago is of the Gallego and Haxall brands, which, because of their dryness and strength, are worth from twenty to fifty per cent. more in the market than other flour. California may not be able to supply the West Indian islands, but she certainly has peculiar advantages for supplying the tropical islands and shores of the Pacific. The flintiness of our wheat is undoubtedly owing to the dryness of the climate, and it is about the same in all the wheat-growing districts of the state. There is no noteworthy difference in this respect between that of the Sacramento valley and that grown on the immediate coast. It is all so dry as to keep well in any climate. Millers in New York and Liverpool make some objection to our wheat, that it is too hard for their millstones; but this is their misfortune, not our fault. The difficulty is remedied by moistening the wheat before grinding.

Most of the wheat of this state is white, but it is not equal in whiteness to that of the Genesee valley, Oregon, Washington, and some other districts of the United States; yet is superior to the wheat of England and of most European countries. The fogs give a dark color to the wheat grown at Half-Moon Bay, in the Pajaro and Petaluma valleys, and on the Santa Rosa plain; but in the other districts a uniform whiteness prevails.

Our wheat generally has a thin skin, and does not make much bran; but in the same districts where the skin is darkened by the fogs, there also it is thick.

Most of the Californian wheat is not well cleaned. It is sent to the market containing oats, barley, chess, alfalfa-seed, and dirt; and when shipped to New York, must usually be cleaned there before it can be ground. Our farmers, however, are gradually becoming more careful in cleaning their wheat.

In the plumpness and size of the berry, our wheat compares well with that of Europe and the Atlantic states, but can perhaps claim no decided superiority. Comparing the different districts of the state with one another on this point, Suscol probably deserves the first place, and Napa the next. In the Sacramento and San Joaquin valleys, the wheat is often shrivelled by hot winds, which blow for three or four successive days while the grain is in the milk, and seem to blast it. Great differences are observed, however, according to the season.

The weight of Californian wheat is usually sixty pounds per bushel, seldom less—frequently sixty-two, and sometimes sixty-five; thus entitling our state to a high position in that respect.

The average yield of Californian wheat-fields is from twenty to twenty-five bushels per acre, which is about thirty-three per cent. more than in the states on the Atlantic slope. An old Spanish book of records, of the mission of San Diego, states that, in 1778, twelve *fanegas* (a fanega is about two bushels) of wheat were sown, and three hundred and fifty fanegas were harvested—an increase of thirty-fold. The next year, sixteen fanegas were sown, and the yield was one hundred and sixty fanegas. In 1780, twenty-four fanegas were sown, and eight hundred harvested—an increase of thirty-three-fold. San Diego is far inferior for wheat-growing to the coast valleys about San Francisco Bay; and previous to the coming of the Americans the ground was not ploughed, but only scratched, and the limb of a tree was used for a harrow.

Colton, in his "Three Years in California" (page 442), states that while the priests still had sole control of the missions and mission-lands, previous to 1833, the *mayordomo* or steward of the Mission of San José harvested 4,300 fanegas of wheat from 40 fanegas of seed; and at the next harvest he had a volunteer crop of 2,600 fanegas on the same land. The first year, according to this report, the increase was 107-fold, and the next year 65-fold. At the Mission of Soledad, according to the same author (page 445), 1,700 fanegas were harvested from 19 sown—an increase of 89-fold; and in 1827, an increase of 58-fold was obtained at San Luis Obispo by scratching the seed in with a harrow upon land unploughed, and not even touched by the thing called a plough in those days. Not less than half a fanega is sown to the acre; so we may suppose that the figures which indicate the increase of the crop over the seed also indicate the number of bushels to the acre. Now, a tenfold increase is considered a fair crop. Crops of 80 bushels to the acre have often been grown in California. Mr. Hill harvested 82½ bushels from an acre in Pajaro valley in 1853, and obtained 660 bushels from 10 acres. In 1851, Mr. P. M. Scooffy harvested 88 bushels; and Mr. N. Carriger 80 bushels in Sonoma valley. In 1853, J. M. Horner harvested 1,000 acres of wheat near the Mission of San José, with an average of 40 bushels, some of it producing 60 bushels to the acre. The next year he had 2,000 acres, with an average of 40 bushels. Large fields of wheat in Eel River valley, according to the report of the assessor of Humboldt county, averaged 73 bushels to the acre in 1857.

In the best wheat districts of the Mississippi valley, the farmers generally believe, or did believe a few years ago, that not more than 45 bushels of wheat ever had been or ever could be grown upon an acre; and when, on a visit from California, I spoke to experienced and intelligent men among them of 60 bushels, I was told that not more than 50 bushels could possibly stand upon the ground. In 1856, the average wheat-crop per acre in California, according to the county assessors' reports, was—25 bushels in Amador and Santa Cruz counties, 30 in Marin, 28 in San Francisco, 19 in Sacramento, 20 in San Joaquin, 15 in Sonoma, and 28 in Tuolumne. The next year it was 35 in Amador, 40 in Del Norte, 20 in Alameda, Santa Cruz, San Joaquin, and Tuolumne, 19 in Sacramento, and 30 in Sonoma. In 1859, the average was 30 bushels in San Mateo, Santa Cruz, Siskiyou, Sonoma, and Yuba, 32 in Butte, 25 in Napa and Santa Clara, 20 in Contra Costa and Solano, 15 in

San Joaquin, and 14 in Sacramento. In 1860, the reported average was 45 bushels in San Luis Obispo, 35 in Yolo and Calaveras, 34 in Placer, 30 in Sonoma, Stanislaus, Yuba, and Amador, 27 in Santa Cruz, 26 in Fresno and Tulare, 25 in Tehama, Butte, Humboldt, and Napa, 24 in Nevada, 20 in San Diego, Santa Clara, and Shasta, 18 in San Joaquin, and 15 in Sacramento.

In 1855, the worst year for wheat we have ever known in California, when both smut and rust raged from Siskiyou to San Diego, the average crop of the state was put down as 15 bushels per acre. Of 12,233 acres sown in Sonoma county, only 3,500 were harvested; and of 2,490 sown in Marin, all but 462 went untouched by the reaper.

In Ohio, the average wheat-crop is about sixteen bushels per acre; and in England, with all their manuring and careful ploughing, twenty-one bushels. In California, no manure is applied; the soil is ploughed but once in most fields, and there is little rest for the land by rotation of crops.

It is a singular fact that where wheat is sown under oak-trees, the stalks are usually thicker and taller, and the grain more abundant, than in other places. This may be owing to the facts that the trees protect the ground under them from the frost, and also retain the moisture; and that while the country was in the hands of the Mexicans, the cattle had possession of the valleys, and, collecting under the trees in the summer-time, their manure enriched the soil there. The roots of the oak-trees in the valleys do not run along the surface of the ground, but go deeper for moisture, and thus the plough can run up to the trunk, and put all the land in order for grain.

The principal wheat-growing counties in the state are San Joaquin, which in 1860 produced 895,000 bushels; Napa, with 652,000; Yolo, with 459,000; Alameda, with 440,000; Santa Clara, with 400,000; Yuba, with 223,000; Santa Cruz, with 243,000; Sonoma, with 275,000; and Contra Costa, with 450,000.

It is almost impossible that there should ever be an entire

failure of the wheat-crop in California, unless the rain should completely fail. After wet winters, the dry lands and hills will produce the best crops; in seasons of light rain-fall, the low, moist lands will take the lead. There are so many soils and so many climates in the state, that some must be favorable. There is no danger that the grain, when nearly ripe, will be beaten down by the hail, as has happened in Europe and the Atlantic states. On only one occasion, within my knowledge or reading, has it happened that the grain has been "lodged" or beaten down by rain, and that was at Suscol and Napa in 1860; and the damage then was slight, for most of the grain recovered, and all of it, if I remember rightly, was reaped by machines.

Wheat is sown from the first of November to the first of April. The most certain crops are those sown early; the largest are those sown late in favorable years. If the amount of rain is small or moderate, the earliest-sown fields are the best; but if the spring be wet, the early-sown fields are surpassed by those sown about the first of February. Wheat is usually sown after barley and oats. The best farmers prefer to sow between New-Year's Day and the middle of February. Most of the sowing is done broadcast, but drills are used to a considerable extent. One ploughing is, by most farmers, considered sufficient. The harvest comes from the middle of June to the middle of July. The expenses of sowing, harvesting, and threshing, are the same as with barley.

§ 137. *Oats.*—The principal varieties of oats cultivated in California are the Australian, English, Bare, Feather, and Tucker. The Bare and Tucker oats thrive best on a heavy soil; the Feather oat prefers a sandy loam. The indigenous wild oat of California is never cultivated; for, although it produces large and tall stalks, they do not contain so much weight or bear so much grain as the domesticated oat. The average crop is from 30 to 40 bushels to the acre, 30 per cent. greater than in the Atlantic states. The Crescent City *Herald* reported in October, 1857, that Rigg and Reid, in Del Norte

8*

county, had grown 125 bushels of oats to the acre; and that John A. Brown, of Crescent City, had a crop of 157½ bushels to the acre. According to the assessors' returns, the average crop per acre of 1860 was—50 bushels in Alameda and Yuba counties; 40 in Butte, Placer, and Santa Cruz; 35 in Napa; 30 in Amador, Sacramento, and Yolo; 28 in Humboldt; 25 in San Joaquin; and 20 in Klamath, Santa Clara, and Sonoma. The largest oat-growing counties in the state are—Alameda, which in 1861 produced 449,000 bushels; Contra Costa, 300,000; Santa Cruz, 262,000; Sonoma, 187,000, and Marin, 174,000.

§ 138. *Maize.*—Maize can be grown to advantage in only a few places in California. Most of the land is too dry and the summer nights too cool for it. The principal maize districts are in the valleys of the upper coast, from Russian River to Humboldt Bay; in Yuba county, upon the moist bottom-lands of the Sacramento River; and at the Monte, in Los Angeles county, where the San Gabriel River sinks and fills the plain with moisture. Sixty bushels to the acre is considered a large crop; the average is not over thirty. Corn can be grown wherever the land can be irrigated, but this is a troublesome and expensive mode of cultivation, though it is not uncommon in gardens near San Fransisco. Green maize, grown in the open air, is in the market from June to September.

The cultivation of rye and buckwheat differs little from that of the same grains in the Eastern states.

§ 139. *Potatoes.*—The potato thrives wonderfully in a few places in California, particularly at Bodega, Tomales, and in Pajaro valley. The average produce per acre is perhaps not larger than in Ohio or England, but the tubers are larger in size and smoother in skin. The average size of those sold in the San Francisco market is probably fifty if not one hundred per cent larger than of those sold in New York. Potatoes six inches long by three inches through, and weighing a pound, are not uncommon; many have been seen to weigh four pounds; and one grew to weigh seven pounds. I saw a cluster that had grown together, eight inches long, six wide, and four deep, that weighed eight pounds.

The soil at Bodega and Tomales, the chief potato district, is a light, sandy loam, and the mists from the ocean supply the abundant moisture which the plant loves. In 1860, Sonoma produced 314,000 bushels, Sacramento 263,000, Marin 240,000, and Alameda 73,000. The potato district of Sacramento county is on the banks of the sloughs of the Sacramento River, near its junction with the San Joaquin. The soil is a very light, warm, rich loam, and the vegetables grown there are among the earliest in the market. According to the assessors' reports, the average crop of Sacramento county in 1860 was 390 bushels per acre; of Sonoma county, 100 bushels; and of Marin, 80. The Californian potatoes are mealy, sound, and palatable. The potato-disease has never made its appearance in this state.

The immediate coast, at least north of Point Conception, is too cold for the sweet potato, which thrives, however, in the Sacramento valley, especially in the low land about the head of Suisun Bay. The true sweet potato has grown here to weigh fifteen pounds—much larger than any I have ever seen in the states east of the Mississippi. The flavor is not equal to those grown at the East. They lack the mealiness and delicate taste which make the Eastern sweet potato so palatable in its season.

§ 140. *Hay.*—In 1860, California had 150,000 head of horses and 1,100,000 head of neat cattle, and cut 200,000 tons of hay, or one ton for six head of large stock. In 1849, Ohio had 463,000 horses and 1,350,000 cattle, and cut 1,500,000 tons of hay, or five tons for six head of stock. Ohio, therefore, cuts five times as much hay, in proportion to the number of her horses and cattle, as does California; and if we suppose that she exports one-fourth of her hay to the slave states, she still makes three times as much in proportion for home use as this state. The cause is, that there every horse and cow must have hay throughout the winter, and many of them through the summer; while here very few cattle are fed with hay at any season of the year, and horses not employed are usually turned out into the open plain. The hay of Ohio is cut in cultivated fields, from tame grasses; that of California is made of wild oats and indigenous grasses, which are grown in the open valleys.

The haying season comes about the first of May. The old adage, that "you must make hay while the sun shines," does not apply in California, for here the sun shines all the time, and the haymaker has ordinarily no fear of rain. It happened, however, in 1860, that a considerable amount of hay was spoilt by the late rains in June. The whole process of haymaking in California is managed by machinery. It is cut with the machine-mower, raked together with horse-rakes into cocks or windrows, and finally the cocks are hauled together on hay-sleds which load themselves by slipping under the cocks. The hay is not turned by hand, nor is the field raked by hand. The hand must be used, however, when wagons are to be loaded or stacks built. Hay is usually cured in the cock or windrow. It is not necessary to turn it by hand, as is customary in the Eastern states. One turning and one day in the sun are enough, when it is raked together and is ready for the stack or the mow.

In Ohio, a good field of timothy will yield four tons of hay to the acre; in California, the wild oat stands so thick in a few places as to yield as much, but the average crop is not over a ton to the acre. The principal hay counties of the state are—San Joaquin, which in 1860 made 37,000 tons; Santa Clara and Yolo, 18,000 each; Sonoma, 17,000; Yuba, 14,000; Sacramento, 13,000; and Contra Costa, 11,000. Very little maize-fodder is used in the state.

Tame grasses occupy, at the present time, a very small place in the agriculture of California. Not one-tenth of the farms in the state have an acre of cultivated pasture; and even in the largest farms, containing from three hundred to a thousand acres under plough, it is rare to find a field of timothy, clover, or alfalfa. The last-mentioned will probably become the principal grass grown in the state, since it is peculiarly fitted to thrive in a climate and soil so dry as ours.

§ 141. *Tobacco, Cotton, Rice.*—California produces tobacco of a fine quality, but the amount grown is small; and the experience of its cultivation is too brief to furnish much information. It requires a moist soil, and most of the attempts to cultivate it in dry places in the Sacramento valley and in the vicinity of Los Angeles have failed. The best crops have been grown near the coast, north of San Pablo Bay and about the head of Suisun Bay. The tobacco-plant has been converted into a perennial at San Francisco; one specimen of it growing up eight or ten feet high, like a tree.

A little cotton of a good quality has been grown, but I think its cultivation can never be extensive. The cotton states have three times as much rain as California, and I presume that only our moistest lands could produce a good crop of it—such, for instance, as the tule-lands in the valley of the San Joaquin.

The question whether rice can be cultivated in the tule-lands has been much discussed, but is not yet decided, though it is the general opinion that some of the tule-lands will produce large and profitable crops.

§ 142. *Hop.*—The hop grows luxuriantly and produces abundantly in California; and indeed there is good reason to doubt whether any country has a climate and soil more favorable to it than ours. We have no heavy dews or showers in summer to wash off the dust which contains the strength of the flowers, or to cover the plant with blight. The failures of crops, from these causes, so frequent in England and the Atlantic states, would never occur here. Not only is the crop certain, but it can be cured here with more ease and in better condition than in other countries. The moisture of the air in England compels the hop-growers to dry the flowers in the sun or in kilns; and if a rain fall upon them while drying, they are ruined: and they are injured by both the sun and kiln-drying. In California, they may be dried in the open air, under sheds;

and thus prepared, they will be superior to any of the European hops.

The Chinese sugar-cane grows luxuriantly in this state, but it is not extensively cultivated.

Flax, hemp, and the basket-willow, are not cultivated, or in patches so small as to be unworthy of notice.

§ 143. *Kitchen Vegetables.*—The vegetables for the kitchen —such as cabbage, cauliflower, beets, parsnips, carrots, radishes, onions, melons, squashes, pumpkins, green peas, string-beans, tomatoes, asparagus, rhubarb, okra, cucumbers, lettuce, garden-egg, and so forth—thrive in California, many of them beyond example elsewhere. Cabbages weighing fifteen pounds are wonders in the New York market; in San Francisco they are common. Whole fields of cabbage-heads, weighing twenty pounds each, have been grown; and hard, solid heads, with no loose leaves, weighing forty-five and fifty-three pounds each, are on record. One cabbage, which did not make a head, grew to be seven feet wide, throwing out leaves three and a half feet long on each side. In many cases the cabbage has been converted into a perennial, evergreen, tree-like plant, by preventing it from going to seed. Several of these are now growing in the state, with stalks from two to six feet high, and a foliage that grows through winter and summer.

The largest squash or soft-skin pumpkin produced in California weighed two hundred and sixty pounds, and the vine which bore it had several others weighing over one hundred pounds each; the total weight of its fruit being more than eight hundred pounds! Elsewhere, sixty pounds is a very large pumpkin or squash; and there is scarcely a record in the Atlantic states of a greater weight than one hundred pounds, which has been frequently surpassed here. In 1857, one squash-vine on the ranch of James Simmons, in Yuba county, produced one hundred and thirty squashes, weighing in all twenty-six hundred and four pounds! In the same year, J. Q. A. Ballou, at San José, grew two squashes, weighing two hundred and ten and two hundred and four pounds respectively.

The largest Californian onion weighed forty-seven ounces avoirdupois, and measured twenty-two inches in circumference. Our onions generally excel those of the Eastern states in size and weight.

Our largest red beet weighed one hundred and eighteen pounds—was five feet long, and a foot in diameter. It was three years old. The first year it grew to weigh forty-eight pounds, and because of its large size was reserved for seed; but it disappointed its owner, and, instead of producing seed the next year, merely kept on growing, and reached the size of eighty-six pounds; and the following year got to a hundred and eighteen. Such beets can be grown in abundance. A beet of twenty pounds is a wonder in New York or London; here it is too common to attract more than a glance. Beets frequently are three feet long, so that it requires no little trouble to dig them out.

Our largest common white turnip weighed, I believe, twenty-six pounds; our largest carrot, ten pounds; our largest water-melon, sixty-five pounds.

Our largest tomato measured twenty-six inches in circumference.

Our kitchen vegetables, grown in the open air, are in the market during a greater part of the year than in any state east of the Mississippi. We have cabbage, cauliflower, lettuce, turnips, beets, carrots, parsnips, radishes, horseradish, celery, green onions, leeks, salsify, and parsley, throughout the year; green peas, string-beans, water-melons, cantaloupes, and nutmeg-melons, from June to November inclusive; tomatoes from May to October; garden-eggs, green okra, Lima-beans, and Californian sweet potatoes, from July to September; asparagus from March to June; and rhubarb from April to July—the months being meant inclusively in every instance. These seasons for the different species of vegetables are, on an average, twice as long as the seasons on the Atlantic slope of the continent in the same latitude. Our tables are thus supplied with a great variety of fresh and wholesome vegetables throughout the year. Another advantage of our climate is, that garden vegetables may be left in the ground all winter. Potatoes are sometimes not dug until the first of January, and turnips and beets are usually left in their beds until they are to be sent to market; there is never enough cold to freeze them. Potatoes are never buried, but after they are dug are piled up in bags under a shed, or are placed in a storehouse.

The cabbage likes a moist air and soil, and thrives best along the coast, from Bodega to Santa Cruz. The melons and tomatoes like a warm climate, and thrive best in the Sacramento valley—and Putah valley, which is tributary to it—where many of the early vegetables for the San Francisco market are grown.

§ 144. *Fruit.*—As a fruit-growing state, California takes a high position. In this particular, as in so many others, her climate gives her great advantages. In no part of the world do fruit-trees grow so rapidly, bear so early, so regularly, and so abundantly, and produce fruit of such large size. Nor is there any other country where so great a variety of fruit can be produced in high excellence. In the matter of flavor, our apples, peaches, and strawberries, or most of them, are inferior to Eastern fruit; in the flavor of other species we are at least equal to other countries. The pear, the plum, the apricot, the grape, and the olive, are peculiarly thrifty, healthy, and productive, as compared with the same kinds of fruit elsewhere.

The Californian orchards are trained low, the lower limbs being within a foot or at most two feet of the ground. All kinds of fruit-trees are trained on the same principle. Men, therefore, do not walk under the trees in an orchard, or climb after the fruit. It would be as absurd to try to walk under or to climb a bearing apple-tree in California as to walk under or climb a gooseberry-bush. One fruit-tree in a hundred may be trained high, not more. The advantages of low training are, that the trees bear fruit earlier—a matter of the greatest importance in California, where the interest of money is so high, and the price of fruit rapidly falling from year to year; the trunk is shaded, and protected against the disease called the sun-scald; the earth about the roots is kept moist; and the trees are protected against the wind.

The trees are planted from one-sixth to one-half nearer together in the orchards than in the Eastern states. This is an additional protection against sun and wind. The ground is ploughed several times every summer, and kept clean; whereas in the Eastern orchards it is common to sow grass or cultivate vegetables. Our apple-trees are free from the borers after the first year, and our plum and cherry trees from the curculio, though the plum suffers from the aphis or louse.

Fruit-trees in California are generally as large at two years old as they are in New York at three and four years. The instances of unusually rapid growth here are without parallel elsewhere. Cherry-trees have grown to be fourteen feet high in one year; pear-trees ten feet high; peach-trees to have trunks from two to three inches in diameter. These were all from buds on yearling stocks, and were well provided with branches—not trimmed to gain height. These specimens of rapid growth were observed on an island near the junction of the Sacramento and San Joaquin Rivers. At Petaluma, a cherry-tree two years old from the graft, and three from the seed, had a trunk seven inches and three-quarters round; a plum-tree, three years from the seed, was eleven feet high, and had a trunk seven inches in circumference; and a peach-tree, one year from the bud, was eight feet high and eight and a half inches round.

Mr. E. B. Crocker, of Sacramento, wrote thus in December, 1858: "In January, 1855, I planted a small almond-tree, with a stem little larger than a goosequill, and which I cut down within a few inches of the ground. It is now a tree twenty feet high, sixteen feet through the top, with branches starting from the surface of the earth. The body below the branches is twenty-four inches in circumference. . . . . A Glout Morceau dwarf pear-tree, planted in 1855, when it had grown one year from the bud, is now ten feet high, four feet through the top, and measures ten inches round the body at the ground, branching about one foot from the surface. A Beurre Diel dwarf, planted in January, 1856, is now seven feet high, three feet through the top, and ten inches in circumference at the ground. A dwarf May Duke cherry, planted in 1856, is now thirteen feet high, and thirteen and a half inches in circumference at the ground. An Old Mixon peach, planted in 1855, and cut down within a few inches of the ground, is now twenty feet high, twenty-two feet through the top, and the trunk twenty-eight inches in circumference. A seedling peach, seed planted in January, 1858, is now eight feet high and well branched, and the trunk four and a half inches in circumference at the ground. The growth of trees, vines, and shrubs, is about double that of similar kinds on the rich prairie-soils of Northern Indiana."

In 1858, a sprig of a peach-tree, a foot long, was stuck into the ground on the Bay-state ranch; the next year it bore fruit. It may be set down as a general rule that, previous to the time of bearing fruit, trees in California make twice as much wood in a year as they do in the middle states.

In Alameda county, plum-trees have grown twelve feet in one year from the bud.

The trees commence to bear fruit at about half the age at which they bear in the Atlantic states. An apple-orchard in New York begins to bear in its fifth or sixth year; in California, in its second or third.

The variety of climates, and the freedom from frosts, severe cold, and furious storms, protect us against a failure of the fruit-crop.

Our apples, pears, apricots, and plums, are larger than the same varieties usually are elsewhere; other fruits are about the same in size.

§ 145. *Apples.*—The Spanish Californians had a few apple-trees, but they were seedlings of a poor class. The first good apples were imported from Oregon in 1849; but the varieties were few, and the trees did not thrive. Either the stock was not the best, or the change of climate had an injurious influence on them. In 1852, a few trees were imported by way of the isthmus of Panama; other importations followed very rapidly; and now the state has millions of trees in nursery, and about eight hundred thousand bearing trees in orchard, including two hundred varieties, the best of Europe and the Atlantic states, both standard and dwarf trees.

Apple-trees are usually planted from twelve to thirty feet apart, fourteen or sixteen being the more common distances. This is much closer together than is customary in the Atlantic states; the reasons for the denser planting here being to prevent injury by the wind, and to keep the earth moist by shading it against the sun. The apple-tree comes into bearing in the third year in California, about two years earlier than in the Eastern states. It also grows more rapidly, a yearling tree here being as large as a two-year-old tree in Ohio. Grafts on yearling stocks have been known to grow six and eight feet in a season—twice as long as similar grafts will grow in the middle states. The fruit usually grows larger here than elsewhere. The *Gloria Mundi* apple, which elsewhere seldom exceeds fourteen ounces in weight, in California frequently reaches twenty ounces, and some have attained the great size of two and even two and a half pounds.

The climate seems to have a tendency to ripen apples more thoroughly here than in other states. Those varieties which are grown for winter use elsewhere, are here generally converted into autumn apples, and only a few will keep to New-Year's Day. A fruit-grower in Alameda has succeeded in keeping several kinds until June. Our list of winter apples is very short, and some years will pass before we can in this respect equal the middle states. Some varieties have been introduced here from Georgia and other Southern states, but we do not yet know how they will succeed.

The flavor of our apples is not equal, as a general rule, to that of the apples grown on the Atlantic slope. They are less juicy, and more mealy. Some varieties, however, are better here than in the Eastern states. Great variations are observed in different parts of the state; an apple may be excellent when the tree grows in the hot summer and cold winter high up on the Sierra Nevada; and be of poor quality if grown in the equable temperature of the coast.

The best varieties, so far as ascertained, about the bay of San Francisco, are the Summer Pearmain, Red Astrakhan, Red June, and Early Harvest, for early apples; the Porter, Gravenstein, and Summer Queen, for late summer apples; the Baldwin, Roxbury Russet, and Rhode Island Greening, for fall apples; the Golden Russet, the Northern Spy, the Yellow Newtown Pippin, the White Winter Pearmain, and the Spitzenberg, for winter apples. The best cider apple is the Smith's Cider.

The leading counties in the production of apples are Santa Clara, Sacramento, Alameda, Sonoma, Napa, Marin, Yola, Yuba, and El Dorado.

The trees grow so rapidly and bear so abundantly, that some persons suppose our orchards must be short-lived; and the oldest American orchard in the state—at the Mission of San José—is cited as proof of this theory. That orchard is evidently dying, though only eleven years old; but its unhealthiness is owing to some influences peculiar to that spot. It has been gradually dying for four or five years, while other orchards six and eight years old are in perfect health. Besides, the fruit-trees of the old missions, many of them thirty years old, are still in excellent health and full bearing, and have not failed at any season during the last score of years to produce a good crop. The indigenous trees in our valleys have a thriftiness of growth and a precocity of development similar to our cultivated fruit-trees, and yet have a longevity equal to that of the similar species east of the Mississippi, where the summers are shorter, the winters colder, the annual growth less, and the development of the reproductive power later.

§ 146. *Peaches.*—The peach-tree grows very rapidly, comes into bearing very early, and produces abundantly, in California;

but nearly all varieties suffer with "the curl," which has given so much trouble during the last two years, that many of the orchards have been cut down. The varieties most free from the curl are the Late and Early Crawford, the Late Admirable, and the Smock. In the valleys and near the ocean, the peaches are not equal, either in size or flavor, to the same varieties on the Atlantic slope; but in the Sierra Nevada they are fully equal to the Eastern fruit. The peach does not thrive in the high winds which prevail about San Francisco Bay. The trees are usually set out in orchard when one year old from the graft or bud; in the second year after that, they begin to bear.

§ 147. *Pears.*—The pear is the most productive and healthy of the fruit-trees of California. It thrives in all parts of the state, and everywhere its fruit is delicate in flavor and large in size. There are pear-trees at San José which produce twenty-five hundred pounds or forty bushels each of fruit annually. The pear was more cultivated by the Spanish Californians than any other fruit; but their varieties were not good, and most of the old trees have been grafted with varieties brought from the Atlantic states during the last eight years. The varieties most prized are the Madeline, Bloodgood, Diane d'été, Dearborn's Seedling, and Bartlett, for summer pears; and the Winter Nelis, Glout Morceau, Easter Beurre, and Pear d'Albert, for winter.

Neither tree nor fruit is troubled by any bug, fire-blight, sun, or rain.

§ 148. *Apricots and Plums.*—The apricot thrives well and bears abundantly, especially in the warmer parts of the state. The fruit, however, in some places is much eaten by bugs and bees. The bugs—some of them of the kind commonly called "Lady-bug," and others similar in appearance and size—eat holes in the apricots before they are ripe; and the bees, which never break the skin, eat at the holes which the bugs have commenced. The apricot-tree is more healthy than the peach, and produces more abundantly; and its fruit supplies the place of the peach in many districts.

The nectarine is affected by the curl, and is not much cultivated.

If the curculio should not be introduced, the plum will occupy a very prominent place in the horticulture of the state. The tree is healthy; and the fruit is large, finely-flavored, and abundant. The climate is very favorable to drying fruit, and prunes might be made here with a profit.

The cherry and the soft-shell almond thrive in all the valleys of the state.

The fig-tree is cultivated from Shasta to San Diego, but does not produce abundantly north of latitude 35°. In the vicinity of Los Angeles it is a very thrifty and productive tree. It produces two crops of fruit annually. North of 37° the second crop is usually killed by the frost or stunted by the cold.

The almond and the English walnut both grow well and produce abundantly about Los Angeles, but do not thrive so well north of Santa Barbara. The almond suffers, and loses its fruit, with a slight touch of frost.

The pomegranate is a healthy and productive bush in California, but its fruit is not profitable. It is cultivated to a small extent in all the large fruit-gardens.

§ 149. *Olives.*—For the cultivation of the olive, California has great advantages. The tree is very healthy, and always bears abundantly; whereas in Italy and Greece, whence most of our olive-oil comes, the crop is frequently destroyed by summer rains, blight, and insects, all of which causes of trouble are unknown here. There it is expected that the crop will fail one year in three, whereas here no failure has ever been known. The number of our olive-trees is small, most of those in bearing having been planted half a century ago. Nor is it likely that there will be a rapid increase. The tree does not come into bearing until ten years of age, at least not in Europe; and although it may live and continue in bearing for five or six centuries, the possibility of a steady income to our remote posterity will not pay Californians for investing their money in a business that will yield no income for a decade of years.

Most of the bearing olive-trees are in the town of Los Angeles, and at the Missions of San Fernando, San Gabriel, and San Juan Sapistrano.

The olive-tree resembles a willow in the form and color of its bark, the shape and proportions of its trunk and branches, and the size, color, and distribution of its leaves. The trees are grown from cuttings or shoots, which latter frequently sprout from the large trees near the surface of the ground. A large olive-orchard in full bearing would prove an excellent income, for the fruit and the oil are in demand.

§ 150. *Oranges.*—The orange is cultivated in Los Angeles; and, although the trees now there are covered with insects to such an extent, that most of them bear no fruit, yet I think there is reason to hope that the fruit will, at no distant day, be cultivated extensively and profitably.

A warmer clime than that of this state is undoubtedly more congenial to the orange than ours; but in those lands where the climate is warmer, the men are less industrious and intelligent. Cultivation, which is the first element in the development of every species of fruit, is wanting there, while here there is no lack. Not that our climate is so cold as to make it doubtful whether we can cultivate the orange in the open air: long experience has settled the fact that the orange-tree will thrive and produce well from Santa Barbara southward.

We have no exact information as to the time when the orange was introduced into California, nor from what stock the old orange-trees came. Probably the first missionaries brought orange-seeds with them from Lower California, that stock having come from the indigenous trees along the western coast of Mexico. The seeds were planted at various old missions, such as San Diego, San Fernando, San Juan Capistrano, and so forth. The trees grew, were planted out, bore well, received little attention or cultivation, and some of them are still standing as monuments to the industry and enterprise of the old priests.

There are now, so far as I can learn, about twenty-five hundred orange-trees set out in orchard in the state, more than two-thirds of them being in the orchard of William Wolfskill, in the town of Los Angeles. About four hundred of the orange-trees in the state are old—from ten to fifty years of age; the remainder are young, from six to eight years old, at which age they begin to come into bearing.

The proper way to raise orange-trees is to make a bed about three feet wide and twenty long, with the earth in it well pulverized; and in January or February plant this bed with seeds, in rows a foot apart, and the seeds six inches apart in the rows, and about six inches deep. The bed should be weeded carefully, and kept constantly moist. If in the dry sand of Los Angeles county, the bed should be irrigated once a week. At the end of three years the trees will be four feet high and an inch and a half thick in the trunk. They should then be set out in the orchard where they are to stand, and be planted twenty-five or thirty feet apart each way. The transplanting should be done in any of the spring months, the earlier the better, and should be immediately followed by irrigation. The transplanting should not take place when the young trees are growing, and therefore the trees should not be irrigated before transplanting, especially if the weather be warm; for warmth and irrigation would have a tendency to start the shoots.

The trees begin to bear in their seventh year, when they are about ten feet high, and the trunks from three to five inches thick. At fourteen years they are in full bearing, and they continue to bear till they are at least fifty years of age, probably much longer. In full bearing, every tree will produce at least one thousand oranges a year, and some trees will regularly produce two or three thousand. The tree grows to be thirty feet high, the top spreading out thirty feet wide. It blossoms early in the spring, and the fruit is ripe in the following February, although it looks ripe in December. The oranges will keep well until May, if left on the tree. The fruit is always in demand, and always commands a high price; and previous to 1857, Mr. Wolfskill made more than one hundred dollars apiece annually from his bearing trees; but since that time the bugs have injured the crop seriously.

The tree is very beautiful, and grows continuously. The wood is hard and valuable. The tops grow very bushy, and frequently branches have to be cut out to allow the air to have access to the fruit and leaves; and sometimes the trees have to be supported, to save them from breaking down under the weight of their fruit. The trees, after setting out in the orchard, should be irrigated thrice every summer, and, unless the land is rich, should be manured.

The bug, a species of Aphis, has fixed itself in most of the bearing trees in the state; and unless some remedy not now in use be applied, it will probably kill all the trees. Many devices to drive away the pest have been tried in vain. But there must be a bane for this bug: when that bane is once found, the cultivation of the orange will take an important place in the horticulture of the southern part of the state, and therefore every good citizen is interested in finding it.

§ 151. *The Grape.*—California is a favorite land of the grape; and indeed many of our vine-growers suppose it to be the best grape country in the world.

The grape region of California extends from the southern boundary, at latitude 32° 30′, to 41°, a distance of five hundred and ninety-five miles from north to south, with an average breadth from east to west of about one hundred miles. The Los Angeles grape district is in an open plain about seventy miles long, and reaching back thirty miles from the ocean—bounded on the east by barren, rugged mountains. The Sonoma, Napa, and Santa Clara grape districts, are in flat, narrow valleys, shut in by steep, rugged ridges of the Coast Mountains, between latitude 37° 30′ and 39°. The Sacramento grape district is in a flat valley, about half way between mountain-ranges fifty miles apart. The grape districts of the Sierra Nevada are situated on the western slopes of those high mountains, usually in very small dales.

The soil of the vineyards at Los Angeles and Anaheim is a

deep, light, warm sand, which, to the inexperieneed eye, looks as though it were too poor to produce any valuable vegetable growth. In those places where water runs through it for a few days, all the mould is dissolved and carried off, leaving a white and almost pure sand. The soil is so dry, that cultivation is possible only with the assistance of irrigation. In Sonora and Napa valleys the vineyards are planted in a red, gravelly clay near the foot of the mountains, or in a light, sandy loam in the centre of the valley. Of late, the vine-growers of these valleys have done without irrigation. In Santa Clara valley most of the vines have been placed in a rich, black loam, but their vineyards are unhealthy. The Sacramento vineyards are planted in sandy loam; those of the Sierra Nevada in sandy loam or in gravelly clay.

The vine was brought to California by the Spanish missionaries about the year 1770. So far as is known, only one variety—that now known as the Los Angeles grape—was brought by them in the last century. It is the vine found in all the old vineyards and in most of the new ones south of the bay of San Francisco. It fills three-fourths of the vineyards in the state. The berry is round, reddish-brown while ripening, and nearly black when fully ripe, about five-eighths of an inch in diametar at its largest size, covered by a strong skin, possessing an abundance of thick and very sweet juice, with little meat, but with no fruitiness of flavor. It has been asserted that this grape is of the Malaga variety; but if so, it has changed so much—perhaps while under cultivation in Mexico, whence the first cuttings that came to California were probably obtained—that it no longer resembles its parent stock.

About 1820, when the missions were established north of the bay of San Francisco, a new variety, now called the Sonoma grape, and said by General Vallejo to be of the Madeira stock, was introduced. It is now extensively cultivated in Sonoma and Napa counties and in the Sacramento valley, and is also found in a few vineyards south of the bay of San Francisco. The berry is bluish-black in color; is covered, when ripe, with a grayish dust, which brushes off, leaving a glossy, smooth skin; is about half an inch in diameter at its largest size; has a thin, sweet juice, with more meat and a little fruitiness of a flavor.

The Sonoma grape makes a light wine, resembling claret; the Los Angeles grape makes a strong wine, resembling port and sherry. The two grapes are classed together as the "Mission," "Native," or "Californian" grapes, and were the only varieties cultivated here previous to 1853. In that year the importation of foreign grapes commenced, and now about two hundred varieties are cultivated. The Mission grapes are hardy, healthy, long-lived, productive, and early in coming into bearing; but they are surpassed in flavor, hardiness, productiveness, earliness of ripening, and earliness of bearing, by many foreign varieties, which, so far as is known, are not inferior in any respect. The latter have been tried, however, only three or four years, and therefore we cannot speak positively whether they will prove so long-lived, or whether they will be equal in some other points to the Mission grapes.

Still, the superiority of the foreign grapes is so great, that no reasonable man, acquainted with the subject, doubts that they will drive the Mission grapes out of the market. Flavor is a matter of vast importance in fresh fruit, and the want of it is the great defect of the Mission grape, which will not command more than six or eight cents per pound in the San Francisco market, at the very time that fine foreign varieties bring twenty-five and thirty-seven cents. Cuttings of the Mission grapes can now be had for ten dollars per thousand, a price that will not more than pay for preparing them for market; while those of the foreign cost from forty to one hundred and fifty dollars per thousand. For wine, the foreign grape has an equal or still greater advantage. Flavor and fruitiness are not less needed there than in fruit to be eaten fresh at the table. The lack of fruitiness is the great misfortune of the wine made from the Californian grape, and the evil can only be remedied by the use of the foreign grape.

For raisins the Mission grape is unsuited, because it is all juice and lacks meat. Again, we want various kinds of grapes to make different kinds of wine, and to give variety to our tables; and we wish also to have early and late grapes, so that our wine-making may extend through a long season, and that our tables may have grapes upon them from August to December. The foreign grapes, it has been observed, are stronger to resist the frost than the Mission grape. The latter is therefore doomed, if not to destruction, at least to a subordinate position.

About two hundred varieties of grape are cultivated in California, including the most noted stocks of Spain, France, Germany, Hungary, and the Eastern states. Nearly all of them thrive, and it can scarcely be said authoritatively that any one of them has proved a failure. The Catawba and Isabella do well; the latter furnishing our finest table-grape for some tastes, while others prefer some of the Muscatels.

The total number of grape-vines planted in vineyard in the state is about three and a half millions, or four thousand five hundred acres, of which more than one-third are in Los Angeles county. One-fifteenth of these may be foreign vines, of which one-half are in Sonoma county. There were probably two hundred thousand bearing vines in the state in 1848, and they still continue productive. Very little was done to increase their number until 1856, and then the business of grape-growing and making wine for the market was commenced. The new vineyards then set out were planted with Mission grapes, the only variety of which cuttings in large quantities could be obtained. A few foreign vines had been imported in 1853, '54, and '55, by nurserymen, but there was little demand for them. When it became clear that California would produce wine largely, the foreign varieties came into demand. It was not until 1859 that the superiority of the foreign grapes as a class over the Mission grape was established by trial.

The advantages of California for the cultivation of the grape are the following:

1. Californian vineyards produce ordinarily twice as much as the vineyards of any other grape district, if general report be true. Here, twelve thousand pounds of grapes per acre is a crop as common as six thousand in France, Germany, or Ohio. Why our vineyards should produce so much more than those elsewhere I know not, but the fact is indubitable. Crops of twenty thousand pounds per acre have been seen here, but never elsewhere, if witnesses, generally considered credible, are to be believed.

2. The grape-crop never fails, as it does in every other country. This is owing partly to the fact that we have no severe frosts, no hail, and no storms of rain and electricity from the time the vine buds until the grape is gathered, each of which often causes a total loss of the crops in Europe. There is abundant time for gathering the grape, while in other vine countries the rain and frost destroy the fruit after it is ripe. The *oidium*—the disease which has done such great damage in France—appeared in 1859, but has done no injury as yet save in a few small, young vineyards. I have heard of it only in Santa Clara, Sonoma, and Alameda counties, where the vines are planted in a wet, black loam, or stiff clay. Bugs and insects, which do much harm in European vineyards, have as yet done no injury worthy of note in California.

3. Vineyards in other countries require more labor than in California. In Europe, the vine is trained with a stalk four feet high, and supported by a pole, which has to be set down every year, and to which the vine is tied. Here the stalk stands alone.

4. The equability and warmth of the climate render it easy to make wine by fermentation without artificial heat during the winter; whereas in other grape countries fires must be kept up in the cellars through the winter.

5. The great variety of grapes which thrive here as compared with every other grape country.

The disadvantages of California consist in the high price of labor (three times as high as in Ohio, and four times as high as in Europe), the ignorance of the people of the arts of vine-growing and wine-making, the dearness of casks (costing from five to twenty cents per gallon), and the necessity of irrigation.

Land suitable for vineyards costs from twenty to one hundred dollars per acre, whereas it is worth several times as much in France; but there is a counterbalancing difference in the interest of money, so that the French vine-land at four hundred dollars per acre, bought with money borrowed at six per cent. a year, costs little more than a Californian vineyard bought for one hundred dollars, with money at twenty-seven per cent.

The vine likes a sandy or gravelly (not very moist) soil, and never thrives in wet, loamy, or stiff clay soil. In California, nearly all the vineyards are planted on flat land; in Europe, hills are preferred, and in Germany the name for a vineyard is "weinberg"—a vine-hill.

Vineyards are planted with cuttings or with rooted vines. The cuttings are obtained at the annual pruning in January or February, are about thirty inches long, and are all of wood less than a year old. They should be taken from vines not less than four years old. The rooted vines are cuttings which are planted in the nursery and allowed to grow there through one season. These latter may be planted out from November to March, inclusive; cuttings from January to March. It is not usual to plough more than once before planting, but several ploughings would be better. The vines are planted either six and a half or eight feet apart each way: the former distance giving one thousand vines to the acre, is customary at Los Angeles; the latter, giving six hundred and eighty vines to the acre, is preferred in Sonoma and Napa. The vines are planted about two feet deep, perpendicularly, leaving about three or four inches with two buds above the surface. The holes are usually made with a crowbar, and after the vine is thrust down into it, a little loose sand or pulverized dirt is poured in to fill up the hole. Sometimes holes are dug with the spade. Unless the ground is very moist, the newly-planted vineyard is irrigated; for the vine, when taking root, likes water. During the first year after planting, the vine-grower has nothing to do save to irrigate twice, to plough several times, and to hoe down such weeds as cannot be reached with the plough.

There is very little growth of wood the first year, but it frequently happens that cuttings bear grapes—one bunch, it may be, to a dozen vines. Rooted vines do not bear the first year. The next year the ground should be kept loose and clean by ploughing and hoeing twice or thrice. Any suckers springing out from buds beneath the surface must be broken off, and a little pruning is done. In pruning, regard is had to the form which the stalk is to have.

The vine bears fruit on new wood; that is, on twigs produced in the same season with the grape. All the twigs are cut off every year, leaving a bare stalk. In the old vineyards of California the stalks are from three to five feet high. Of late, the more general custom is to make the stalks about fifteen inches high. It is observed that the nearer the grapes to the ground, the earlier they ripen, and the less liable they are to injury from frost and wind. The strongest shoot is selected to make the stalk, and it is tied to a little stake stuck into the ground at its side, and the other shoots are cut off. It is a matter of importance to use the stake so that the vines may grow straight up. Vineyards planted with cuttings bear no grapes the second year; those planted with rooted vines may bear a few. In the southern part of the state the vineyard must be irrigated at least twice every summer; in many localities in the northern and middle districts, irrigation is considered unnecessary, though it would undoubtedly be beneficial during the first year.

The third year, the ploughing and hoeing is the same as the second. More attention must be given to the pruning. All the twigs are cut off save two or three, which sprout from the top of the stalk, and these are pruned so as to leave but two buds on each, which are to produce all the wood and fruit of the season. This year the vines should produce three or four pounds of grapes each; some vineyards have averaged twelve pounds to the stalk the third year.

The fourth year, the five or six twigs all starting from the top of the stalk are left with two eyes each; and this year the yield should be six or eight pounds per vine. The fifth year, there should be seven or eight twigs, with two eyes each, and the grape-yield should be ten pounds per vine. The sixth year, the vine is in full growth, and there should be eight or ten twigs, and from ten to fifteen pounds of fruit per vine. About the fortieth year the vine begins to decay. After the third or fourth year, if the vine has been well trained, it needs no stake for support, but stands alone.

The towns most notable for the cultivation of the vine in California are Los Angeles, Anaheim, and Sonoma—all grape-towns, and the only towns which depend chiefly on the grape for their revenue. Los Angeles has about 900,000 vines, Anaheim 400,000, and Sonoma 500,000. Many of the vines of Sonoma are not yet old enough to bear; those of Anaheim are all now (1862) in their fourth year; while more than one-half of those of Los Angeles have been in full bearing for several years. The Cocomongo ranch, in San Bernardino county, owned by J. Rains, has a vineyard of 165,000 vines, most of them very young yet. This ranch is one thousand feet above the sea-level, and has the reputation of producing the best wine in the state. Hock farm, in Sutter county, has a large vineyard, owned by Emil Sutter, a son of Captain J. A. Sutter, and the wine produced there has an excellent reputation.

Among the largest vineyards in the state are the following:

A. Haraszthy, Sonoma	290,000 vines.
John Rains, Cocomongo	165,000 vines.
B. D. Wilson, San Gabriel	100,000 vines.
William Wolfskill, Los Angeles	85,000 vines.
C. H. S. Williams, Sonoma	68,000 vines.
Matthew Keller, Los Angeles	61,000 vines.
Corbett and Dibblee, Santa Anita	60,000 vines.
T. J. White, Los Angeles	50,000 vines.
J. R. Scott, Los Angeles	50,000 vines.

There are many vineyards in the mining counties, but they are small and young. In 1858, an acre of bearing vineyard was worth a thousand dollars; but since then the supply of grapes and native wines has increased to such an extent, that the vineyards have depreciated fifty per cent. The profits of wine-making several years ago, induced the vine-growers to make their wine hastily and carelessly, and much of it is poor stuff, that has brought all native wines into discredit. The wine-business, just now depressed, will in a year or two become better, and then vine-planting will take a new start, and vineyards will rise in value.

§ 152. *Wine-making.*—The making of wine is considered a a branch of agriculture. In 1861, California probably made about a million gallons of wine, and the amount will increase within five years to three million gallons. The best wines are made from foreign grapes, of which, however, not many are as yet produced in the state; so that the Mission grapes yield the chief supply. The principal classifications of wine are into red and white, light and heavy, still and sparkling.

Wine-making commences with the ripening of the grapes, about the middle of September. The berry is considered to be fully ripe when the heart has taken a tinge resembling the darkness of the skin; when the berry is perfectly sweet, and comes off easily from the stem, leaving no juice upon it; and when, on holding a bunch up to the sun, the fibres running from the stem into the berry are nearly or quite invisible.

The branches are cut off with a knife, after the dew or fog (if any) has been dispelled, put into a basket, and carried to the press. Here the rotten and unripe berries are carefully picked out, and the bunches are then thrown upon a coarse wire sieve. A man presses the bunches upon this sieve, through which the grapes fall, some broken and others unbroken, while the large stems and leaves will not pass, and are thrown away. Below the sieve is the masher, composed of two rollers, ten inches in diameter and three feet long, made of iron or wood. These rollers, turning toward each other, crush the berries, but do not bruise the seeds, which, if crushed, would give a bitter taste to the wine. If the wine is to be white, the pulp is pressed as it comes from the masher; if it is to be red, the pulp is left to stand for six or eight days, so that the red color of the skins may be communicated to the juice. This is the only mode in which wines are colored. The juice as it comes from the fresh berry is never red, but some varieties of grape make a yellowish juice.

After the pressing, the red and white wines are treated in the same manner. The juice is put into large casks, usually those of one hundred and forty gallons each, and about one hundred and fifteen gallons are put in each. The casks are thus not filled entirely, but a considerable surface of the wine is left exposed to the air. This is to favor fermentation, to which the atmosphere is necessary. The cask lies upon its side, the bunghole is left open, and in three or four days the fermentation begins; in three or four more its period of greatest activity has passed. The temperature is a matter of the utmost importance to fermentation, the proper degree being about 65° Fahrenheit; and if the liquid be kept either warmer or colder than that figure, it will be in great danger of spoiling. The fermentation is accompanied by a rising of little air-bubbles to the surface, where they burst, making a noise that may be heard by applying the ear to the bunghole, and which is sometimes so loud as to be heard in the cellar at a distance of ten or twenty feet from the barrel.

After the fermentation has been in progress three or four days, the wine-maker pours in six or eight gallons of fresh juice every day, until the cask is full; and for several days after that he leaves the bunghole still open, and throws out all scum that rises to the surface there. When the scum has ceased to rise, the barrel is closed, and not disturbed for a period which should not be less than three weeks nor more than three months. After this, comes the "racking off." All the liquor, except about four inches at the bottom, containing sediment, is drawn off through a siphon, or a cock placed above the level of the sediment. The remainder is filtered through a doubled cotton cloth, and is then poured in with the clear liquor, or used in making brandy. The sediment deposited in the bottom of the cask within the first three months is about one-twentieth in weight of the juice as it comes from the press. After the first racking, the new cask is filled up, the bung is put in, and the wine is not disturbed till March or April, when it begins to feel a more lively fermentation, for that process never ceases entirely.

It is said that the wine sympathizes with the vine, and that whenever the latter is in active development, the former feels a peculiar impulse also. Thus, the periods when the vine sprouts in March or April, when it blossoms in June, and when the grape ripens in September, are also the times when the fermentation is the most active. At those seasons the bungs must be taken off, or at least loosened, and the barrels must not be moved.

It is an important point with wine-makers to avoid disturbing the process of fermentation. Between times, when the wine is at rest, it should be racked off, and placed in a clean cask. At the end of a year and a half the wine has become clear, but it continues to grow better with age for about a score of years, about the expiration of which period it has acquired a mellowness and delicacy of flavor and an oiliness of consistency which neither gain nor lose by longer preservation.

Many kinds of wine are made in California. The light wines come in this state, as in other parts of the world, from the northern, and the strong wines from the southern districts. The wines of Los Angelica have a body like those of Spain, and the wines of Sonoma and the upper Sacramento valley resemble claret and hock. A wine similar to port is made in the southern part of the state, by leaving the grapes on the vines until they are "dead ripe," and somewhat shrivelled by the sun. The juice is then very strong, and, being left with the pulp ten days or two weeks, takes a strong, dark red color.

Wine is defined to be "the fermented juice of the grape," and therefore "Angelica" is not properly a wine, though it is usually classed under that title. It is made by mixing brandy, in proportions varying from one-fourth to one-half, with the grape-juice fresh from the press, and then adding some sugar. The brandy prevents fermentation. Angelica is a sweet liquor, and is usually considered a proper drink for women; but it is really stronger than the fermented wines.

The wine made of the juice that drips from the masher, or of the first drainings from the press, is considered superior to that obtained by severe pressure from the pulp. After the pulp has been pressed, it is sometimes covered with water, allowed to stand a few days, and pressed again; and the wine made of this liquor is called "Piquet," a very light wine, and generally sourish in taste.

Some sparkling wine has been made in California, but thus far without great success. The only house which has engaged extensively in the business is that of Sainsevain Brothers, who brought an experienced workman from the champagne district of France. They own a large vineyard at Los Angeles, and another at San José, and have a large capital invested in the wine-business. Their failure in making the "sparkling California," as they call their effervescent wine, was owing to the strength of the Los Angeles grape, and the "earthiness" of its taste, which earthiness is stronger in the sparkling than in the still liquid. Effervescent wines should not be strong; and the grapes grown at Los Angeles, or at least the Mission grapes grown there, are too strong. I speak of the experiment as a "failure," meaning thereby that the Sainsevains have not made an article equal to the best brands of imported champagnes; but, for all that, it is a passably good wine. Attempts are to be made this year, and for several years to come, with the lighter wines made in the middle and northern parts of the state, and there is good reason to believe that the experiment will prove entirely successful.

Sparkling wine is treated like still wine until after the first fermentation, except that it receives much more attention, and is made from a more careful selection of berries, than any other kind of wine. After the first fermentation, the wine is put in bottles, and these are placed in racks with their necks down, the racks being made so that the bottles can be raised and lowered; and the position of the bottles is changed from time to time, to assist fermentation, which continues, though in a suppressed form—the carbonic acid gas being retained in the wine, instead of escaping as it does during the fermentation of still wines in the open barrel. The management of sparkling wines is very complicated. It varies greatly in different places, and is usually kept as secret as possible. Thus, the Sainsevains keep their process to themselves.

White grapes will not make a red wine; but the skins, if left to ferment with the juice, will give it a dirty-yellow or light-brown color.

The general custom, in making wine, is to use the pure juice of the grape, but wine-makers consider it not unwholesome or disreputable to put sugar, water, or brandy, into certain kinds of wines; all of which, however, are unnecessary and injurious to the finer kinds of still wines. Sugar and water endanger the keeping qualities of the wine, and brandy spoils the flavor. In France, it is common to put sugar or rock-candy into wine intended for the sweet tastes of the Americans; and in bad years, when the grapes are sourish, they sweeten a little for home consumption. Brandy is sometimes used to prevent wine from turning into vinegar; but the mixture, if strong enough to have the desired effect, deserves rather the name of adulterated brandy than of wine. Different kinds of pure wines may be mixed without impropriety, but the label should not misrepresent the nature of the mixture. It is a fraud to mix a bad wine with a fine article, and then sell it by the name of the latter. If a wine-maker sends his wine into the market under his own name, no other person can honestly mix any thing else with it, and still preserve the name of the original maker.

Most of the wines hitherto made in California have been pure, but not fine in flavor. The Mission grape lacks delicacy and fruitiness of taste, and gives an earthiness or harshness to the wine. These defects will probably be remedied by the use of the foreign grapes, and their mixture with the native Mission grape. Still wines, equal to the best still wines of France, have been made from foreign grapes in California; and we presume that we can make equally as good wine, in very large quantities, so soon as we have the grapes. Only about one-tenth of the foreign vines planted in the state are now in bearing.

The skin of the grape probably contains tannin, for the red wines have an astringent taste not common to the white.

The cellar is a matter of great importance to the wine-maker. From the moment when the grape-juice comes from the press until the wine is brought upon the table to be drunk, it should be kept in a cellar; and it is only in a cellar that the equability and coolness of temperature proper to favor fermentation can be obtained. In France and Germany, it is often necessary to have fires in the cellars; and it would be well to have them occasionally in California. Indeed, wine-makers generally have no cellars, but only houses. In Los Angeles county, most of the wine is kept in adobe houses. The sandiness of the land, the frequent irrigation, and the proximity of the vines to the places where the wine is stored, would lead to the filling of deep cellars with water; so the cellars are dug only three or four feet into the ground; and an adobe wall three feet thick, and a thick covering, render the cellars pretty cool. In Sonoma, Colonel Haraszthy has dug a wine-cellar in the side of a hill of magnesian limestone. The wine-cellar should be used for wine alone, because the presence of other things—especially salt meat, leather, and putrefying vegetables—may spoil the flavor of the wine.

It is probable that, in many of the vineyards, the soil will not produce a first-rate wine. In Europe, the wines from the flat lands are generally of an inferior quality. To what extent this rule prevails here, cannot be ascertained until we have given the finer foreign grapes a fair trial. Certainly the Mission grape takes up in most of the vineyards an earthiness of taste which must never be found in wines of the best quality. We cannot yet tell what are our best grape-soils, or how they differ from one another in their influences on the wine. It is certainly no easy matter to make fine wine out of the Mission grape, and most of our wine-makers have little experience in the business. Again, they send their wine to the market too soon after it is made. They often use old barrels and bottles, which may give a taste to the wine. They have also been too careless in pressing grapes before they were fully ripe, and without picking out the green and rotten fruit.

§ 153. *Berries.*—Alameda county cultivates, chiefly for the San Francisco market, four hundred and fifty acres of strawberries, one hundred of raspberries, and thirty of blackberries —more than all the remainder of the state. The varieties of strawberries most prized are the British Queen and Longworth's Prolific. They are planted in rows, thirty inches apart, and the plants are a foot apart in the rows. The strawberry comes into the market in April, and continues abundant till July, but it may be obtained in any month in the year; and the only reason why large quantities are not grown from August to October inclusive, is, that they are not in demand, because of the abundance of cheaper fruits. It must always be costly as compared with the tree-fruits, because it is more perishable, requires greater cultivation, costs more for picking, and produces less to the acre. The picking alone costs about two cents a pound, being done by Chinamen, who pick forty pounds in a day, and are paid seventy-five cents a day, they providing their own food. The average yield per acre is about one thousand pounds, and the average wholesale price in 1861, during the season of their abundance, nine cents per pound, making a gross yield of one hundred and twenty dollars to the acre. The largest field of strawberries contains eighty acres, the second seventy, and the third sixteen. Very few strawberries are preserved, or used in any way save to be eaten fresh on the table.

Most of the raspberry-fields of California are in Oakland and its immediate vicinity. The land about a mile northward from the Oakland wharf is said to be peculiarly favorable to the raspberry. The average yield is about fifteen hundred pounds per acre, and the wholesale price in 1861 was ten cents per pound. The picking is done by Chinamen, and costs from three to four cents a pound. The bushes are planted in rows five feet apart, and one foot apart in the row. The berries are produced on wood a year old, and at the end of every season all the wood that has borne is cut away. It is not known how long the raspberry-bush will continue to bear, but probably ten years at the least. The largest raspberry field in the state contains twenty acres.

An excellent wine is made of raspberries, and some of the Oakland people have gone into the business, hoping to derive a large profit from it. The berries are bruised, a gallon of water is poured on for three pounds of berries, and after three days the berries are pressed; three pounds of sugar are added for every gallon of the liquid, which is then put into barrels, and allowed to stand, with the bung open, until active fermentation has ceased—a period varying from five weeks to two months—and the wine is ready for use, though the longer it stands the better it becomes. Soft water is better than hard for putting on the berries, and hard water should not be used without boiling. An acre of raspberries will yield six hundred gallons of wine, which, if well made, is worth a dollar or a dollar and a half a gallon at wholesale. The raspberry cultivated at Oakland is the Fastolf variety. The fruit ripens during June and July.

The Lawton blackberry is also cultivated about Oakland. The rows are from six to eight feet apart, and the plants from four to six feet apart in the row. The yield of the blackberry and the cost of picking are about the same with the raspberry, but the price is much higher, the blackberry selling at wholesale in 1861 at twenty-five cents per pound. It is in the market from the first of July to the middle of September. Wine is made from the blackberry in the same manner as from the raspberry, and sometimes the two berries are combined together.

§ 154. *Ornamental Shrubs.*—Professional gardeners say that California is better fitted by Nature than any part of Europe or the Atlantic slopes to have beautiful ornamental gardens. Our shrubs are more numerous, grow larger, remain green longer, and have a longer blooming season, than those of other states. The mayo and malva trees, the rose, the daisy, the pansy, the œlyssum, the clyanthus punceus, the flowering verbena, the hollyhock, and the calla or Ethiopian lily, bloom here in the open air every month in the year. The honeysuckle, metrosideros, and myrtle, bloom from March to December; the geranium and snow-ball from April to October; the violet from October to May; the pittosporum from November to March; the spireas and flowering almond from March to June; and the camelia japonica from January to May, all in the open air. Persons at all familiar with the cultivation of these flowers in New York will observe that the blooming season here is, on an average, fully double its length there. Not only do they bloom in the open air, but they retain their leaves through most of the winter months, so that our gardens are never bare and cheerless as they are in the Atlantic winters. I have seen a rosebush bearing twenty full-blown roses in January, and that in the open air, with no assistance from artificial heat, and no protection save that of clambering up a brick wall on the southern side of an unoccupied house. Our roses are larger as well as more abundant than in the Eastern states, but their perfume is not so strong.

A marked feature of our ornamental gardening is our ability to cultivate in the open air many plants which can only be preserved in this latitude east of the Rocky Mountains under glass and with the aid of artificial heat. These plants are too numerous to be all specially named here; but some of the more important are the orange, camelia japonica, laurastinus, myoporum, ericas, casuarina, daphne, eucalyptus, metrosideros, and thirty varieties of acacia, twenty of them from Australia. It might be almost said that we have no hot-houses in the state, but only green-houses, for it is scarcely ever necessary to make a fire, even to protect the most delicate of tropical plants.

Our climate is very favorable to the growth of evergreens, especially to those strange and beautiful ones from Australia, with the graceful growth and the brilliant, feathery foliage. Two of the most striking features of our ornamental horticulture are the malva and mayo trees. The former, a native of the southern part of this state, grows to be about fifteen feet high, continues green throughout the year, and is always covered with abundant foliage and a wreath of large crimson flowers, resembling the flowers of the crimson hollyhock in size, shape, and color. The mayo-tree is an evergreen, originally from Chili, always brilliant with abundant yellow flowers.

Among the most common and beautiful creeping vines grown in California is the Australian bean, which has a dense, bright, evergreen foliage, and abundant flowers throughout the year. It climbs strings, and is therefore well suited to shade verandas and to grow in the front of porticoes.

The rose, the honeysuckle, the veronica, the oleander, the laurastinus, the euonimus japonica, and the verbenas—especially the lemon verbena—may safely be said to make twice as much wood in a year as they do on the Atlantic coast. The geraniums in San Francisco are almost trees. Rose-sprouts often grow twenty feet in a season, and other plants in proportion. There is scarcely any tree or shrub cultivated in the Atlantic states which does not thrive equally as well here, except the weeping willow.

California has thus far furnished very little for our gardens. There are many singular plants in our mountains, but few have found favor with our gardeners. The malva and the ceanothus are the chief ornamental shrubs, indigenous in California, adopted for cultivation.

We have not yet had time to produce many ornamental gardens; but I think the time is not far distant when no place in the world will, within so small a district, have so many fine gardens as the valleys round San Francisco Bay.

§ 155. *Pests of the Farmer.*—Certain "pests" of the farmer must be mentioned here, among which are the spermophile, gopher, grasshopper, locust, grape-bug, orange-bug, army-worm, Canada thistle, mullen, dock, fern, and so forth. Of the spermophiles and their habits I have spoken in the chapter on the zoölogy of the state. The amount of mischief which they do is very great. The most effective means of driving them off are poisons, chiefly strychnine and phosphorus. About a drachm of strychnine is dissolved in a quart of whiskey, and then the solution is poured over dry wheat in such quantity, that the surface of the liquid is just on a level with the top of the grain. In the course of twelve hours the wheat absorbs all the liquor, and a few grains may then be thrown in front of every squirrel-hole. When phosphorus is to be used, the wheat is soaked in boiling water until it is soft, when the water is drawn off, and the wheat in a pan is put in or over boiling water to keep it near the boiling heat. A stick of phosphorus three inches long is put into the hot wheat, melts in ten minutes, and the wheat is stirred about well, so that the melted phosphorus will touch every grain. The wheat is then poured upon some bran in which it is rolled so that every kernel may be covered, and the grain is ready for its purposes of destruction. A couple of kernels will kill a squirrel; and if a cat eats the squirrel, it will kill him; and if a raven picks out the eyes of the cat, he will die too: and such a progressive destruction has been observed more than once in California.

The gopher is more readily caught with traps than the spermophile. In the chapter on zoölogy I have described the trench used for keeping gophers out of orchards and gardens, and for catching them. Several traps are in common use, but it is not easy to describe them; so I will not attempt it.

The grasshoppers are the greatest pests of the farmer in

California, and I fear that we have not yet seen the worst of them, though several times during the last fifteen years they have eaten every green thing within large districts. They come in millions upon millions, and darken the air, moving forward at the rate of a mile or two a day, and leaving no grass or leaf behind them. Grains, grass, weeds, kitchen vegetables, and fruit-trees, are alike eaten bare of every green particle. Grasshoppers are abundant in countries where the summers are dry, the winters warm, and the vegetation vigorous; and if a large extent of land be uncultivated, they will occasionally be so numerous as to destroy every green thing. They are bred in the hills of California, and after dry winters descend into the valleys, usually content to eat the wild grasses, but sometimes attack the cultivated fields. There is no known method of killing them after they have entered a field, or of driving them away from it; but they may be kept out by digging a trench, putting straw in it, with some moist straw on top, and then setting fire to it. The grasshoppers do not like the fire and smoke, and will try to avoid them.

Under the head of the grape and the orange, I have spoken of the bugs which infest them. The army-worm has been seen in California, but has done little damage as yet. The curculio and weevil are not known in the state. The Canada thistle, the mullen, and the dock, have been introduced, but have not yet given much trouble.

§ 156. *Neat Cattle.*—California has 1,100,000 neat cattle, 900,000 sheep, and 150,000 horses, nearly all bred in the open air and open plains, and fed only on wild grasses. The system of breeding live-stock differs much in California from that which prevails in the Atlantic states, where cattle are kept in fields and stables all the time, and fed with cultivated food. Here domestic animals grow more rapidly, and reach their full development earlier, than east of the Sierra Nevada.

§ 157. *Spanish Cattle.*—Most of our neat cattle are of the old Californian breed, brought hither by the Spanish missionaries from Mexico, about 1770. At what time their stock came originally to Mexico is not precisely known, but without doubt it was in the seventeenth century, soon after the conquest by Cortes, and they must have been imported from Spain. They are called "Spanish cattle." In Mexico, as subsequently in California, they were allowed to run almost wild, and they took something of the appearance of wild animals. They have nearly the same range of colors as the neat cattle of Europe; but mouse, dun, and brindle colors—almost infallible signs of "scrub" blood—are more frequent; and the deep red, fine cream-color, and delicate mottling of deep red and white, found only in animals of high blood, are entirely wanting. Their legs are long and thin, their noses sharp, their forms graceful, their heads high, their horns long, slender, and wide-spread; and they have a duskiness about the eyes and nostrils similar to that of the deer, between which animal and a young Spanish cow there are many points of resemblance. The general carriage of the Spanish cow is like that of a wild animal: she is quick, uneasy, restless, frequently on the lookout for danger, snuffing the air, moving with a high and elastic trot, and excited at the sight of a man, particularly if afoot, when she will often attack him. In some counties it is, for this reason, unsafe to go about on foot. The native Californians are always mounted, and to these the cattle are accustomed; but a man afoot is considered to be a dangerous animal, deserving of the same treatment with wolves and coyotes. The Spanish cow is small, does not fatten readily, produces little milk, and her meat is not so tender and juicy as that of American cattle.

The breeding of neat cattle was almost the only business of the country previous to the American conquest, and they were killed for their hides and tallow, which were the chief exports. The meat went to enrich the land; there was too much of it to be eaten. The breeding of cattle, being the chief occupation of the Californians, determined their mode of life, the structure of their society, and the size of their ranches. Nobody wanted to own less than a square league (four thousand four hundred and thirty-eight acres) of land; and the government granted it away without charge, in tracts varying from one to eleven leagues, to anybody who would undertake to erect a house and put a hundred head of cattle on the place. It was common for one man to own five thousand head of cattle. The cows were kept for breeding, and the steers were regularly killed as they reached the age of three or four years. All had the freedom of the country, and ranged where they pleased, except that several times a year every man collected his own upon his ranch. There was about one bull to fifty cows. No attempt was made to improve the breed, nor was any profit to be obtained from an improvement. Most of the calves were born about the beginning of the year, and in March the first *rodeo* was held.

§ 158. *Rodeos.*—The word *rodeo* comes from the same root with "rotate," and means a surrounding, a gathering of all the cattle on a ranch, and the separation and removal of those belonging to other ranches. There are general and special *rodeos*. A rodeo may be for one ranch, or for several; but every ranchero owning a large ranch and many cattle has his own rodeo. Every large ranchero must have at least one rodeo in the spring, and another in the fall. The general rodeo is held for the benefit of all the cattle-owners in the neighborhood; the special rodeo is held for the benefit of some particular person or persons who desire an opportunity to remove their cattle from a ranch. Every owner of a rancho is required by law to give a general rodeo every spring.

When a general rodeo is to be held, the ranchero sends notice several days or weeks in advance to the cattle-owners in the vicinity; and in the cattle-districts the neighborhood extends forty or fifty miles, for cattle will stray that distance. On the day appointed, the ranchero having selected some place where the cattle are to be collected, sends out his mounted *vaqueros* or herdsmen at daylight to drive the cattle to the appointed place, where they are gathered at ten o'clock. By that time, the interested rancheros with their vaqueros have made their appearance, and are on the ground, all mounted, and prepared for the day's work.

The ranchero who gives the rodeo is present to entertain his visitors, and his men are instructed to keep the cattle together. The herd may be very large. I have seen eight thousand head of cattle in a rodeo, forming a solid body about a quarter of a mile in diameter in every direction. The visiting rancheros who have come from the greatest distance are permitted to enter the mass first, select their cattle, and drive them out. Each man has a position chosen at a distance of half a mile or a mile, whither he drives his cattle; and there are several men there mounted, to prevent them from returning to the main herd. When a ranchero sees one of his cows in the herd, he calls to a friend, and the two chase her out. She does not wish to go, and tries to hide herself among the other cattle. The horses, accustomed to the rodeo, soon recognize the cow that is to be parted out, and enjoy the work. They turn with every turn of hers, and she is soon tired and compelled to go out. If the cow be accompanied by a large unmarked calf, the latter is often caught with the lasso, thrown down, and then marked with the knife. While these rancheros are riding about among the herd and seeking their own, the cattle are driven by a few vaqueros belonging to the ranch so as to move about in a circular manner. As the cattle are thus moving round in one direction, the rancheros of the immediate neighborhood, whose time has not yet come for entering the centre of the rodeo, ride round in a direction contrary to the course of the herd, and thus are enabled to see them to more advantage than if they were standing still. After the rancheros from a distance have parted out all their cattle, those of the vicinity ride in, and the whole day is thus spent in racing and chasing after cattle.

The man who gives the rodeo does not attempt to examine the cattle which are taken away. He takes it for granted that every one will drive off only his own animals. Sometimes several days are necessary to complete the general rodeo of a ranch, and the work is continued from day to day until finished. All the rodeos of a neighborhood are usually held in a regular and close connection. The rancheros from a distance, therefore, stay until they have attended all the rodeos in a district to which they suppose that any of their cattle have strayed; and they are usually the guests of the man upon whose ranch the rodeo is given.

When a cow is driven out, her calf follows. Every ranchero knows his cattle by the brand, which law and custom require him to use. Of course, when a man has four or five thousand head of cattle, he cannot recognize them all by sight; he can only distinguish them by marks. He knows his cows by their brands, and his calves by their following the cows.

The spring rodeos are the busiest seasons of the rancheros, and are for them the chief occasions of general meeting, exciting adventure, conversation, and festivity, in the course of the year. Frequently three or four hundred men will meet at these places, mounted on their best horses, and ready for fun. All the work of the rodeo is exciting. Lively scenes are enacting at every moment, and in every direction. Calves will try to get away from the herd, and escape to the hills. Cows which have been driven out will endeavor to get back. These must be chased by the horsemen. Frequently the lasso must be used. Many of the vaqueros are fond of showing their skill before so many spectators, and astonishing feats of horsemanship are performed.

When a ranchero returns from a rodeo, with his cattle which had strayed away, he drives them into his corral, and brands and marks his calves; so that if they should return to their former range, he will know them the next year. If those that have been on other ranches are too numerous to he branded and marked in one day, some of his vaqueros stay with them on horseback, and herd them until all can be marked. When a cow has become accustomed to a ranch, she likes to return to it. After all the calves are marked, the owner does not care much whither they go, provided that they do not stray beyond the limits of the ranches, the rodeos of which he attends. It is only in times of extraordinary scarcity of grass, that the rancheros are particular to drive the cattle of other owners off their lands.

The rodeo season being over—that is, when the ranchero has all his cattle on his own ranch, and his alone—he commences the work of branding. His vaqueros drive about two hundred cows with their calves into the corral every morning, and two or three good vaqueros will brand these calves in a day. The vaqueros enter the corral with their horses, which they need when the calves are large and strong, for many of them are three and four months old. If the calf be small, the vaquero may be afoot to lasso him. One vaquero throws a reata over the calf's head, and another catches him by the leg; they throw him down, and one holds him, while the other gets a hot branding-iron and burns the owner's mark upon its hip. Thus the work goes on from day to day, and from week to week, until every calf on the ranch is marked.

§ 159. *Brands.*—The law requires that every horse and cow shall be branded, with a brand belonging to their owner. The brand is made of iron, sometimes representing one or two letters, sometimes other arbitrary signs, such as a cross, a circle, a triangle, or any other design. The brand may be six inches long by four wide, and the thickness of the iron is about a third of an inch. There is an iron handle, with a wooden cross-piece at the end, so that the brand can be handled when hot, and held down firmly upon the prostrate calf, until the figure is indelibly burned into the skin. A copy of every brand must be burned upon leather, and deposited in the county recorder's office. Every minor and servant on a ranch must use the brand of the owner of the ranch. The brand must be burned, under penalty, upon all horses and neat cattle, before the age of eighteen months. The brand is burned upon the hip, and indicates ownership; when the animal is sold, the brand is burned upon the shoulder, and indicates sale. The purchaser then puts his brand upon the hip; and thus the skin of a Californian horse or cow contains the history of its ownership. Many of the brands are well known to the rancheros over a large portion of the state; and by looking at the animal, they will tell where it was born, and who have owned it at different times. The hips and shoulders on both sides are often covered with brands. Sometimes the brands grow with the animals; in other cases they remain nearly of their original size. A brand well burned into the skin is perceptible as long as the animal lives, though it grows less and less distinct with the advance of years.

In the fall there is another season of rodeos, to brand such calves as may have escaped notice at the spring rodeos, or may have been too small to be branded.

The rancheros sometimes have a mark in addition to their brand, such as slitting the ear or cutting a notch in the dewlap. A drawing of the mark must be deposited in the county recorder's office. It is contrary to law to cut off the end of the ear, or to cut it on both sides so as to bring it to a point; for those modes of marking would give opportunities to cut away the marks of other people. The bull-calves are usually altered at the rodeos, as well as branded and marked. The cattle on many ranches are touched only twice in their lives by the hand of man—first, when they are branded; and next, when they are slaughtered.

§ 160. *Early Maturity of Californian Cows.*—The cows calve almost invariably before they are two years old, frequently before they are eighteen months, and sometimes before fourteen months. They generally arrive at maternity a year sooner than in the Atlantic states. It is said that suckling heifers have been seen to take the bull. The Spanish rancheros have eight or ten bulls to a hundred cows; the Americans usually four or five. The calves suckle from six to ten months; that is, from January or February, when they are born, until November, when the pasturage is very scanty. The Spanish cows have small udders, and yield little milk; and notwithstanding their great number in the country, butter, milk, and cheese were very rarely seen on the table previous to the coming of the Americans. American cows are the only ones used for the dairy, but many of them are now kept also for breeding alone, and, like the Spanish cows, they are never milked.

§ 161. *Corral and Reata.*—The *corral* is an important part of all cattle-ranchos, and on many of them it is the only enclosure. It is a pen, from thirty to fifty yards square, surrounded by a high, strong fence. It is used whenever horses or cattle are to be branded.

The *reata*, used for lassoing, is a rawhide rope, about five-eighths of an inch in diameter and thirty yards long. It is made of four strips of cowhide, from which the hair has been scraped; and after plaiting, it is greased and dragged along on the ground after a saddle, to render it pliable. Rawhide is better than any other material, because it has just the proper weight and stiffness for the purpose. A running noose, which slips very easily, is arranged at one end. When the reata is to be used, the noose is made from four to six feet long; one side of the noose and the reata just outside are taken in the right hand, so that while in the hand the noose will not slip; the remainder of the reata is held coiled up in the left hand, ready to be let go. The vaquero swings the noose round his head, in such a way as to keep it open; and when he has a good swing, he lets go, and away it will fly, its whole length. If it catches the object aimed at, the noose draws tight. It is not an uncommon thing for a vaquero to catch a cow at a distance of thirty feet, while she and his horse are both running rapidly; but usually he will get within fifteen or twenty feet, if he can, before throwing his reata. A good vaquero, standing in front of another man, can push the latter back, and the moment his foot leaves the ground, throw a reata under it, and thus lasso him by the leg. When cattle or horses are to be branded, they must be thrown down; and this is generally accomplished by catching the head with one reata, and a hind-leg with another.

§ 162. *Occasional Starvation.*—Nineteen out of twenty of the cattle of California never get any food, save such as grows indigenously in the open country, and they always suffer for it. From March to July the pasture is abundant and excellent, and the cattle are fat; from July to October, in ordinary years, the grasses and clovers, though dry and brown, are nutritious, and the cattle still remain in good condition; but from October to January they grow lean rapidly, and almost every year a considerable number of them die by starvation. Either the grass may be all consumed, or it may be deprived of its nutriment. The first case happens when the grass is very scanty, because of a small fall of rain during the winter; the second occurs when a heavy rain, lasting a day or two, comes before New Year's day, and is followed by cold dry weather. The rain takes away the palatable and nutritious qualities of the old grass, and the cold and dry weather prevents the starting of the new grass, and between the two the cattle suffer. In 1856, seventy thousand head of cattle died in Los Angeles county alone by starvation, one-third of the entire number in the county. The state has not one large cattle-ranch surrounded by fence, and therefore, if a man owns good pasture land, the cattle of other people come and eat the grass, and later in the season his own must suffer. It is impossible for the *vaqueros* to drive away the strange cattle, because they enter the ranch on every side every day, and if the grass be much better than on adjoining ranches, they will, if driven away one day, return the next. The proper and profitable method of managing an extensive cattle-ranch, is to have it all fenced in and divided off into a few large fields, in which the cattle could be pastured at different seasons. It would also be a source of profit to have hay for them or green alfalfa in the early winter, so that there would be no danger of a reduction to skin and bone: for it costs thrice as much to replace as to preserve a pound of flesh. Spanish cattle, when slaughtered between September and February, are usually very thin, and in the Atlantic States it would be grossly impolitic to send such animals to the market.

§ 163. *Imported Cattle.*—The great majority of the cattle

in the state are of pure Spanish blood, but there is a considerable number of American cows and half-bloods, which are constantly increasing in number, while the Spanish cattle are decreasing. Most of the American cattle are in the Sacramento basin and the coast valleys north of Monterey, while along the southern coast they are very few in number.

During the last four or five years about seventy-five pure-blood Durhams, twenty-five Devons, and four Ayrshire bulls and cows, have been imported across the Isthmus of Panama. The freight on the steamer and railroad, from New York to San Francisco, for a bull or cow, is two hundred and seventy-five dollars. The food is carried free of charge, but the owner must find a person to feed and take care of the animal during the voyage. Some of these imported animals are of the best English stock. The better the blood the larger the beeves grow, and the more readily and rapidly they fatten. The Spanish cattle are too uneasy; they run about so much that they lose flesh. The half breed American cattle are more quiet, the full-blood American still more quiet, and the Durham the best of all. But as our beef cattle get no food save such as they pick up in the open country, I doubt whether it will pay to import these pure-blood Durhams to improve the breed. The excellence of the Durham is caused by care in breeding, protection against the weather, and an abundance of good food; and the excellence cannot be maintained without a continuance of the same system of management. The Durham will, in a few generations, cease to be a Durham, if he gets no food save such as he can pick up in the valleys and hills; but if carefully fed, he will, of course, do as well here as elsewhere. It is said that the Durham needs succulent food; if so, the blood will soon degenerate on the dry grasses of California. The Devon stock, which has been praised by the importers of it as peculiarly fitted to thrive on our indigenous grasses, has not found much favor. Our dairy cows are the only ones which are well taken care of; and therefore the Ayrshire blood is really more needed, and likely to be better preserved, than either the Durham or Devon. Roots are seldom cultivated for cattle; hay, barley, and wheat-bran, are used for feeding them when kept in the yard. Beeves are never stall-fed in California.

§ 164. *Dairies.*—The chief dairy districts of the state are the valleys in the vicinity of the bay of San Francisco. The business is very profitable, but requires a considerable capital.

The climate near the coast is very favorable for making butter and cheese.

In 1860, according to the assessors' reports, Santa Clara county produced 220,000 lbs. of butter and 300,000 lbs. of cheese; Marin, 226,000 lbs. of butter and 170,000 lbs. of cheese; Sonoma, 220,000 lbs. of butter and 103,000 lbs. of cheese; Sacramento, 148,000 lbs. of butter and 122,000 lbs. of cheese; Yuba, 92,000 lbs. of butter and 5,745 lbs. of cheese; and Alameda, 79,000 lbs. of butter and 103,000 lbs. of cheese.

§ 165. *Spanish Horses.*—California has about one hundred and fifty thousand horses, of which about one-third are American; one-third wild Spanish; and one-third tame Spanish. The Spanish horses are of the old stock imported, sent early in the sixteenth century from Spain to Mexico, and thence brought to California about eighty years ago. Like the neat cattle, the Spanish horses run wild, and partake, to some extent, of the wild nature. They show their base blood by their colors—mouse color, dull duns of various shades, and calico color, or mixtures of white with red or black, in numerous large spots or blotches, are common; while chestnut, bright sorrel, blood-bay, and dappled gray, are very rare among them. They are quick, tough, healthy, and unsurpassable for the uses of the rider and the *vaquero;* but small, lacking in weight, strength, and beauty, and unfitted for the heavy, steady work of the plough, cart, or wagon. They are wanting in the docility, kindly disposition and steadiness of the well-bred horse; and they have little of that kind of sense which leads an American horse to be quiet and gentle, even in circumstances strange to him. For California, as it was in 1845, there were no better horses than the Spanish-Mexican. They have a wonderful toughness, and some of their exploits in the way of travelling are unsurpassed in the annals of the turf. A number of instances are on record where Californian horses have carried a rider one hundred miles in a day, and that with no food save grass. Sixty miles a day is not an uncommon ride, nor is it considered a severe one. Fremont, on one occasion, rode four hundred miles in four days, riding different horses, but driving them before him from the beginning to the end of the journey.

More than half of the brood-mares of the state are wild Spanish; that is, they live entirely in the open plain, are unbroken, and many of them have never been touched save when they were to be branded. They are in bands called *manadas*, numbering from thirty to sixty mares, which are under the guidance of one stallion or garañon. He knows every one of his band, keeps them together, conducts them to what he considers the best pastures, and drives away geldings, stallions, mules, and whatever animals he may dislike. When a vaquero tries to drive the manada into a corral for the purpose of catching some of the band, the garañon will frequently divide them and scatter them about, and render it impossible for the vaquero to get them together; for while he drives in one place, the stallion is equally busy at another, and the mares fear his teeth and heels as much as the swinging reata of the horseman. The garañon is usually from five to nine years of age. He guards his manada with the most jealous care. It sometimes happens that one garañon tries to take away a mare from the band of another, and then a fight ensues, in which the weaker has to suffer a severe biting and kicking, and then lose the object of the battle too. The manada keeps together for year after year, but when it gets too large, the vaquero will divide it and give a portion to the charge of another garañon. All the mares foal before they are three years old, whereas in the Atlantic States they seldom foal until a year later. They also breed more regularly than elsewhere, for when mares are kept in stables, they frequently pass seasons without breeding. The foals are branded at the age of three or four months, and are weaned at the age of eight or ten months. The fillies continue to run with the manada, and become part of it. The colts are altered when branded, and continue to run with the manada until they are three or four years of age, when they are broken and put into the *caballada*, or herd of broken horses. The Mexicans never broke their mares, and considered it discreditable and a mark of great poverty to ride one.

§ 166. *Horse-breaking.*—The Mexican system of breaking horses is peculiar. They are broken only to the saddle; for horses were never used before wagons by the Spanish-Californians. The horse having run free all his life, is too wild to be caught without a lasso, or to be approached at first while he is on his feet. He is therefore caught by the neck with one reata, by a hind-foot with another, and then thrown down by pulling the reatas in different directions. A vaquero goes to the horse as he lies down and puts a *jáquima*, or a kind of halter, on his head. The jáquima is provided with a piece of leather, which can be pulled over the eyes so as to blind the horse, but it is first lifted to let him see. The reatas are taken from the neck and leg of the prostrate horse, and a long rope having been fastened to the jáquima, he is allowed to rise. This is the first time he has been haltered and he dislikes the restraint, but he exerts himself in vain to get loose. After he has tired himself in vain efforts, the vaquero goes up for the purpose of pulling down the blind. The horse is terrified at so near an approach of a man, and trembles with excitement. If he pulls back hard the vaquero sees there is little danger, and slowly advancing and putting his hand to the blind pulls it down over the horse's eyes. But if the horse stand up, it is probable that he will rear up when the vaquero comes near, or strike at him with a fore-foot. Such animals are very dangerous, and are usually allowed to tire themselves by standing without food, or else they are drawn up to a fence or tree, behind which the vaquero can protect himself. So soon as the blind is down, the horse is perfectly quiet. He can be rubbed, and saddled, and mounted without a motion. The first thing after blinding is to put a saddle on him. It is fastened well, the blind is raised, and the full length of the rope given to him. He does not understand the saddle. It may be a carnivorous beast for all he knows. He is terrified at it. He jumps, and snorts, and kicks, rears and pitches, throws himself down, worries himself out, and falls into an agony of despair. After an hour or so of such work, the vaquero advances again, puts down the blind, and the horse stands trembling with fear and exhaustion. He is now to be ridden. The rope is fastened under the chin, so that it can be used for a bridle-rein; the horse's ears are pushed down under the upper part of the jáquima, so that he shall be deaf as well as blind; the saddle-girth is tightened, and the rider mounts. Over the saddle he has a second girth, which is loose enough to allow him to get the point of his knee, bent at right angles, under it. This girth ties him upon the horse. The more he presses the knee outward, the tighter the girth holds him; whereas by turning his knees inward and straightening his legs, he can be free in an instant. Having put his knees under the girth (he does not care for the stirrups and cannot use them) he reaches forward, takes the ears out from under the jáquima, and raises the blind. The horse, as soon as he sees the man on his back, is stricken with a new terror. He immediately commences to jump stiff-legged. He springs up into the air and comes down on his fore-feet with his legs stiff. This is the way in which horses try to shake off panthers, and they resort to the same method with men. The shock would be severe if the man were not tied down to the saddle, but he moves with the horse and is not hurt with the shock. Sometimes a horse will jump thus for hour after hour, and the rider is very well satisfied, for there is no danger in the jumping, and it is very tiresome to the animal. Some horses, after jumping for a few minutes, will commence to run. To this the rider makes no opposition, but practises the horse with the reins to accustom him to guidance. The most dangerous horses are those which rear up and fall backward. In such case the vaquero must be ready to throw off the girth from his knees, and alight upon his feet as the horse falls. After riding for a couple of hours, the vaquero reaches forward, pushes the blind down over the horse's eyes and dismounts. To dismount without first putting down the blind would be very dangerous, for the horse would probably kick him. He takes off the saddle, hoists the blind, ties him to a fence or a tree with his head up, and gives him neither food nor water. The next morning he is very stiff and hungry. The vaquero does not feed him, but pulls down the blind, puts on the saddle, mounts him again, and the scenes of the previous day are re-enacted, though the jumping is less furious. After a couple of hours of exercise, the horse is tied where he can get water and grass. Every day he is ridden. In five or six days he quits jumping. In three months the blind is laid aside. In four months a bit is put in his mouth. This is strange to him, and he jumps stiff-legged. If the first bit used is American, he will jump again when the harsher Spanish bit is used. When any thing is wrong he jumps stiff-legged. During the first month or two his nose will be very sore where the jáquima or halter crosses it, caused by the pulling of the halter in holding and guiding the animal. At the end of a couple of months he learns to follow the guidance of the jáquima almost as readily as afterward the bit. After he has been ridden daily for six months, he has become tame and quiet, and he commences to fatten up again; for during the first three or four weeks he worries himself so much, with his vain plungings, that he loses flesh rapidly. The Californian horse, when once broken, is kindly in disposition. He rarely bites or kicks, no matter how roughly he may be used.

After having been broken to the saddle, he must be taught the uses of the reata. The vaquero always carries his reata with him, and the horse soon learns to see it swinging about the rider's head. The reata is first thrown at small calves and then at larger ones, and the horse gradually learns that he can best hold a lassoed animal by presenting his head toward it, and bracing himself back with his fore-feet. The reata is fastened to the horn of a saddle, strong enough to hold a bull.

The horse learns to watch the reata; if it catches, he slackens his pace, or stops suddenly; if it does not catch, he continues at full speed, while the vaquero pulls up the reata and prepares to throw it again. The saddle and bridle are both peculiar, and necessary to the trade of the vaquero. The saddle-tree, or *fuste*, is made of four pieces of wood, two of which are longitudinal and rest on the sides of the horse; one forms a high back; and the fourth is a fork, which rises in a large strong horn. The pieces are strongly fastened together, and the whole framework is covered with wet rawhide, which shrinks when dry, and contributes much to its strength. A good fuste is stout enough to hold the strongest bull. The girth is four inches wide, and is made of a number of little ropes of horse-hair, connecting two iron rings four inches in diameter. One of these rings is fastened on the right side of the saddle, by straps running over the front and back of the fuste; and a similar ring is fastened in like manner on the left side of the fuste. In this latter ring is a rawhide strap three feet long. When the saddle is to be fastened, this strap runs through the ring at the loose end of the girth, then through the upper ring, then down to the girth ring again; and the vaquero pulls, and usually draws so tightly that the wide girth cuts into the horse's belly and evidently displaces its contents. When a vaquero is preparing for a day of lassoing, there is no danger that the saddle will slip. The girth has no buckle about it, and is made wide to give it strength and to prevent it from hurting the horse. It is placed farther back than in American saddles, and rarely cuts the skin, just behind the fore-legs. Sometimes in lassoing, the animal caught will get off sideways from the horse, and the saddle must be so firmly in its place, that it cannot move; and sometimes the vaquero will drop his reata after having caught his cow or horse, and then while his horse is going at full speed he must reach down, supporting himself with one hand on the horn of the saddle and with the other seize the reata. For such feats, the girth must not be loose. The stirrups are of wood, about three inches wide, and are covered in front with pieces of thick leather, called *tapaderas*, which prevent the feet from slipping through. The fuste is usually covered by large flaps of leather called *mochilas;* and along the stirrup-straps are pieces of leather called *sudaderas*, to protect the legs of the rider from the sweat of the horse's side. Cruppers and martingales are never used by the Spanish-Californians. The bridle is so made that a hard pull on the rein hurts the horse, and a severe jerk will throw him back on his haunches. The bit has an arm projecting about two inches up in the mouth. On ordinary occasions this arm lies flat on the tongue, but when the rein is pulled it rises and presses against the roof of the mouth. The slightest pull, therefore, on a Spanish bridle is felt by the horse, and he will stop instantaneously, though at full speed, if the reins be jerked severely. It may be cruel to the horse, but it is very convenient to the rider, and necessary to the vaquero.

The common gait of the Californian horse under the saddle is the gallop. He never paces or racks, and rarely tries a sharp trot, but rests himself with a walk or a slow trot. His gallop has an easy motion which does not tire the rider in fifty miles. He has an excellent speed for a race of a quarter a mile, but he cannot run two-mile heats with the thoroughbred, though for a gallop of a hundred miles he has probably no superior.

Many of the Californian horses have of late years been broken to the wagon and the plough, and they do very well for farm work, though not equal to American horses.

§ 167. *Blood-horses.*—The American horses, that is the common stock of horses brought from the Atlantic States within the last twelve years, and their offspring, are large, fine animals, not so healthy and tough as the Californian horses, but larger, more active, stronger and more handsome in shape and color.

A large number of stallions and mares of fine blood have been imported, including about fifty thoroughbreds or English racers, two dozen Morgans or American trotters, and a dozen Clydesdale and Flemish, or heavy cart and truck horses. Some of these horses are valued as high as ten thousand dollars apiece. The Clydesdale and Flemish are considered the most valuable for crossing with the Californian mares, the offspring being large, strong farm-horses, worth twice as much in the market as the pure Californian. The thoroughbred horses in California are of the purest blood, and some of them have few superiors in speed in any country.

§ 168. *Mules.*—Nearly all the farm work of California, where draught animals are necessary, is done with horses. Mules are too dear and oxen are too slow. A great number of mules and horses are used in packing merchandise in those districts where there are no good wagon-roads. For the ordinary uses of the farm the mule is preferable to the horse, being longer-lived, more healthy, not so much injured physically or morally by ill-treatment, and contented with much cheaper and simpler food. But the mule is not considered handsome, and the small farmer wants a horse which he can ride, and with which he can take his family out, so he never gets a mule. Mules are now used on very few farms, but I think they will gradually gain in favor. Few mules are bred in the state at present, but there are some excellent jacks in Contra Costa county and in the San Joaquin valley.

§ 169. *Camels.*—Some camels have been introduced into California, but they are as yet few in number, and have been here but a short time. Our experience with them is therefore small, and we do not know what value they are to have in the future. They are healthy, increase, thrive on our wild pastures, are strong and active, climb our rugged mountains as well as horses or mules, carry loads of one thousand pounds each, go three or four days at a time without water, and are readily managed; but they have not been tried in a regular business way, and such trial is necessary before we can tell what they are to be.

§ 170. *Sheep.*—The climate of California is peculiarly favorable to the growth, increase, and health of the sheep. Our mild winters permit them to grow throughout the year; and it is an accepted principle among those familiar with the subject, that a sheep, born and bred in California, is, at two years of age, usually as large and heavy as one of three years, born and bred in the Atlantic States. The ewes produce twins and triplets more frequently here than east of the Rocky Mountains. The health of the herds is better. No fatal disease has ever prevailed to any serious extent. The "scab" exists in many herds, but in a mild form, and few have died of it. It is the general opinion of sheep-breeders that the sheep bred in California will produce more wool than those of other states. The heaviest unwashed fleece on record, is that of "Grizzly," a French Merino buck. It was fourteen months old, weighed forty-two pounds, and was sheared by Flint, Bixby & Co., in Monterey county, in 1859.

Sheep in California are never kept under shelter, and except a few of fine blood, seldom get any food save such as they can pick up on the open hills and plains. Sometimes lambs are lost with cold, but this is very rare when they are well managed. At night the herds are driven into corrals or pens, to protect them against the coyotes, and to keep them from being lost. On the large sheep ranches, one herdsman is employed for a thousand sheep. There are a few shepherd-dogs in the state, some brought from Australia, others from Scotland. The word "corral" is understood by these dogs, and when they hear it, they immediately drive the herd to the corral. At the sight of a wolf, they hastily collect the sheep into a dense body, with their tails out and the lambs in the centre. If a sheep turns his head out, the dog bites his knees and makes him turn about. The dog seems to understand that the wolf cannot do much harm by biting the rump of a sheep, but would soon kill it after catching its throat.

In most other sheep countries, the sheep-breeder is at great disadvantages as compared with California; the land is dear; it must be cultivated; the sheep must be fed by hand every day during a considerable part of the year; the herds must be under shelter in the winter; four or five men are required, on an average, to attend to a thousand sheep; the herds are not so healthy, do not increase so rapidly, do not grow so large within the first two years, and do not produce so much wool. The land of the sheep ranches in California is not worth more than five dollars per acre, on an average probably not more than three dollars. It follows that sheep-breeding should be very profitable here, and so it is. The ewes, when properly taken care of, have lambs before they are a year old—increase one hundred per cent. every year. The cost of keeping large herds is variously estimated from thirty-seven to fifty cents per head annually, exclusive of the interest of the land used for pasturage. The wool of a good sheep will pay twice the cost of keeping it; and the wool and lamb together, of a fine-blood ewe, are worth eight or ten times the cost. It is the present custom to sell the wethers for mutton when a year old, but this is bad policy, save with the poorest sheep.

The old missions had large herds of sheep, but after the management of those large establishments was taken from the priests and given to civil officers, in 1833, the sheep were neglected and most of them were killed. Twenty years later very few were left in the state, but there was a demand for mutton, so large herds were driven from New Mexico. These were a very poor stock, but they were for a long time the only sheep that could be had, and they now form the great majority of the sheep in the state. The first attempt to breed sheep as an exclusive business in California, since the American conquest, was commenced in 1853, by a poor man who had nothing save nine hundred ewes; and they increased so rapidly and proved so profitable, that now, if report be true, he has ten thousand sheep, sixteen thousand acres of land, and other property to the value of one hundred thousand dollars. Within the last three years many sheep of fine blood have been imported, and these will gradually swallow up the Mexican stock. The imported kinds are American, Southdown, Australian Merino, French Merino, and Spanish Merino. Of the two latter varieties there are few save bucks. The prices of sheep fluctuate, but the relative prices of the different breeds remain about the same. Thus when a Mexican ewe is worth about three dollars, an American is worth five dollars, a half-Merino six dollars, a Southdown six dollars, and an Australian Merino twelve dollars. The Mexican sheep produces on an average two pounds of wool per year, worth from five to seven cents per pound; the American four pounds, worth from fifteen to twenty cents; the half-Merino six pounds, worth from eighteen to twenty-four cents; the Southdown five pounds, worth twenty or twenty-one cents; the Australian Merino seven pounds, worth twenty or twenty-one cents. These weights indicate the weights of the unwashed fleeces, and the prices paid in this market for unwashed wool. The Californian wool, especially that grown in the southern part of the state, is filled with grease, dust, and sand. In one case, a fleece weighing sixteen pounds was reduced by washing to six pounds. The finer the wool, and the farther south it is grown, the greater the proportion of dirt. The wool grown in the northern part of the Sacramento valley, is cleaner than that of Alameda county, and that of the latter place is superior to the wool of San Luis Obispo. There are a few Chinese sheep in the state, and much value was for a time attached to them, because the ewes very frequently produce triplets, but it requires a good ewe to suckle two lambs well, and twins are sufficiently abundant among American sheep. Sheep-growers are divided in opinion as to whether the French or Spanish Merino be the best sheep for the state. The French Merino grows large, and averages more wool to the sheep than any other kind, but it is said that the Spanish Merino, though smaller, will produce more wool to the acre. About a thousand sheep are kept in a herd. One sheep-owner in Monterey county has 30,000 head; and others have 15,000 and 20,000 head each. The largest sheep county is Monterey, which has about 150,000; Solano has nearly as many, and after these come Los Angeles, San Bernardino, and San Luis Obispo. There are about 900,000 sheep in the state, and their number is rapidly increasing. One of the drawbacks of wool-growing in California is, that in the summer the wool gets full of the little burs of the bur-clover, which gives much trouble in washing and carding. By sheering early this trouble is to some extent avoided. Of course the sheep grow lean in the late fall, as do the horses and neat cattle, and some of them die of starvation. The same remarks may apply to high-blood sheep as to other high-blood animals—they must degenerate if they get no cultivated food; but the stock may be kept up by crossing with high-fed bucks of pure blood.

§ 171. *Swine.*—Swine are not favorite animals in California. They increase rapidly and are healthy, and their meat commands a high price, but they do not thrive upon the dry pastures; they are not permitted to run at large in many counties; the mast is scanty in the agricultural counties, and grain suitable for feed is dear. It is probable that in a few years great numbers of swine will be bred in the tules, the roots of which they like to eat; but the tule-lands at present are in wide undivided tracts, and the swine which have access to them soon get lost. The present number of swine in the state is about six hundred thousand.

§ 172. *Poultry.*—Poultry command very high prices in this state, but all attempts to breed them on a large scale, have proved unprofitable. Hens are worth from fifty to seventy-five cents each, and eggs from twenty-five to fifty cents per dozen. Chickens are healthy and increase rapidly in small poultry-yards or farms; but when more than five hundred are collected a fatal epidemic appears, and they die off. The disease seems to be a kind of apoplexy, for it attacks the fattest chickens, and they die suddenly. One large hennery, on the French plan, has been established about eight miles from Oakland, and it contains one thousand five hundred hens, with accommodations for five thousand. The poultry-yard covers four acres of ground, and one acre of it is separated from the remainder. The hens lay in the lower story of a frame house, which is open on one side. The nests are in a long trough, a foot square, open on top, and separated into nests by partitions.

Each nest has some hay in it and a mock egg of porcelain. Several times in the course of a day, the eggs are taken out and placed in a covered box near at hand. Four or five hens may use the same nest in the course of a day, and if the eggs were left in the nest the warmth might start the development of the chick, and injure the egg for either hatching or eating. It is considered bad policy to let a hen sit on eggs while laying, even if she is to hatch them herself, for some will be farther advanced in incubation than others, and then the proportion of loss will be great. Over the laying department is the roosting place, which is eighteen feet high, and has the perches so fixed that the droppings of one hen do not fall on another A stairway leads up from the ground to the roosting chamber

In the lower story of another house is the hatching department, with nests for six hundred hens. The nests are about a foot square, with a door in front, opening on a level with the floor. They are numbered and divided into sections, each of which has one door, and has hens which commenced sitting at the same date. The hens are fed by sections; ten, twenty or thirty being let out at a time, and called to eat. When first called they do not understand it, and after they have eaten, they have difficulty in finding their places; but in three or four days they come out immediately as soon as the door is opened, and when the signal for closing is given, they go to their places without the least confusion. About a dozen eggs, usually not more than four, and never more than ten days old, are given to each sitting hen, and of this dozen, nine or ten are hatched on an average. Every nest has snuff in the bottom of it to keep out the lice. When the hen has hatched out her brood, she and they are transferred to the "young-chick room," over the hatching room, where every hen is put in a pen. During the first twenty-four hours the little chicks get nothing to eat; then they are fed twice on fine bread, after that on boiled rice and corn-meal. They are fed four times a day. When five days old, the chicks, with their mother, are placed in the smaller enclosure of the poultry-yard, which has several streams of clear water running through it, bushes and mustard-plants for shade, and boxes into which the hens can retire for protection against the cold and rain. When the young brood enters the yard, the older occupants, or some of them, make war on it, and one chick in a dozen is slain in these hostilities. After the hen has proved by fighting her right to be there, peace is restored, and she and her little ones are ready to make war on any subsequent intruders. After the chickens are three months old, they are turned into the large yard, where they have to struggle for their food and lives, with all the old hens and cocks on the place. In the large yard the chickens are fed twice a day, and four ounces to each chicken per day. The food is wheat, rice, oats, barley, raw meat, cabbage-leaves, sorrel, chalk, oyster-shells, and green mustard. Regularity in the time of feeding is considered a matter of much importance. There should be one cock to a dozen hens. The hens commence laying when about eight months old and lay most in their second year. They will continue laying till their fifth or sixth year, but they are usually killed about the end of the fourth year. Hens that eat eggs and that crow are killed. Hens which want to sit when their services are not needed in that way, are shut up in the callaboose, kept there one or two days, with no food save green vegetables, ducked several times in cold water and then let out cured. In eight or ten days they are ready to commence laying again. A hen will lay fifteen or eighteen eggs before wanting to sit.

One man near Oakland, devotes himself entirely to the business of breeding rabbits, and according to rumor, finds it very profitable.

Goats, pigeons, and pheasants are bred in California, but on only a small scale.

§ 173. *Bees.*—There were no bees in California until within the last seven years, and it was supposed they could not live here, because of the dryness of the vegetation during the last half of the year; but for these insects, as for larger animals, it was found on trial, that our climate is peculiarly favorable, and that they thrive better here than east of the mountains. A good hive here will make two hundred pounds of honey, or may be made to produce twenty swarms in a season, an increase ten times as great, and a production of honey five times as large as that in the Eastern States. Mr. H. Hamilton, a bee-keeper of Stockton, reports that he had thirty-five swarms of bees on the first of February, 1860; and by the first of October, they had increased to five hundred hives, and produced twenty thousand and seventy-five pounds of honey; a production said to be without parallel. Bees are not idle during six months of the year, as in New York, but busy during nine or ten months. They find their food in wild and cultivated flowers, in the blossoms of manzanita bushes, fruit-trees, grasses, clovers, and grains, in grapes, fruits, and honey-dew. They seem to thrive in the driest portions of the state, where there are no cultivated fields and no flowers or green herbage. They are very fond of apricots, which they eat in places where the skin has been previously cut through by bugs. When the latter have made a hole, the bees come and eat side by side with the bugs, which are of the "lady-bug" kind, and other similar species. Many of the bees lose their lives in consequence of their fondness for the apricot. Either they eat too much, or they eat the meat after it has passed into the alcoholic fermentation; but whether intoxicated or surfeited, they are unable to get home, and they perish during the night. In places where the honey-dew is abundant, especially in the mountains on the eastern border of the Tulare valley, the bees make honey very rapidly. When the food for bees was becoming scarce in the midsummer of 1860, in the Santa Clara valley, a man owning seventy hives sent them to the vicinity of Visalia, so that they could get honey-dew. Indeed it is the custom of several bee-keepers in California, to move their bees about from place to place, according to the pasture and the season. Hitherto little honey has been sold in the market, the chief object of bee-keepers being to produce swarms, which for a time were worth one hundred dollars each. This business yielded so large a profit that a dozen men devoted themselves to it exclusively, and some of them, who commenced five years ago with only three or four hives, now have comfortable fortunes. The hives have increased so rapidly, however, that they have fallen greatly in value, and are now worth from ten to thirty dollars each; and honey, which is worth one dollar per pound, will come into the market. The honey made in the mountains is very similar in quality to that of the Eastern states; that made in the valleys is not so good. Many swarms have gone off and found homes for themselves in the woods, so it is not rare to find "bee-trees." Most of the bees in the Sacramento basin during 1861, were destroyed by the flood of 1862.

§ 174. *Silkworms.*—A few silkworms have been hatched in California, and have been found to thrive extremely well, but the high price of labor has prevented any extensive experiment in the production of silk. Our climate is very favorable to them in three respects, equability of temperature, exemption from electrical convulsions, and dryness in summer. The silkworms should be kept at a temperature of about 75° Fahrenheit, and this is very near the summer temperature of some of the valleys near the coast. Extreme heat and extreme cold are both very prejudicial to them. A considerable proportion of them will die if the thermometer falls to 45° or rises to 100°; thunderstorms kill a large portion of the worms every year in France, Italy, and China; in California such storms are unknown. In all the countries where silk is now produced extensively, there are showers which wet the mulberry leaves, and this moisture gives a diarrhœa to the worms whereby many are killed. These are important advantages, and may enable us to compete soon with Europe and China in the production of silk.

NOTE.—I owe acknowledgments for information about grain to J. W. Osborn, of Napa; about fruit to A. A. Cohen, of Alameda; about the grape to Charles Kohler; about the orange to John Frohling; and about the quantity of wheat from different districts to Isaac Friedlander, of San Francisco.

CHAPTER VIII.

MINING.

§ 175. *Chief Industry.*—Mining is the chief industry of California. It employs more men and pays larger average wages, than any other branch of physical labor. Although it has been gradually decreasing in the amount of its production, in the profits to the individuals engaged in it, and in its relative importance in the business of the state, it is yet and will long continue to be the largest source of our wealth, and the basis to support the other kinds of occupation.

§ 176. *Metals obtained.*—Our mines now wrought are of gold, silver, quicksilver, copper, and coal. Ores of tin, lead, and antimony in large veins, beds of sulphur, alum, and asphaltum; lakes of borax and springs of sulphate of magnesia, are also found in the state, but they are not wrought at the present time, though they will probably all become valuable in a few years. Platinum, iridium, and osmium are obtained with the gold in some of the placer mines, but are never found alone, nor are they ever the main object sought by the miner. The annual yield of our gold mines is about forty millions of dollars, of our quicksilver two millions of dollars. Our silver, copper, and coal mines have been opened within a year, and their value is yet unknown. All our other mining is of little importance as compared with the gold.

§ 177. *Gold Mines.*—Our gold mines are divided into placer and quartz. In the former, the metal is found imbedded in layers of earthy matter, such as clay, sand, and gravel; in the latter it is incased in veins of rock. The methods of mining must be adapted to the size of the particles of gold, and the nature of the material in which they are found. In placer mining, the earthy matter containing the gold, called the "pay-dirt," is washed in water, which dissolves the clay and carries it off in solution, and the current sweeps away the sand, gravel, and stones, while the gold, by reason of the higher specific gravity, remains in the channel or is caught with quicksilver. In quartz mining the auriferous rock is ground to a very fine powder, the gold in which is caught in quicksilver, or on the rough surface of a blanket, over which the fine material is borne by a stream of water. About two-thirds of our gold is obtained from the placers, and one-third from the quartz.

A mine is defined and generally understood to mean "a subterraneous work or excavation for obtaining metals, metallic ores or mineral substances;" but this definition does not apply to our placer mines, which are places where gold is taken from diluvial or alluvial deposits. Most of the work is not subterraneous; it is done in the full light of day. In some of the claims the pay-dirt lies within two feet of the surface; in others it lies much deeper, but all the superincumbent matter is swept away.

Water is the great agent of the placer miner; it is the element of his power; its amount is the measure of his work, and its cost is the measure of his profit. With an abundance of water he can wash every thing; without water he can do little or nothing. Placer mining is almost entirely mechanical, and of such a kind that no accuracy of workmanship or scientific or literary education is necessary to mastery in it. Amalgamation is a chemical process it is true, but it is so simple that after a few days' experience, the rudest laborer will manage it as well as the most thorough chemist.

It is impossible to ascertain the amount of gold which has been taken from the mines of California. Records have been kept of the sums manifested at the San Francisco Custom House, for exportation, and deposited for coinage in the mints of the United States; and there is also some knowledge of the amounts sent in bars and dust to England; but we have

no account of the sums carried by passengers to foreign countries and coined elsewhere than at London, or used as jewelry, or of the amount now in circulation in this state. According to the books of the custom-house of San Francisco, the sums manifested for export were as follows:

In 1849, $4,921,250; in 1850, $27,676,346; in 1851, $42,-582,695; in 1852, $46,586,134; in 1853, $57,331,034; in 1854, $51,328,653; in 1855, $45,182,631; in 1856, $48,887,543; in 1857, $48,976,697; in 1858, $47,548,025; in 1859, $47,640,462; in 1860, $42,303,345; in 1861, $40,639,089; a total of $551,603,-904 in twelve years.

The exportation of gold commenced in 1848, but we have no record of the sums sent away in that year. Previous to 1854 very large sums were carried away by passengers, who gave no statement at the custom-house; since that year, the manifests show the exportation correctly within a few millions. I am entirely satisfied that the total gold yield of California has been not less than seven hundred millions of dollars; but I have not room here to state the reasons for this opinion. My estimate is considerably less than that of most business men of the state, and less than that made by Hunt's *Merchants' Magazine.* There was undoubtedly a regular increase in the annual yield of the mines from 1848 to the end of 1853; and there has been a gradual decrease since the beginning of 1854—a decrease perhaps not very regular but still certain. Since 1854 considerable sums exported from San Francisco and included in our tables, came from mines beyond the limits of California, such as the mines in Southern Oregon, in the eastern part of Washington Territory, in British Columbia, and in Nevada Territory; and while the Californian gold yield has been decreasing, these extraneous supplies have been increasing. Several millions must be deducted from the annual shipments since 1858, for foreign gold. The gold yield will undoubtedly continue to fall, but to what point and at what rate no one can know. I believe that in 1870, the yield will not exceed thirty millions of dollars.

§ 178. *Placer Mines.*—Placer mines are divided into many classifications. The first and most important is into deep and shallow. In the former the pay-dirt is found deep, twenty feet or more beneath the surface; in the latter, near the surface. The shallow or surface diggings are chiefly found in the beds of ravines and gullies, in the bars of rivers, and in shallow flats; the deep diggings are in hills and deep flats. The pay-dirt is usually covered by layers of barren dirt, which is sometimes washed, and sometimes left undisturbed, while the pay-dirt is taken out from beneath it through tunnels or shafts. So far as our present information goes, we have reason to believe that no gold country ever possessed so large an extent of paying placer mines, with the pay-dirt so near the surface, and with so many facilities for working them as California. In Australia the diggings are very deep and spotted, that is, the gold is unevenly distributed, and the supply of water for mining is scanty. In Siberia the winter is terribly cold during six months of the year. In Brazil the diggings were not so extensive nor so rich as in this state. Here we have numerous large streams coming down through the mining districts, very large bodies of pay-dirt, and a mild climate.

After dividing placers into deep and shallow, the next classification will be according to their topographical position, as into hill, flat, bench, bar, river-bed, ancient river-bed, and gulch mines. Hill diggings are those where the pay-dirt is in or under a hill. Flat diggings are in a flat. Bench diggings are in a "bench" or narrow table on the side of a hill above a river. Benches of this kind are not uncommon in California, and they often indicate the place where the stream ran in some very remote age. Bars are low collections of sand and gravel at the side of a river and above its surface at low water. River-bed claims are those beneath the surface of the river at low water, and access is obtained to them only by removing the water from the bed by flumes or ditches. Ancient river-bed claims are those of which the gold was deposited by streams in places where no streams now exist. Gulch claims are those in gullies which have no water, save during a small part of the year. A "claim" is the mining land owned or held by one man or a company.

The placer mines are again classified according to the manner in which, or the instruments with which they are wrought. There are sluice claims, hydraulic claims, tunnel claims, dry washing, dry digging, and knife claims. In 1849 and 1850, the main classification of the placers was into wet diggings and dry diggings, the former meaning mines in the bars and beds of rivers, and dry diggings were those in gullies and flats where water could be obtained only part of the year or not at all. That classification was made while nearly all the mining was done near the surface, before the great deposits of pay-dirt in the hills had been discovered, and before ditches, sluices, and the hydraulic process had been introduced. The class of mines then known as the "dry diggings," and which for several years furnished nearly half of the gold yield of the state, are now, with a few unimportant exceptions, exhausted, or left to the attention of the Chinamen.

The purpose of all placer miners is not to catch all the gold in the dirt which they wash, but to catch the greatest possible quantity within a given time. It is not supposed that any process used in gold mining catches all the metal. Part of it is lost; in some processes a considerable proportion. The general estimate in California is, that one-twentieth of the gold in the dirt which is washed is lost. Many of the particles are so very small as to be invisible to the naked eye, and so light that their specific gravity does not avail to prevent them from being carried away by the water like sand. The larger pieces will sink to the bottom and resist the force of the water; the smaller the particles, the greater the danger that it will be borne away. Many devices have been tried to catch all the gold, but none have succeeded perfectly, and some which have caught a portion of what escaped from the ordinary modes of mining, have been found to cost more than their yield. The miner does not grieve about that which he cannot catch. He is not careful to catch all that he could. His purpose is to draw the largest possible revenue per day from his claim. He does not intend to spend many years in mining, or if he does he has become thriftless and improvident. In either case, he wishes to derive the utmost immediate profit from his mine. If his claim contain a dollar to the ton, and he can save five dollars by slowly washing only six tons in a day, while he might make ten dollars by rapidly washing fifteen tons in a day, he will prefer the latter result, though he will lose twice as much of the precious metal by the fast as by the slow mode of working. The object of the miner is the practical dispatch of work, and his success will depend to a great extent upon the amount of dirt which he can wash within a given space of time. He regrets that any of the gold should be wasted, but his regret is because it escapes from his sluice and his pocket, rather than because it is lost to industry and commerce.

§ 179. *The Sluice.*—The board-sluice is a long wooden trough, through which a constant stream of water runs, and into which the auriferous dirt is thrown. The water carries away the clay, sand, gravel, and stones, and leaves the gold in the bottom of the sluice, where it is caught by its gravity and by quicksilver. The board-sluice is the great washing machine, and the most important instrument used in the placer mining of California. It washes nearly all the dirt and catches nearly all the placer gold of the country. It was invented here, although it had previously been used elsewhere; it has been more extensively employed here than in any other country, and it can be used here to more advantage than elsewhere. It is not less than fifty feet long, nor less than a foot wide, made of boards. The width is usually sixteen or eighteen inches; and never exceeds five feet. The length is ordinarily several hundred and sometimes several thousand feet. It is made in sections or "boxes" twelve or fourteen feet long. The boards are an inch and a half thick, and are sawn for that special purpose, the bottom boards being four inches wider at one end than the other. The narrow end of one box therefore fits in the wide end of another, and in that way the sluice is put together, a long succession of boxes, the lower end of each resting in the upper end of another, and not fastened together otherwise. These boxes stand upon trestles, with a descent varying from eight to eighteen inches in twelve feet. It is therefore an easy matter to put up or take down a sluice after the boxes are made, and it is not uncommon for the miners to haul their boxes from one claim to another. The descent of a sluice is usually the same throughout its length, and is called its "grade." If there be a fall of eight inches in twelve feet, the sluice has an "eight-inch grade," and if the fall be twice as great, it is a "sixteen-inch grade." The grade depends upon the character of the pay-dirt, the length of the sluice, and its position. The steeper the descent, the more rapidly the dirt is dissolved, but the greater the danger also that the fine particles of gold will be carried away by the water. The tougher the dirt, that is, the greater its resistance to the dissolving power of the water, the steeper, other things being equal, should be the sluice. A slow current does not dissolve tough clay, and that is the greater part of the pay-dirt, so rapidly as a swift one. The shorter the sluice, other things being equal, the smaller the grade should be. There is more danger that the fine particles of gold will be lost by a short sluice than by a longer one, and to diminish this danger, the rapidity of the current must be reduced by a small grade. The greater the amount of dirt to be washed, other things being equal, the steeper should be the grade; for a swift current will wash more dirt than a slow one. In many claims the pay-dirt is full of large stones and boulders, weighing from one hundred to five hundred pounds each, all of which must be carried away through the sluice. Some are sent down whole, and others are broken into pieces with sledge-hammers before they are thrown into the box. These require a swift current and a large body of water. The larger the supply of water, the steeper the sluice is made, other things being equal. Of course economy and convenience of working require that the sluice should be near the level of the ground, and as that may be steep or level below the claim, the grade of the sluice must to some extent conform to it. There are thus a multitude of points to be taken into consideration in fixing the grade of a sluice; but a fall of less than eight or more than twenty inches, in a box of twelve feet, would be considered as unsuitable for the board-sluice. Sometimes the upper part of the sluice is made steeper so as to dissolve the dirt, and the lower part has a small grade to catch the gold. The clayey matter of ordinary pay-dirt is fully dissolved in a sluice two hundred feet long with a low grade, so the use of the boxes beyond that length is merely to catch the gold. There are claims however in which the clay is so extremely tough that it will roll in large balls more than a quarter of a mile through a steep sluice with a large head of water, and come out at the lower end scarcely diminished in size.

The gold is caught in the sluice-boxes by false bottoms of various kinds. It would not do to leave the smooth boards, for the water would sweep all the gold away, and the boards themselves would soon be worn through. The most common false bottom is the longitudinal riffle-bar, which is from two to four inches thick, from three to seven inches wide, and six feet long. Two sets of these riffle-bars go into each sluice-box, the box being twice as long as the bar. A set of riffle-bars is as many as fill one-half of a box. They are wedged in, from an inch to two inches apart; the wedging being used, because the bars can more readily be fastened in their places, and more easily taken up, than if nails were used. Before the work of sluicing commences, all the boxes are fitted with riffle-bars, and the bottom of the sluice is therefore full of holes from one to two inches wide, from three to seven inches deep, and six feet long. These are the places in which the gold, quicksilver, and amalgam are caught. Quicksilver is used now in nearly all the sluices, and is the more necessary the smaller the particles of gold. The large pieces of the metal would all be caught by their specific gravity without the aid of amalgamation.

The sluice-boxes having been made, and set up with the proper grade, the water is turned in. The boxes are made of the rough boards as they come from the saw, and the joints are not waterproof, but the leaks are soon stopped by the swelling of the wood, or by the dirt. The stream of water in the sluice is at least two inches deep over the bottom. The height of the sides of the boxes is from eight inches to two feet. The sluice usually runs through the claim, and the auriferous dirt is thrown in with shovels, of which from four to twenty are constantly at work. A man will throw in from two to five cubic yards of dirt in one day. The water rushing over the dirt as it lies in the box, rapidly dissolves the clay and loam, and then sweeps the sand, gravel, and stones down. The first dirt in the box goes to fill the spaces between the riffle-bars. After the sluicing has been in progress a couple of hours, some quicksilver is put in at the head of the sluice, and it gradually finds its way downward, most of it stopping, however, near where it is put in.

§ 180. *Amalgamation.*—There are a few metals, including gold, silver, copper, and tin, which, with quicksilver, form a peculiar chemical union called amalgamation, a process of great importance to the gold miner. When a piece of gold or silver is placed in mercury, the latter metal gradually penetrates through it, destroys the coherence of its particles, and forms with it a mass like dough. A lump of gold as large as a bean will be soaked through in three or four days; with silver and copper the process is slower, but they are affected in the same manner. Amalgamation, though a union of a solid with a liquid, differs much from a solution. In the latter the union is mechanical; in the former it its chemical. In the latter the solid is reduced to particles of impalpable fineness; in the former it is not. An ounce of salt will be dissolved in, and nearly equally diffused through, a pint of water; but if an ounce of gold be thrown into a pint of quicksilver, it will, after forming an amalgam with the quicksilver, remain at the bottom. We have no texture so fine that it will strain salt out of water; but the particles of gold are so coarse in amalgam that they can easily be strained out by means of buckskin or tight cloths. However a little gold will remain in the quicksilver—about the fiftieth part of an ounce of gold in every pound of quicksilver; and the only method of obtaining this gold is by retorting.

Quicksilver is used in gold mining for catching the small particles of metal; the large ones are caught by their weight. But many of the particles are so small that they are almost invisible to the naked eye, and when in moving water they float. Miners frequently show visitors the fineness of their gold by putting some of the dust in a vial with water; and upon shaking, the particles of metal can be seen floating about in the clear water. Riffles, and all the devices to get the benefit of specific gravity, are of little use to arrest this "float-gold," so amalgamation is employed. If a bit of quicksilver is put in the way of the fine gold, the two metals unite at once and make a larger bulk, which can be caught.

There is no such attraction between gold and quicksilver as there is between the magnet and iron; but when the two former metals once touch, an amalgam is immediately formed, and if the proportions of the metals be about even, they in time make a hard mass. Some gold does not amalgamate readily; in various diggings of Siskiyou county, the gold has a reddish coating, which prevents amalgamation. Grease or resin in the water used for washing, is also unfavorable. So is cold. Heat is favorable, and therefore less gold is lost in summer than in winter. Quicksilver that has been once used is considered better than that fresh from the flask.

No tinned iron or copper vessel should be used for holding or panning out amalgam, or dirt containing amalgam; since quicksilver forms an amalgam with tin and copper, and will stick to the sides of a tinned or copper pan.

In most sluices, the quicksilver is put in above the riffle-bars at various places along in the boxes, with a confidence that the great specific gravity of the metal will prevent it from being lost. The greater the quantity and proportion of fine gold, the greater the importance of the quicksilver.

The best method of catching very fine gold by amalgamation is to cover a large copper plate with mercury, and let the dirt and water, in a thickness of not more than a quarter of an inch, pass over it slowly. There are various methods of covering copper plates with quicksilver. The first thing, in every case, is to wash the copper with diluted nitric acid, so as to remove all dirt and grease. The quicksilver may then be rubbed on with a rag ; or, still better, it may be dissolved in nitric acid, and the liquid nitrate of quicksilver may be applied with a rag. The nitric acid will attack the copper, and leave the quicksilver as an amalgam on the surface of the copper. This is the most common process, but the nitrate of copper continues for a long time to come up through the quicksilver and interfere with the catching of the gold. When the nitrate of copper appears—it is a green slime—it should be scraped off and the place rubbed over with quicksilver. When a plate is once covered with mercury, the operation need never be repeated; but more mercury must be sprinkled on as the gold collects and forms a solid amalgam. The plate is usually three feet wide and six feet long, and is set nearly level. In very large sluices the stream should be divided so as to run over different plates. The slowness of the current and the shallowness of the water are important, for with a swift current or deep water many of the particles of float-gold may escape without touching the quicksilver. Wherever a speck of gold has fixed itself on the plate, there others will collect about it, evidently preferring to fix themselves in a neighborhood rather than in a waste place. The more gold there is on a plate, the better it is considered to be. The seasons for cleaning up are usually determined by the danger of theft. Miners do not like to leave their gold out in quantities so large as to attract thieves. The amalgam is sometimes half an inch thick, and is usually, at cleaning-up time, a hard mass, which must be loosened by heat. The plate is put on a fire, and when it gets so warm that the hand can scarcely bear it, the amalgam is softened and loosened, so that it can be scraped off readily. The plate is then sprinkled anew with quicksilver, and is ready for use again. Mercury does not amalgamate with copper so readily as with gold or silver. A copper plate, the sixteenth of an inch thick, may be used for at least five years, and perhaps for ten; whereas a gold plate of equal thickness would, if exposed to the action of quicksilver in the same manner, fall to pieces in a few weeks. After a time the quicksilver pervades the copper, and gives it a silvery whiteness all through on the under side. It is said that a solution of cyanuret or prussiate of potash, is used instead of nitric acid in applying mercury to copper plates, and that it is still better, there being then no trouble with the green spots of nitrate of copper.

A good amalgamated copper plate is considered as serviceable as a bed of quicksilver of equal size, and it is very much cheaper and more convenient to manage.

The dirt and water should be admitted to the copper plate, by falling first through a sheet-iron plate, pierced with holes half an inch long and a sixteenth of an inch wide. Some miners place this sheet-iron plate immediately over the copper.

Very soon after the water and dirt commence to run in the sluice, all the spaces between the riffle-bars are filled with sand, gravel, and dirt; which, however, present many little inequalities of surface, sufficient to catch all the particles of gold larger than a pin-head. The largest gold is caught near the head of the sluice; and the farther down the sluice, the finer the gold. In some sluices, where the pay-dirt contains much coarse gold, the quicksilver is introduced from thirty to sixty yards below the head, so as to catch only the fine particles of metal.

§ 181. *Cleaning up.*—The separation of the gold, amalgam, and quicksilver, from the dirt in the bottom of the sluice, is called "cleaning up;" and the period between one "cleaning" up and another is called a "run." A run in a common board-sluice usually lasts from six to ten days. Ordinarily the sluice runs only during daylight, but in some claims the work continues night and day. Cleaning up occupies from half a day to a day, and therefore must not be repeated too often, because it consumes too much time. In some sluices the cleaning up does not occur until the riffle-bars have been worn out or much bruised by the wear of the stones and gravel. Cleaning up is considered light and pleasant work as compared with other sluicing, and is often reserved for Sunday. At the time fixed, the throwing in of dirt ceases, and the water runs until it becomes clear. Five or six sets of riffle-bars, a distance of thirty or thirty-five feet, are taken up at the head of the sluice, and the dirt between the bars is washed down, while the gold and amalgam lodge above the first remaining set of riffle-bars, whence it is taken out with a scoop or large spoon, and put into a pan. Five or six more sets of bars are taken up, and so on down. Sometimes all the riffle-bars are taken up at once, save one set in every thirty-six feet, and then the work of cleaning up is dispatched much more rapidly.

The quicksilver and amalgam taken from the sluice are put into a buckskin or cloth, and pressed, so that the liquid metal passes through, and the amalgam is retained. The amalgam is then heated, to drive off the mercury. This may be done either in an open pan or in a close retort. In the former, the quicksilver is lost; in the latter, it is saved. The pan is generally preferred. Often a shovel or plate of iron is used. Three pounds of amalgam, from which the liquid metal has been carefully pressed out, will yield one pound of gold. The gold remaining after the quicksilver has been driven off by heat from the amalgam, is a porous mass, somewhat resembling sponge-cake in appearance.

§ 182. *Riffle-Bars.*—The riffle-bars are usually sawn longitudinally with the grain of the wood, but "block riffle-bars" are considered preferable; the latter are cut across the tree, and the grain stands upright in the sluice-box. The block riffle-bars are three times more durable than the longitudinal; and as the latter kind are worn out in a week in some large sluices, there is a considerable saving in using the former. The block riffle-bars are only two or three feet long.

In some small sluices the riffle-bars are not placed in the boxes longitudinally, nor in sets; but one bar near the head runs downward at an angle of forty-five degrees to the course of the box, not touching its lower end to the side of the box, but leaving an open space of an inch there. Just below this open space another bar starts from the side of the box and runs downward at right angles to the course of the first bar, and an open space is again left at the end of this bar; and so on down to near the lower end of the sluice, where there are longitudinal riffle-bars in sets as described in the preceding paragraphs. The consequence of using this kind of riffle-bar is, that though much of the water and light dirt runs straight over the bars, the heavier material runs down from side to side in a zigzag course. Near the head of the sluice is a vessel, from which quicksilver falls by drops into the box; and it follows the course of riffle-bars, overtaking the gold which takes the same route. These zigzag riffle-bars are nailed down. In all sluices, men must keep watch to see that the boxes do not choke; that is, that the dirt and stones do not collect in one place, so as to make a dam, and cause the water to run over the sides, and thus waste the gold.

There are small sluices, from which all stones as large as a doubled fist are thrown out. For this purpose the miner uses a sluice-fork, which is like a large manure-fork or garden-fork, but has tines which are blunt and of equal width all the way down; the bluntness being intended to prevent the tines from catching in the wood, and the equality of width to prevent the stones from getting fast in the fork.

In some sluices, the "block riffle-bars"—that is, bars cut across the grain of the tree—are set transversely in the boxes, and about two inches apart.

Another device is, to fill the pores of such riffle-bars with quicksilver. This is done by driving an iron cylinder with a sharp edge into the surface of the bar, then putting mercury into the cylinder, and pressing it into the wood. The quicksilver, thus fastened in the wood, catches particles of gold, which must be scraped off when the time for "cleaning up" comes.

§ 183. *Double Sluices.*—Sluices are sometimes made double—that is, with a longitudinal division through the middle, so that there are two distinct sluice-boxes side by side. Two companies may be working side by side, so that it will be cheaper for them to build their sluices jointly. In some places the amount of water varies greatly; so that in the winter there is enough to run two sluices, and in the summer only one. And there are companies which wish to continue washing without interruption; so they wash first on one side and then on the other, and clean up without any interruption to the process of washing.

Another device for saving gold in sluices is the "under-current box." There is a grating of iron bars in the bottom of a box, near the lower end of a sluice; and under this grating is another sluice, with an additional supply of clean water, and with a lower grade. The grating allows only the fine material to fall through; and the current of water being moderate, many particles of gold, that would otherwise be lost, are saved. Sometimes the matter from the under-current box is led back to the main sluice.

§ 184. *Rock-Sluices.*—Large sluices are frequently paved with stone, which makes a more durable false bottom than wood, and catches fine gold better than riffle-bars. The stone bottoms have another advantage—that it is not so easy for thieves to come and clean up at night, as is often done in riffle-bar sluices. But, on the other hand, cleaning up is more difficult and tedious in a rock-sluice, and so is the putting down of the false bottom after cleaning up. The stones used are cobbles, six or eight inches through at the greatest diameter, and usually flattish. A good workman will pave eight hundred square feet of sluice-box with them in a day; and after the water and dirt have run over them for an hour, they are fastened very tightly by the sand collected between them. In large sluices, wooden riffle-bars are worn away very rapidly—the expense amounting sometimes, in very large and long sluices, to twenty or thirty dollars a day; and in this point there is an important saving by using the stone bottoms. They are used only in large sluices, and they generally have a grade of twelve or fourteen inches to the box of twelve feet.

§ 185. *Hydraulic Mining.*—After the board-sluice, with its various adjuncts of riffle-bars, stone bottoms, copper plates, and so forth, the next instrument of importance in the gold-mining of California is the hydraulic hose, used to let water down from a considerable height, and throw it under the pressure of its own weight against the pay-dirt, which is thus torn down, broken up, dissolved, and carried into the sluice below. The sluice is a necessary part of hydraulic mining. The hose is used, not to wash the dirt, but to save digging with shovels, and to carry it to the sluice.

The hydraulic process is applied only in claims where the dirt is deep and where the water is abundant. If the dirt were shallow in the claim and its vicinity, the necessary head of water could not be obtained. Hydraulic claims are usually in hills. The water is led along on the hill at a height varying from fifty to two hundred feet above the bed-rock, to the claim at the end or side of the hill, where the water, playing against the dirt, soon cuts a large hole, with perpendicular or at least steep banks. At the top of the bank is a little reservoir, containing perhaps not more than a hundred gallons, into which the water runs constantly, and from which the hose extends down to the bottom of the claim. The hose is of heavy duck, sometimes double, sewn by machine. This hose when full is from four to ten inches in diameter, and will bear a perpendicular column of water fifty feet high; but a greater height will burst it. Now, as the force of the stream increases with the height of the water, it is a matter of great importance to have the hose as strong as possible; and for this purpose, in some claims, it is surrounded by iron bands, which are about two inches wide, and are connected by four ropes which run perpendicularly down. The rings are about three inches apart. The "crinoline hose," thus made, is very flexible, and will support a column of water one hundred and fifty or two hundred feet high. The pipe at the end of the hose is like the pipe of a fire-engine hose, though usually larger. Sometimes the pipe will be eight inches in diameter where it connects with the hose, and not more than two inches at the mouth; and the force with which the stream rushes from it is so great, that it will kill a man instantaneously, and tear down a hill more rapidly than could a hundred men with shovels.

One or two men are required to hold the pipe. They usually turn the stream upon the bank near its bottom until a large mass of dirt tumbles down, and then they wash this all away into the sluice; when they commence at the bottom of the bank again, and so on. If the bank is one hundred and fifty feet high, the mass of earth that tumbles down is of course immense, and the pipemen must stand far off, for fear that they will be caught in the avalanche Such accidents are of daily occurrence, and the deaths from this cause probably are not less than threescore every year in the state. Often legs are broken; still more frequently the pipemen have warning, and escape in time. When men are buried in the falling dirt, the water is used to wash them out. In some claims, the pipe will tear down more dirt than the sluice can wash; in other claims, the sluice always demands more dirt than the pipe can bring down. In the latter case, blasting may be used to loosen the dirt, or the miners may undermine the bank, leaving a few columns of dirt for support; and then these being washed away by the pipe, the whole bank comes tumbling down.

In hydraulic claims, all the dirt is washed; in all other kinds of claims, such dirt as contains no gold is thrown to one side, or "stripped off."—"Hydraulic mining" is the highest branch of placer mining; it washes more dirt, and requires more water, and a larger sluice, than any other kind of mining.

The number of men employed in a hydraulic claim, however, is usually small, from three to six, the water doing nearly all the work. In some claims a man is constantly employed with a heavy sledge-hammer in breaking up large stones, so that the pieces may be sent down the sluice. One man attends to the sluice, and sees that the dirt does not choke up in the sluice, or in the claim above it.

The quantity of dirt that can be washed with a hydraulic pipe depends upon various circumstances—such as the supply of water, the height of its fall, the toughness of the dirt, and the amount of moisture in it. More can be washed in winter than in summer, because the dirt is then moister, and requires less water to loosen and dissolve it. The quantity of water used in a hydraulic claim is from forty to two hundred inches. With one hundred inches, at least thirty cubic yards can be washed in ten hours, on an average; and three men can do all the work. If there were a cent's worth of gold in each cubic foot, the thirty cubic yards would yield eight dollars and ten cents per day, or two dollars and seventy cents to the man, exclusive of the cost of water. But, as a matter of fact, nearly all the hydraulic claims pay more than that, and they will average at least three cents to the cubic foot, and many of them yield five cents. The water usually costs twenty cents an inch per day, so that one hundred inches would cost twenty dollars. Allowing for the water at that ráte, a claim in which thirty cubic yards could be washed in a day with one hundred inches of water, and in which the dirt contained five cents to the cubic foot, would leave a net pay of six dollars and sixty-six cents to each man per day.

One hydraulic company, of whose labors I have a note, washed two hundred and twenty-four thousand cubic feet of dirt in six days, using two hundred inches of water, and employing ten men. The wages of the men amounted, at four dollars per day each, to two hundred and forty dollars; the water cost three hundred dollars; and the waste of quicksilver, and wear of sluice, perhaps one hundred dollars more, making a total expenditure of six hundred and forty dollars: and the gold obtained was three thousand dollars, leaving a clear profit of twenty-three hundred and fifty dollars. The dirt contained one cent and a fifth of gold in a cubic foot. The greater the amount of water used, the greater the proportionate amount of dirt that can be washed, and the greater the proportionate profits. It is far more profitable to have a large sluice than a little one, if the water and dirt can be obtained in abundance.

Usually, in a hydraulic claim, the dirt is washed down to the bed-rock; but in some places the washing stops far above the bed-rock, because there is no outlet for the water.

§ 186. *Blasting.*—In some hydraulic claims, the dirt, in dry seasons, is blasted, so as to loosen it. A drift or hole is cut into the bottom of a hill one or two hundred feet high, and a number of kegs of powder (from twenty to two hundred) are introduced, and they are fired with a slow match. The explosion makes an earthquake in the vicinity; and the ground is loosened to such an extent that there is a great saving of labor. The breaking up of the dirt and the exposure to the air are supposed to facilitate the washing greatly.

More water is required for piping down banks than for washing the dirt; and often the sluice is almost idle for want of dirt, while the water, after being thrown against the hill-side, runs away without doing any service at washing. Blasting, therefore, by loosening the earth, enables the hydraulic miner to have an abundant and regular supply of dirt in his sluice, at an expense much less than the cost of manual labor to dig the bank down with pick and shovel.

§ 187. *Tail-Sluice.*—The tail-sluice is a large sluice made for rewashing the tailings or dirt which has previously passed through other sluices. It is placed ordinarily in the bed of a ravine or creek through which tailings run, and it receives no attention for weeks or months at a time, save to keep it from choking. The sluices emptying into it furnish both dirt and water, and in the dirt there is always a large amount of fine gold, as is plainly proved by the fact that some of the tail-sluices have paid large profits to their owners. Tail-sluices are always large, long, and paved with stones; and sometimes they are double, so that one side may be cleaned up while the other continues washing. In a branch of the Yuba there is, or was not long since, a tail-sluice twenty feet wide.

§ 188. *Tunnel-Sluice.*—A tunnel-sluice is a sluice in a tunnel. It sometimes happens that a considerable body of water runs out through a tunnel; and in such case, a sluice at the bottom of the tunnel offers the easiest method of getting out and washing the dirt. The tunnels are never cut level, but with a slightly-ascending grade, so that the water will always run out. The grade is so low, that transverse riffle-bars must be used; for with longitudinal riffle-bars or stones, there would be too much danger of choking. These tunnel-sluices, because of their low grades, require much more attention than any other kind of sluices.

§ 189. *Ground-Sluice.*—All the sluices hitherto mentioned and described have wooden boxes, but the ground-sluice has no box: the water runs on the ground. The place selected for the ground-sluice is some spot where there is a considerable supply of water, a steep descent for it, and much poor dirt. The stream is turned through a little ditch, which the miners labor to deepen and enlarge, and when it is deep they prize off the high banks so that the dirt may fall down into the ditch. This is a very cheap and expeditious way of washing, but it is not applied extensively. It is used to the most advantage for washing where the water is abundant for only a few weeks after heavy rains, and where it would not pay to erect large sluices. A few cobble-stones should be left or thrown at intervals in the bed of the ground-sluice to arrest the gold, for if the bed were smooth clay, the precious metal might all be carried off. Quicksilver is not used in the ground-sluice. After the dirt has all been put through the ground-sluice, it is cleaned up in a short board-sluice, or a tom.

190. *Long Tom.*—The tom or long tom, an instrument extensively used in the Californian mines in 1851 and 1852, but

now rarely seen, is a wooden trough about twelve feet long, eighteen inches wide at the upper end, and widening at the lower to thirty inches, with sides eight inches high. It is used like a board-sluice, but has no riffle-bars, and at the lower end its bottom is of sheet-iron, perforated with holes half an inch in diameter. This sheet-iron is turned up at the lower end, so that the water never runs over there, but always drops down through the perforated sheet-iron or riddle, into a little riffle-box, containing transverse riffle-bars. A stream of water of about ten inches makes a "tom-head"—or the amount considered necessary for a tom—through the tom, which has a grade similar to that of a board-sluice. The dirt is thrown in at the head of the tom, and a man is constantly employed in moving the dirt with a shovel, throwing back such pieces of clay as are not dissolved, to the head of the tom, and throwing out stones. From two to four men can work with a tom; but the amount of dirt that can be washed is not half that of a sluice. The tom may be used to advantage in diggings where the amount of pay-dirt is small and the gold coarse. The riffle-box contains quicksilver, and as the dirt in it is kept loose by the water falling down on it from the riddle above, a large part of the gold is caught; but where the particles are fine, much must be lost.

§ 191. *Cradle.*—The rocker or cradle is still less than the tom and inferior in capacity. It bears some resemblance in shape and size to a child's cradle, and rests upon similar rockers. The cradle-box is about forty inches long, twenty wide, and four high, and it stands with the upper end about two feet higher than the lower end, which is open so that the tailings can run out. On the upper end of the cradle-box stands a hopper or riddle box twenty inches square with sides four inches high. The bottom of this riddle-box is of sheet-iron, perforated with holes half an inch in diameter. The riddle-box is not nailed to the cradle-box, but can be lifted off without difficulty. Under the riddle is an "apron" of wood or cloth, fastened to the sides of the cradle-box and sloping down to the upper end of it. Across the bottom of the cradle-box are two riffle-bars about an inch square, one in the middle, the other at the end of the box. The dirt is shovelled into the hopper, the "cradler" sits down beside his machine, and while with one hand with a ladle he pours water from a pool at his side upon the dirt, with the other he rocks the cradle. With the water and the motion the dirt is dissolved, and carried down through the riddle, falling upon the apron which carries it to the head of the cradle-box, whence it runs downward and out, leaving its gold, black sand, and heavier particles of sand and gravel behind the riffle-bars. The man who rocks a cradle learns to appreciate the fact, that the "golden sands" of California are not pure sand, but are often extremely tough clay, a hopperful of which must be shaken about for ten minutes before it will dissolve under a constant pouring of water. Many large stones are found in the pay-dirt. Such as give an unpleasant shock to the cradle, as they roll from side to side of the riddle-box are pitched out by hand, and after a glance to see that no gold sticks to their sides, are thrown away; but the smaller ones are left until the hopperful has been washed, so that nothing but clean stones remain in the riddle, and then the cradler rises from his seat, lifts up his hopper, and with a jerk throws all the stones out. The water and the rocking are both necessary. Without the water, the dirt could not be washed; and without the rocking, the dirt would dissolve very slowly, and the gold would most of it be lost. The rocking keeps the dirt in the bottom of the cradle more or less loose, so that the particles of gold can sink down in it, whereas if the cradle stood still the sand there would almost immediately pack down into a hard floor, over which the gold would run almost as readily as over a board. The whole business of washing with a cradle is a repetition of the process already described—some dirt, about one-third or one-fourth of what the hopper would hold, if full, is put into the hopper, and while the cradle is rocked with one hand, the other pours in the water. The cradle is cleaned up two or four times in a day. The cleaning up is done by lifting the hopper, taking out the apron, scraping up all the dirt in the bottom of the cradle with an iron spoon, putting it into a pan and washing out the dirt, so that only the gold will be left. This last process is called panning out, and will be described in the next section. Most of the gold collects above the upper riffle-bar, including all the larger lumps. If the apron be of rough woollen cloth, some of the fine gold will be caught there. In diggings where the gold is very fine, the hopper is sometimes placed over the lower end of the cradle, and the apron is made twice as long, and with a lower inclination than in the more common form of the rocker. The water for the cradle should be supplied by a little ditch, with a reservoir at the head of the cradle, to contain five or six gallons. The dipper should be of tin, shaped like a basin, hold about a gallon when full, and have a handle an inch and a half in diameter, and eight inches long. The difference of height between the upper and lower ends of the cradle should not be more than two inches: a steeper inclination will make the current running through it too strong, and the gold will be carried off; and, on the other hand, if the cradle be nearer a level it will be hard to rock, and the dirt in the bottom will pack more rapidly. The amount of dirt that can be washed in a day with a cradle, varies from one to three cubic yards. The dirt is usually shovelled into a pan or bucket, from which it is thrown into the hopper. The miners usually measure the amount of dirt washed by the number of "pans." One man working alone with a cradle ought to wash from seventy-five to one hundred and fifty pans in a day, and two men will wash twice as much. A pan may contain one-third or one-half of a cubic foot. Two men can work more conveniently with the rocker than one. There is enough work to give constant employment to a cradler and a shoveller. The latter has a couple of buckets or pans, which he fills alternately, always keeping one full and near the cradler, so that without moving his feet he can pick it up and empty it into the riddle-box. If the rocker have only one man, he must stop rocking after washing every pan and get more dirt. This delay is injurious to the process of washing, because it allows the dirt in the bottom of the cradle to harden and pack, and some gold is always lost as a consequence. If the dirt and water be convenient not more than two men can work to a profit with a rocker. But sometimes it happens that water cannot be led to the claim, and in such case the dirt must be carried to the water, a greater weight of which is used than of dirt. At least three times as much water as dirt is required for washing. If the distance from the hole to the water be not over ten or twenty feet, the miners will usually carry the dirt in buckets; if farther they will use wheelbarrows; and sometimes for greater distances pack-mules or wagons. The greater the distance, the more the men required for carrying the dirt. Sometimes, too, it happens that the claim is troubled by water, and then one man may be constantly employed in baling.

It is of great importance in mining with the cradle, to have the cradle placed within four or five feet of the hole from which the pay-dirt is obtained, and to have a good supply of water at the head of the cradle, and then to have a good descent below the cradle, so that the tailings may all be carried away by the water, so as not to accumulate. The rocker washes about one-half the amount of dirt that can be washed by an equal number of men with the tom, one-fourth of what can be washed with the sluice, and one-hundredth of the amount that can be washed with the hydraulic process; but it is peculiarly fitted for some kinds of diggings. Many little gullies, containing coarse gold in their beds, cannot obtain water for washing except during rains, and then only for a few days at a time. In these gullies the cradle can be used to the best advantage, for it can easily be transported, and it is very good for saving coarse gold. While dirt that would pay from ten to twenty-five cents, was abundant at the surface of the earth in the Californian mines, the cradle was extensively used, but now it has been abandoned by the whites, and is left to the Chinamen, who think themselves doing well if they make seventy-five cents or one dollar per day.

The great difficulty in mining with the cradle, is that the sand will "pack," or make a hard mass on a level with the top of the riffle-bars, and the gold then is lost. So long as the cradle is in motion the dirt does not pack, but when the rocking ceases, the mass hardens in a few minutes. If the miner leaves his cradle standing for fifteen minutes, he stirs up the dirt with his spoon before commencing again to wash. One device to prevent packing is to put a little block under each end of the rockers, so that at the end of every motion the cradle receives a shock. Quicksilver is sometimes used in cradles, but not usually.

§ 192. *Pan.*—The pan is used in all branches of gold mining, either as an instrument for washing, or as a receptacle for gold, amalgam, or rich dirt. It is made of stiff tin or sheet-iron, with a flat bottom about a foot across, and with sides six inches high, rising at an angle of forty-five degrees. A little variation in the size or shape of the pan will not injure its value for washing. Sheet-iron is preferable to tin, because it is usually stronger and does not amalgamate with mercury. The pan is the simplest of all instruments used for washing auriferous dirt. Some dirt, not enough to fill it full, is put in, and the pan is then put under water. The water ought to be not more than a foot deep, so that the pan may rest on the bottom, while the miner inserts his fingers in and under the dirt and lifts it up a little, so that the whole mass is wet. If the water be deep, the pan may be held in one hand while the other is used to stir up the dirt, but it is more convenient to take both. The dirt having been filled with water, the miner catches the pan at the sides, raises that part toward his body, and lowers the outer edge a little, and commences to shake the pan from side to side, holding it so that all the dirt is under water, and so that a little of the dirt can escape over the outer edge. The earthy part of the dirt is rapidly dissolved by the water, assisted by the shaking of the pan and the rolling of the gravel from side to side, and forms a mud which runs out while clean water runs in. The light sand flows out with the thin mud, while the lumps of tough clay and the large stones remain. The stones collect on the top of the clay, and they are scraped together with the fingers and thrown out. This process continues, the pan being gradually raised in the water, and its outer edge depressed, until all the earthy matter has been dissolved, and that as well as the stones swept away by the water, while the gold remains at the bottom. Panning is not difficult, but it requires practice to learn the degree of shaking which dissolves the dirt and throws out the stones most rapidly without losing the gold. If the shaking be too mild and slow, the process consumes too much time; whereas if it be too rapid and violent, the gold is carried off with the stones. Sometimes the pan is shaken so that the dirt receives a rotary motion. This is the most rapid method of washing dirt, but also the most dangerous. The pan must always be used in cleaning up the dirt which collects in the cradle, in prospecting, and frequently in washing small quantities of dirt collected in other kinds of placer mining. Amalgam can be separated from dirt by washing, almost as well as gold. In panning out, it frequently happens that considerable amounts of black sand containing fine particles of gold are obtained, and this sand is so heavy that it cannot be separated from the gold by washing, while it is easily separated by that process from gravel, stones, and common dirt. The black sand is dried, and a small quantity of it is placed in a "blower," a shallow tin dish open at one end. The miner then holding the pan with the open end from him, blows out the sand, leaving the particles of gold. He must blow gently, just strong enough to blow out the sand, and no stronger. From time to time he must shake the blower so as to change the position of the particles, and bring all the sand in the range of his breath. The gold cannot be cleansed perfectly in this manner, but the sand contains iron, and the little of it remaining is easily removed by a magnet. The blower should be very smooth, and made of either tin, brass, or copper.

§ 193. *Dry Washing.*—Dry washing is a method of winnowing gold from dirt. In many parts of the mining districts of California, water cannot be obtained during the summer for mining purposes. The miner therefore manages to wash his dirt without water. He takes only rich dirt, and putting it on a rawhide, he pulverizes all the lumps and picks out the large stones. He then with a large flat basin throws the dirt up into the air, catches it as it comes down, throws it up again, and repeats this operation until nothing but the gold remains. Of course a pleasant breeze, that will carry away the dust, is a great assistance to the operation. Sometimes two men have a hide or a blanket, with which they throw up the dirt. The process is very similar to the ancient method of separating grain from chaff. The miner who devotes himself to dry washing must be very particular to take only rich dirt, so he scrapes the bed-rock carefully. He never digs very deep—not more then twenty feet; and when he goes beyond seven or eight feet he "coyotes," or burrows after the pay-dirt. He may coyote into the side of a hill, or sink a shaft and coyote in all directions from it. This style of mining is named from the resemblance of the holes to the burrows of the coyote, or Californian wolf. Coyoting is not confined to the dry washing, but is used also by miners washing with the pan and cradle. One of the Congressmen elected some years ago to represent California at Washington, was a miner at the time of his nomination, and was so fond of coyoting, that he was generally known as "Coyote Joe."

§ 194. *Dry Digging.*—Dry digging is that mining where the miner, after using the shovel to strip off the barren dirt, scrapes the pay-dirt over with a knife, picking out the particles of gold as he comes to them, and throwing away the earthy matter. This is a slow process, but in rich placers may be profitable. The miner is, of course, particular to examine all the crevices in the bed-rock; and if the material be slate, he digs up part of it, to see whether the gold has not found its way into cracks scarcely perceptible on the surface. "Dry digging," as a mode of mining, must not be confounded with "dry diggings," a kind of mining-ground which has been described near the beginning of this chapter.

Knife-mining differs a little from dry digging. In the latter, a shovel is used to strip off the barren dirt; whereas the knife-mining is practised in those places where the gold is deposited in crevices in rocks along the banks of streams, without any covering of barren dirt, so that the knife alone is used in scraping out the dirt; and afterward the dirt, being placed in a pan, may be washed in water, which is never used in dry digging.

§ 195. *Puddling-Box.*—The puddling-box is a rough wooden box, about a foot deep and six feet square, and is used for dissolving very tough clay. The clay is thrown into the box, with water, and a miner stirs the stuff with a hoe until the clay is all thoroughly dissolved, when he takes a plug from an auger-hole about four inches from the bottom, and lets the thin solution of the clay run off, while the heavier material, including the gold, remains at the bottom. He then puts in the plug again, fills up the box with water, throws in more clay, and repeats the process again and again until night, when he cleans up with a cradle or pan. The puddling-box is used only in small mining operations, and never with the sluice, or in hydraulic claims.

§ 196. *Quicksilver-Machine.*—The quicksilver-machine, or Burke rocker, is a cradle about seven feet long, two feet wide, and two feet high. In the bottom are a number of compartments, all containing quicksilver. One man rocks the machine without cessation. A constant stream of water pours into the machine at its head. The riddle extends the whole length of the machine; and the stones, after being washed clean, fall off the riddle at the lower end. One man is employed constantly working with a shovel to keep the dirt on the riddle under the stream of water, and in throwing off the big stones. If the pay-dirt is very convenient, two men can shovel enough to keep the machine in operation. The Burke rocker was extensively used in California eight and ten years ago, but now it is a great rarity.

§ 197. *Tunnel-Mining.*—A tunnel, in Californian mining, is an adit or drift entering a hill-side, or running out from a shaft. Mining-tunnels are usually nearly horizontal—those entering hill-sides having a slight ascent, for the double purpose of draining the mine, and to facilitate the removal of the pay-dirt. In a few hills the tunnels run downward at an angle of twenty degrees or more, to avoid veins or ledges of rock, which would have to be blasted through if the tunnel were cut horizontally; but this can only be done with safety in hills which are drained by older horizontal tunnels.

The mining-tunnel does not run through a hill, but only into it. The length of tunnels varies greatly; the longest are about a mile. The usual height is seven feet, the width five feet. Ordinarily the top must be supported by timbers, to prevent it from falling in, and not unfrequently the sides must also be protected by boards. The cost of cutting a tunnel varies from two to forty dollars a longitudinal foot, according to the nature of the ground, the cost of getting timbers, &c. Tunnels are usually made by companies of eight or ten men, of whom one-half may be merchants, lawyers, physicians, or office-holders, and the remainder laboring miners. The latter class do the work; the former furnish provisions and tools, and a certain amount of cash weekly until the pay-dirt is reached. Two or three men work at a time cutting a tunnel; one or two to dig the dirt; and one or two to haul it out. The dirt of the first fifty yards is hauled out in a wheelbarrow; beyond that distance a little tram-way or railroad is laid down, and the dirt is hauled out in cars, pushed by the miners. It is not customary to use horses. It is common to have two relays of laborers—one set working from noon to midnight, the other from midnight to noon. Work in a tunnel is as pleasant at night as in the daytime. When a company is rich, or has many laborers, it may have three relays, each to work eight hours in the twenty-four.

It is not uncommon for two companies, owning adjacent claims in a hill, to unite and cut a tunnel on joint account along the dividing line. They go in until they reach the pay-dirt, and then a surveyor is employed to run the line between their claims, and the tunnel is continued through the pay-dirt. The dirt from the tunnel is washed for the joint account of the two companies. After the dividing line has been established, each company keeps on its own side, and each has its time to use the tram-way. They may also have a joint-stock sluice at the mouth of the tunnel—one company having the privilege of using the sluice one week, and the other the next. All the dirt brought out in a week can readily be washed in a day. The work of taking out the pay-dirt after the main tunnel has been cut, is called "drifting;" and the holes made by the men engaged in it are termed "drifts." The drifts are usually not so high as the tunnels. The large stones and barren dirt obtained in the drifts are piled up here and there to sustain the earth overhead. Sometimes wooden posts are likewise necessary.

§ 198. *Shafts.*—Shafts are used in prospecting, and also in mining, where the claims are deep and cannot be reached by either the hydraulic process or the tunnel. The prospecting shaft is sometimes sunk into hills supposed to be auriferous, where the shaft is far less expensive than the tunnel. After the shaft demonstrates that the dirt is rich, and precisely the altitude at which it lies, a tunnel is cut to strike it. The shaft may be the cheaper for prospecting, but the tunnel is usually the cheaper if any large amount of dirt is to be taken out.

The shaft is dug by one man in the hole, and one or two are employed at a windlass in hauling up the dirt. Mining-shafts in placer diggings are rarely over one hundred feet deep; but one was dug in Trinity county to the depth of six hundred feet, for the purpose of prospecting, but it found neither pay-dirt nor the bed-rock.

§ 199. *River-Mining.*—River-mining is mining for gold in the beds of rivers, below low-water mark. The only practicable method of doing this is by damming the stream, and taking the water out of its bed in a ditch or flume. It has been proposed by persons who never saw the mines, to get the gold by dredging, or with a diving-bell; but such schemes are absurd in the eyes of miners. The rivers in which the gold is found are mountain-torrents, in which a canoe can scarcely float in summer, much less a dredging-machine; and any large scoop working under water would miss the crevices and corners in the rocks, where most of the gold is found. As the water is very seldom more than a couple of feet deep, a diving-bell would be of little service. The flume, the ditch, and the wing-dam, are the chief tasks of the river-miner. The ditch is rarely used, because the banks of the mining-streams are usually so steep, high, rocky, and crooked, that a flume is cheaper. The wing-dam is not often used, because the river-beds are in most places too narrow. The flume is almost universally employed.

The work of river-mining can be done only during the summer and fall, while the water is low, and while the miner can have confidence that it will not rise. It may be as low in January as in August, but the winter is the season of rains; and when the flood comes, it sweeps dams, flumes, and every thing before it. If the dam and flume be commenced too early in the season, they may be carried off before they are finished; and it frequently happens that they are destroyed in the fall just when the miners are commencing to reap the reward of their summer's labor.

River-mining has many disadvantages, as compared with other branches of mining. The miner cannot work at it more than half the year; he cannot prospect the dirt which is hidden under water; he must erect expensive dams and flumes, which can be used for only a few months; and then he is exposed to floods which may come and destroy all his work before he has commenced to wash. These disadvantages, and the exhaustion of most of the river-diggings in the state, have almost put an end to river-mining in California. In a few cases, extensive fluming enterprises have proved profitable; but, as a general rule, river-mining in this state has cost more than it has produced. A river is seldom flumed for less than three hundred yards, and sometimes for a mile; and the lumber and labor required to make so long a flume, and one large enough to hold all the water of a river, are very expensive. The dam will always leak, and water will run into the bed from the adjacent hills and mountains, and this water must be lifted out by pumps driven by wheels placed in the flume. The river-beds are full of large rocks, weighing from one to ten tons, and these must be moved by machinery, to allow the dirt to be taken out.

River-mining is now never undertaken by an individual, but always by large associations, generally called "fluming companies," sometimes composed of miners exclusively, sometimes of miners and all the principal business-men living near the place where the work is to be done. The lawyers, doctors, and office-holders, pay their assessments in cash; the merchants furnish provisions, the lumbermen supply lumber, and the miners make the dam, and help the carpenters build the flume.

§ 200. *Beach-Mining.*— Beach-mining is the business of washing the sands of the ocean-beach. Between Point Mendocino, in California, and the mouth of the Umpqua River, in Oregon, the beach-sand contains gold, and in some places it is very rich. The beach is narrow, and lies at the foot of a bluff bank of auriferous sand. In times of storm, the waves wash against this bank, undermine it, sweep away the pieces which tumble down, leaving the gold on the beach. The gold is in very fine particles, and it moves with the heavier sand, which alters its position frequently under the influence of the waves and surf. One day, the beach will have six feet depth of sand; the next, there will be nothing save bare rocks. The sand differs greatly in richness at various times: one day, it will be full of golden specks; a few days later, at the same place it will be barren. The sand in the mean time has been moved by the waves, and replaced by other sand.

It is a very difficult matter to know where the sand is rich and where it is not. The companies employed in mining on the beach number about ten men; and there is a foreman who rides out early every morning, following the beach about two miles to the northward and two miles to the southward of the camp, for the purpose of finding where the sand is the best. So changeable is the sand, that a new examination is made every day; and only three or four men are supposed to be good judges of the quality of sand, from its appearance.

When the foreman has selected a place, he orders all the men to it, and they go with twenty pack-mules, which carry the sand in *alforjas*, or rawhide sacks, to the place of washing, which is up on the bluff, probably a mile or more distant from the spot where the sand is obtained. It happens occasionally that the foreman rides long distances on the beach, and sometimes he will order the sand to be obtained ten miles from the washing-place. The sand must, of course, be very rich, to pay for such transportation, but the beach-sand at times in the sunlight is said to be actually dazzling yellow with gold. The purpose of going upon the bluff to wash it is to get fresh water for washing; foı the sea-water is not so good, nor can it be obtained conveniently. The richest dirt is that the farthest down on the beach, so still weather and low tide are the best times for getting it. When a rich place is discovered low down on the beach, great exertions are made to get as much of the sand as possible before the tide rises. When high tide and storm come together, little can be done. The sand, having been separated from all clay and soluble matter by the action of the sea, is very easily washed, and all collected in a month can be washed in two days in a sluice.

§ 201. *Mining-Ditches.*—The placer-mines of California would yield very little gold, were it not for the numerous ditches which supply them with water for washing. The auriferous districts are very dry in summer, and in some places there is not a spring nor a brook within many miles. The artificial ditch supplies the want. The ditches are made by large companies, which sell the water by the "inch." An inch of water is as much as will run out of an orifice an inch square, with the water standing six or seven inches deep in the flume over the orifice. The depth of water over the orifice is called the "head." The orifice is usually two inches high, and as long as necessary to give the amount of water desired. Nobody wants less than ten or twelve inches for mining: a "sluice-head" is about eighteen inches; a "hydraulic head" is from forty to two hundred inches. The water, however, is not measured accurately. Of course, the amount which runs through the orifice will depend to a considerable extent upon the "head," which is usually greater in the morning than at night. At sunrise there may be fifteen inches head, and at sunset only three. The water collects during the night, and is exhausted during the day. The price of water is in no place less than ten cents an inch per day; in some places it is forty cents; the average is about twenty cents.

Many of these ditches are extensive enterprises, and have cost hundreds of thousands of dollars. When they cross ravines and valleys, large flumes—wonders of carpentry—must be built. Some of these are two hundred feet high and a mile long, and so large that a horse and wagon can be driven through them. In all, save length and durability, they are as wonderful as the great Roman aqueducts, whose tall ruins still stand in the Campagna, near the Eternal City. In some cases iron tubes have been used, and, although they are very expensive, yet they may pay for themselves, by preventing evaporation, leaking, and soaking, which take away much of the water from flumes and ditches.

§ 202. *Prospecting.*—"Prospecting" is the search for gold. The instruments used by the prospector for placer-mines are usually the pan, pick, and shovel. He should be familiar with the general laws of the distribution of gold, and then try the dirt in the most favorable places. If there is any gold in a district, he can scarcely fail to find specks of it by washing dirt from the bed-rock in the ravines, and in bars. The existence of gold in a district having been established, close observation will suggest to the prospector where he may reasonably expect to find the best diggings. It is usually found that placer-gold is collected in those places where, if he had been familiar with the ancient topography of the country, he should have had reason to suppose that it would be.

§ 203. *Quartz Mining.*—Quartz mining differs much from placer mining. For the former, more capital, more experience, more complicated machinery and richer material are required than for the latter. The placer miner throws the dirt into the water, which then does the work; whereas the pulverizing of rock is a nice operation, requiring constant attention. Quartz requires a mill and water-power; placer dirt is washed in a simple sluice. Dirt containing ten cents in the cubic yard may pay the hydraulic miner, but the quartz miner must have a hundred times as much in a cubic yard of vein-stone, or he cannot work. The placer gold, when freed from the baser material surrounding it, is much of it in coarse particles, which are easily caught by their specific gravity; the quartz gold must be reduced to a fine powder before it can be set free from its gangue, and with the fineness of the particles increases the difficulty of catching them.

Auriferous quartz lodes are often found by accident. Not unfrequently it happens that a rich streak of pay-dirt in a placer claim is followed up to the quartz vein from which it came. While miners are out walking or hunting, they occasionally will come upon lodes in which the gold is seen sparkling. Some good leads have been found by men employed in making roads and cutting ditches. The quartz might be covered with soil, but the pick and shovel revealed its position and wealth. In Tuolumne county in 1858, a hunter shot a grizzly bear on the side of a steep cañon, and the animal tumbling down, was caught by a projecting point of rock. The hunter followed his game, and while skinning the animal, discovered that the point of rock was auriferous quartz. In Mariposa county in 1855, a robber attacked a miner, and the latter saw the rock behind his assailant sparkle in the sunlight, at a spot where a bullet struck a wall of rock. He killed the robber, and found that the rock was gold-bearing quartz. In Nevada county several years ago, a couple of unfortunate miners who had prepared to leave California, and were out on a drunken frolic, started a large boulder down a steep hill. On its way down, it struck a brown rock and broke a portion of it off—exposing a vein of white quartz which proved to be auriferous, induced the disappointed miners to remain some months longer in the state, and paid them well for remaining. Science and experience do not appear to give much assistance in prospecting for quartz lodes. Chemists, geologists, mineralogists, and old miners, have not done better than ignorant men and new-comers. Most of the best veins have been discovered by poor and ignorant men. Not one has been found by a man of high education as a miner, or geologist. No doubt geological knowledge is valuable to a miner, and it should assist him in prospecting; but it has never yet enabled any body to find a valuable claim.

§ 204. *Distribution of Gold in Quartz.*—The rich quartz-veins of California extend from Kern River to the Siskiyou, are found on hills, in cañons and in vales. They are at least two thousand feet above the level of the sea, and not more than ten thousand feet above it. Their course is generally from north-northwest to south-southeast, and they dip steeply to the eastward, sometimes being nearly perpendicular. They differ in thickness from a line to sixty feet. Quartz veins are very numerous in most of the mining districts, so the task is not to find the veins, but rather to find those which are gold-bearing. It is supposed that nearly all large veins come to the surface of the bed rock or "country;" but many of them are covered with soil and thus are hidden. Hidden veins are called "blind;" those plainly visible on the surface are called "croppings veins," because their position is shown by the out-croppings. Experience has not ascertained whether large or small veins are more likely to contain gold. It is found in both. The porous quartz, or that containing many cavities, is more frequently found auriferous and richly auriferous, than the very compact quartz. The best gold-bearing veins are usually yellowish or brownish in tinge, near the surface at least; but very rich specimens are found in white and bluish-white rock. Most quartz veins in California contain a little gold; the metal seems to have been distributed most lavishly, but unfortunately in nine-tenths of the veins, the proportion of metal is too small to pay. Most of the large veins are supposed to run for miles upon miles, though they can rarely be traced clearly on the surface for more than a furlong. The auriferous veins vary much in richness. No vein is wrought for more than a few hundred feet. Beyond that, it is either too poor to pay, or the vein is hidden. Some persons have supposed that there is one great gold-bearing quartz vein running along the side of the Sierra Nevada, from Mariposa to Plumas county, and that many of the richest claims are really in this one vein; but this is a supposition which cannot be proved now. Sometimes a vein seems to spread out and divide into a number of smaller veins, all of which afterward unite again. These points of junction, and the narrower places in the vein, are usually richer than other parts of it. When two veins cross each other, one may be auriferous on one side of the intersection and not on the other; but in this case the other vein will be auriferous on both sides. It is as though they were streams, one rich, the other barren, and that after meeting, the wealth of the one was divided between them. It is a general rule that metalliferous veins running parallel with the strata of the bed-rock or country are not extensive. In fact they are rather deposits than veins, and though often extremely rich are soon exhausted, while the lodes which run across the stratification run far and deep, and have a regular and straight course and dip. Lodes lying between two different kinds of rock, are usually richer than those which have the same kind of rock on both sides. Thus it is said that the richest veins of auriferous quartz in California, have been discovered at the intersection of trap and serpentine, and the richest places in veins are where they cross from one kind of bed-rock into another. The richest part of a lode of auriferous quartz is almost invariably on the lower side of the vein, near the foot-wall. All these are facts to be remembered by the prospector as a guide, and an assistance to him in his search for a rich gold-bearing vein. If the lode is covered with earthy matter, he may sometimes trace its course by the difference in the color of the dirt and stones over it from that elsewhere. When the prospector finds dirt and stones on a vein, evidently disintegrated portions of it, he should wash some of the dirt in a pan, and if he finds no gold, there is a strong presumption that the vein is barren.

§ 205. *Prospecting Quartz Rock.*—After finding a gold-bearing vein, the question arises whether it will pay. Great sums are lost in gold-mining countries by injudicious investments in mills and machinery to work the auriferous rock, and persons going into the business should be particularly careful not to commit this great error. The business of quartz mining has great profits but also great pecuniary dangers connected with it. It is rarely that all the rock of a vein will pay for working. In some lodes, the vein-stone will average one hundred dollars to the ton, for all the stone found in a certain part of the lode, but beyond that the rock may be poor or worthless. Picked specimens may be worth several thousand dollars to the ton, but perhaps not more than a ton of such specimens has been obtained in the best lode ever opened in the state. The most profitable lodes are those which have a large supply of rock, easily to be obtained, and all of it yielding something above the cost of working. The common method of ascertaining whether rock will pay, is to pulverize a little of it and wash it in a horn spoon. In taking out the quartz rock in large lodes, it is important to take out only that which will pay, and to determine this, the superintendent of the quarry-men must occasionally test the vein-stone. He takes several little pieces of it, average specimens, places them on a hard, smooth, flat stone about a foot square, on which he crushes them with a stone muller four inches square, and then by rubbing with the muller he reduces them to a fine powder. He has a horn spoon, made of a large ox-horn, with a bowl about three inches wide and eight inches long, being merely one-half of the horn in its natural shape. With this spoon he washes out the powder in water, and if he does not find a speck of gold or a "color," as it is called, in a pound of the rock, he infers that it will not pay. The three principal quartz mines in the state are those of Fremont in Mariposa county, of the Allison company in Nevada county, and of the Sierra Butte company in Sierra county. The first has produced $75,000 in a month, the second $60,000, and the third $20,000, but the average is probably thirty per cent. less, and the expenses about thirty per cent. of the total product. The average yield of the Fremont rock is fourteen dollars to the ton, of the Sierra Butte rock eighteen dollars, and that of the Allison company, according to report, has for more than a year at a time been one hundred dollars per ton. The cost of working quartz rock, including quarrying, crushing, and amalgamating, is in the best mills from five to ten dollars per ton. The width of the vein, the softness of the rock, the amount of work done, and the skill and industry of the workmen, all are items of great importance in estimating the cost of quartz-mining. It is a business which the owner of the mill ought to understand. The cost of quarrying common quartz rock is about two dollars per ton, that is, for mill-owners that understand the business and superintend the labor themselves. When given out by the job, it usually costs more. When quartz is crushed in a custom mill, that is, a mill built to crush for all applicants, the cost is rarely less than five dollars per ton, and in Washoe, the price was at one time thirty dollars per ton; but in the large mills, where many tons are crushed every day, is about two dollars per ton.

§ 206. *The Divining Rod.*—In prospecting for auriferous quartz, use is sometimes made of the divining rod, a practice not without credit with some good miners. The rod is a fork of a green hazel-bush, shaped like a V, with the arms about a foot long. The prospector holds the end of an arm in each hand, with the point of the V directed forward horizontally, and as he walks along, the point turns down whenever he comes over a metalliferous vein, metallic body or water. It is supposed that very few persons can use the divining rod effectually; for most men it refuses to turn. It is used in nearly every civilized country, especially by miners, and is generally considered superstitious, because it is employed by ignorant people, and because there has been no generally accepted scientific explanation of the manner in which a stick could be influenced by a metal hidden under ground. A scientific explanation of the principle of the divining rod has been offered to the world, by Baron Reichenbach (see page sixty of his *Odic-Magnetic Letters*, translated by John S. Hittell).

§ 207. *Quarrying Quartz.*—The quarrying of quartz rock differs little from the quarrying of other metalliferous veinstones. The lode descends steeply, and the excavation must follow its course. Sometimes the quartz is so soft that it may easily be loosened with the pick. The harder rock is blasted. Soft quartz is that which is penetrated by numerous cavities, though the lumps between the cavities may be very hard. Some quartz on exposure to the air crumbles into sand, though hard when first taken from the vein. In narrow lodes, some of the wall-rock must be cut away to get room for the workmen. In wide lodes, that part of the vein-stone which does not pay is left. Sometimes the gold from the lode penetrates a little way into the foot-wall, and in that case the quarrying must extend beyond the vein-stone. The quartz loosened in the vein, must either be hoisted perpendicularly in a bucket with a windlass, or be hauled out through a tunnel. The common method is to hoist the rock with a windlass. Most of the veins are in such places that shafts are more easily dug than tunnels. After the excavation has extended twenty or thirty feet below the surface, it is usual to dig a perpendicular shaft, so as to strike the vein sixty or seventy feet below the surface, and from this point the miner or "drifter" works upward, and as he loosens the rock it falls to the bottom of the shaft, where it is put in the bucket to be hoisted to the surface. Our quartz mines are generally in dry hills, so that they are not troubled much by water; but there are a few shafts where steam-pumps are constantly at work to carry off the water.

Occasionally the miners find small quantities of auriferous quartz which are so easily broken up, and the pieces of gold in which are so coarse, that after the rock has been pounded a little in a mortar, the metal can easily be picked out with the fingers.

§ 208. *Arastra.*—Quartz is pulverized either in an arastra, or Chilean mill, or by stamps.

The arastra is the simplest instrument for grinding auriferous quartz. It is a circular bed of stone, from eight to twenty feet in diameter, on which the quartz is ground by a large stone dragged round and round by horse or mule power. There are two kinds of arastras, the rude or improved. The rude arastra is made with a pavement of unhewn flat stones, which are usually laid down in clay. The pavement of the improved arastra is made of hewn stone, cut very accurately and laid down in cement. In the centre of the bed of the arastra is an upright post which turns on a pivot, and running through the post is a horizontal bar, projecting on each side to the outer edge of the pavement. On each arm of this bar is attached by a chain a large flat stone or muller, weighing from three hundred to five hundred pounds. It is so hung that the forward end is about an inch above the bed, and the hind end drags on the bed. A mule hitched to one arm will drag two such mullers. In some arastras there are four mullers and two mules. Outside of the pavement is a wall of stone a foot high to keep the quartz within reach of the mullers. About four hundred pounds of quartz, previously broken into pieces about the size of a pigeon's egg, are called a "charge" for an arastra ten feet in diameter, and are put in at a time. The mule is started and in four or five hours the quartz is pulverized. Water is now poured in until the powder is thoroughly mixed with it, and the mass has the consistence of thick cream. Care is taken that the mixture be not too thin, for the thickness of it is important to the amalgamation. The paste being all right, some quicksilver (an ounce and a quarter of it for every ounce of gold in the quartz, and the amount of gold is guessed at from the appearance of the rock) is scattered over the arastra. The grinding continues for about two hours more, during which time it is supposed the quicksilver is divided up into very fine globules and mixed all through the paste (which is so stiff that the metal does not sink in it to the bottom), and that all the particles of gold are caught and amalgamated. The amalgamation having been completed, some water is let in three or four inches deep over the paste, and the mule is made to move slowly. The paste is thus dissolved in the water, and the gold, quicksilver, and amalgam have an opportunity to fall to the bottom. At the end of half an hour, or sooner, the thin mud of the arastra is allowed to run off, leaving the precious material at the bottom. Another charge of broken quartz is now put in and the process is repeated, and so on. The length of a "run," or the period from one cleaning up to another, varies much in different places. In the rude arastra a run is seldom less than a week, and sometimes three or four. The amalgam having settled down between the paving stones, the bed must be dug up and all the dirt between them carefully washed. In the improved arastra the paving fits so closely together, that the quicksilver and amalgam do not get down between them, but remain on the surface, and can readily be brushed up into a little pan, and therefore cleaning up is much less troublesome and is more frequently repeated than in the rude arastras; besides there is a greater need of frequent cleaning up in the improved arastras, because the amount of work done within a given time is usually greater.

The arastra is a slow instrument, but in some important respects it is superior to any other method of working auriferous quartz. It grinds the quartz well, is unsurpassable as an amalgamator, is very cheap and simple, requires no chemical knowledge or peculiar mechanical skill in the work, requires but little power, and very little water—all of them important considerations. In many places, the scarcity of water alone is enough to enable the arastra to pay a larger profit than any other method. Again, if a miner finds a rich spot in a lode, he may be doubtful as to the amount of paying rock which he can obtain. Such cases very frequently happen in California, and the arastra is just the thing for the case; for then if the amount of paying rock is small, nothing is lost, whereas the erection of a stamping-mill would cost much time and money, and before it could get into smooth operation the rich rock would be exhausted, and the mill perhaps become worthless. No other simple process of amalgamation is equal to that of the arastra; and it has on various occasions happened in California, that Mexicans making from fifty to sixty dollars per ton from quartz, have sold out to Americans who have erected large mills at great expense, with patent amalgamators, and have not been able to get more than ten or fifteen dollars from a ton. The arastra is sometimes used for amalgamating tailings which have passed through stamping-mills.

§ 209. *Chilean Mill.*—The Chilean mill has a circular bed like the arastra, but much smaller, and the quartz is crushed by two large stone wheels which roll round on their edges. In the centre of the bed is an upright post, the top of which serves as a pivot for the axle on which both of the stones revolve. A mule is usually hitched to the end of one of the axles. The methods of managing the rock and amalgamating with the Chilean mill, are very similar to those of the arastra. The Chilean mill, however, is rarely used in California; the arastra being considered far preferable.

§ 210. *Stamps.*—Nine-tenths of the quartz crushed in California is pulverized by stamps, of which there are two kinds, the square and rotary. The square stamp has a perpendicular wooden shaft, six or eight feet long, and six or eight inches square, with an iron shoe, weighing from a hundred to a thousand pounds. The wooden shaft has a mortice in front near the top, and a cam on a revolving horizontal shaft enters this mortice at every revolution. When the cam slips out of the mortice, the stamp falls with all its weight upon the quartz in the "battery" or "stamping-box." The rotary stamp has a shaft of wrought iron about two inches in diameter, and just before falling this shaft receives a whirling motion, which is continued by the shoe as it strikes the quartz. The rotary stamp is considered superior to the square, its advantage being that it crushes more rock with the same power, that it crushes more within the same space, and that it wears away less of the shoe in proportion to the amount of rock crushed. There are usually half a dozen square stamps or more, standing side by side in a square-stamp mill, and these do not all fall at the same moment, but successively, running from the head to the foot of the "battery." The quartz is put in at the head of the battery and is gradually driven to the foot. The rotary stamps sometimes stand side by side, and sometimes in a circle. The battery of both rotary and square stamps is surrounded by wire gauze, or a perforated iron plate, allowing the finely pulverized quartz to escape, and retaining the coarser particles. Quartz is crushed wet and dry. In wet crushing a little stream of water runs into the battery on one side and escapes on the other, carrying all the fine quartz with it.

§ 211. *Separation.*—After pulverization comes the separation of the gold from the rocky portion of the powder. The means of separation are mechanical or chemical. The chemical process is amalgamation ; the mechanical are those wherein the gold is caught on a rough surface with the aid of its specific gravity. The chief reliance is upon amalgamation, and in some large quartz-mills mechanical appliances are not used at all for catching the particles of gold, but only for catching amalgam.

The mechanical appliances used in quartz-mills in separating the gold from the pulverized rock, are the blanket, the sluice, and the rawhide.

The blanket is a coarse, rough, gray blanket, which is laid down in a trough sixteen inches wide and six feet long. The pulverized quartz is carried over this by a stream of water, and the particles of gold are caught in the wool. The blanket is taken up and washed, at intervals depending upon the amount of gold deposited. In some mills where a large amount of rock is crushed, and where the powder is taken over the blanket before trying any other process of separation, the washing takes place every half-hour. In mills where the pulverized quartz is exposed to amalgamation first, the blanket may be washed three or four times a day. The washing is done in a vat, kept for that especial purpose.

The sluice used in quartz-mills is similar to the placer board-sluice, but the amount of matter to be washed is less, and there is no dirt to be dissolved, and there are no large stones, and therefore the sluice is not so large, so strong, or so steep in grade, as the placer-sluice, and the riffle-bars are not so deep. In some quartz-mill sluices there are transverse riffle-bars. If the quartz has much iron or copper pyrites, the sluice is used to collect this material and save it for separation at some future time. The pyrites ordinarily contains, or is accompanied by much gold, which it protects from amalgamation. This separation of the pyrites from the pulverized rock is called "concentrating the tailings," and the material collected is called "concentrated tailings." In the sluices of some quartz-mills cast-iron riffle-bars are used; cast in sections about fifteen inches square, and about an inch deep. Much study has been devoted to the subject of making these riffle-bars in such a manner that the dirt will not pack in them, but will always remain loose, and keep in constant motion under the influence of the water running over them; but the object has never been fully attained. Quicksilver is used in nearly all quartz-mill sluices.

The rawhide used in separating gold from the pulverized

quartz is a common cowhide, laid down in a trough with the hairy side up, and the grain of the hair against the course of the water. The gold is then caught in the hair. Sheep-hides have been used in the same manner, recalling to mind the Golden Fleece. The hides, however, are inferior to the blankets for this purpose, and are never used in the best mills.

The methods of amalgamating are numerous. Among them are amalgamation in the battery, amalgamation with the copper plate, amalgamating bowls, and patent amalgamation of many kinds.

In many mills quicksilver is placed in the battery, two ounces of quicksilver for one of gold; and about two-thirds of the gold is caught thus. The copper plate in quartz-mills is made in the same manner as in placer-sluices, under which head a description of the plate may be found. Some amalgamating bowls or basins are little Chilean mills and arastras, made of cast-iron. One plan of amalgamation is to use a cast-iron bowl about four feet in diameter and a foot deep. Near the bottom are horizontal iron arms, which revolve and stir the quicksilver and pulverized quartz together. Four or five of these bowls sit in a row but at different levels: the bottom of the first bowl being level with the top of the second, and so on. The pulverized quartz passes through them all. Under each bowl a fire is kept up, because heat forms the action of amalgamation. If there be any pyrites in the quartz, some common salt is thrown in to assist in releasing the gold from the embraces of the sulphurets, and preparing it to be seized by the mercury. Another amalgamating bowl revolves on an axis that stands at an angle of about seventy-five degrees to the horizon, so that the material in the bowl is continually moving; and the bottom is divided by little compartments, which make a constant riffle. In other bowls the pulverized quartz is forced with water through the mercury. The methods of amalgamation differ very much, and a book might be filled with a description and discussion of the processes used at different quartz-mills in California.

§ 212. *Sulphurets.*—Many auriferous quartz veins contain considerable quantities of sulphurets or pyrites of iron, copper and lead, and their presence prevents amalgamation, and thus causes a great loss of gold. It is said that on some occasions in good mills, not more than twenty or thirty dollars have been obtained from a ton of vein-stone which had seven or eight hundred dollars of gold in every ton. The best method of treating the quartz containing pyrites, is to roast it, and thus drive off the sulphur, but this process is so expensive that it is seldom used; and the common practice is to crush and amalgamate the rock, and save the concentrated tailings for some future time, when there may be a sale for them, or when it will be cheaper to reduce them. The pulverized sulphurets are decomposed by exposure to the air, and after the tailings have been preserved for a time, they may pay better at the second amalgamation than at the first. A mixture of common salt assists the decomposition of the pyrites.

§ 213. *Chief Quartz-Mills.*—The most productive quartz-mill in the state is the Benton mill, on Fremont's Ranch, in Mariposa county. It is also the largest, having forty-eight stamps. There are four mills on the estate, with ninety-one stamps in all, and their average yield per month is sixty thousand dollars. A railroad four miles long conveys the quartz from the lode to the mills. The Allison quartz mine in Nevada county produces forty thousand dollars per month. The Sierra Buttes quartz-mill, twelve miles from Downieville, yields about fifteen thousand dollars per month. These last mills run night and day, and crush and amalgamate ten thousand tons of rock a year or twenty-eight tons per day. Forty men are employed, twenty-five to quarry the rock, five in the mill to attend to the stamps and amalgamation, one to do carpentry, one for blacksmithing, and eight for getting out timber, transporting quartz, and so forth. The cost of quarrying, crushing, and amalgamating a ton of rock is six dollars. The wages of the men are from fifty to seventy dollars per month with boarding. The average wages is sixty dollars. About ten miles eastward of Sonora, in Tuolumne county, are some rich veins of auriferous quartz, the most prominent of which are the Soulsby and Blakeslee lodes. The Soulsby mill produced forty thousand dollars in three weeks, when it commenced work in 1858, but it has not been so profitable of late.

§ 214. *Silver Mining.*—Silver mining has not yet been established fairly as a business in California. The silver ores of Washoe were discovered in 1859, and mining has been fairly commenced there, but the mines of Esmeralda and Coso, within the limits of this state, were not found until the summer of 1860, and up the present time no mills have been established there.

Silver mining differs much from gold mining. Gold is always found as a metal, never as an ore, and the separation from the accompanying vein-stone with which it is mixed mechanically, is much more simple and easy than the reduction of the argentiferous ores in which the silver is chemically combined with base substances, for which it has a strong affinity. Chemical knowledge and chemical processes are more necessary in mining for silver than for gold; and while all auriferous quartz is of the same kind, and may be treated in the same manner, there are many different kinds of silver ores, each of which requires a peculiar treatment. The reduction of silver ore costs on an average, from three to five times as much as the reduction of auriferous quartz.

The silver ore of Esmeralda and Coso is a sulphuret of silver, nearly all the veins having the same material, though the amount of it scattered through the vein-stone differs greatly in different lodes. In some veins there is much free gold, that is, little specks of metallic gold which can be separated from the other material in the same manner that gold is separated from auriferous quartz. The methods of reducing silver ore are so numerous and complex, and vary so much in different districts and under different circumstances, that it is impossible to know now what process will be used in Esmeralda and Coso, the resources of which places have been so little studied.

Besides it is said that new processes for reducing silver ore have been invented, far superior to all the old methods; and these processes are kept secret. It is therefore unnecessary that I should go into a long description of the various processes practised elsewhere. Silver ore after pulverization is smelted by mixing with it fifty per cent. of lead in metal or ore, and ten per cent. of iron, and exposing the whole to a heat sufficient to melt the silver which runs off. The metal thus obtained is not pure but contains much lead, which is driven off by heat while the silver is kept in a molten condition for a period of four or six hours. The cost of smelting in California at present, is about one hundred and twenty-five dollars per ton. In most of the other methods of reducing silver ore, the ore is roasted to drive off the sulphur. In the barrel amalgamation, which has been used at Washoe, and will probably be used at Esmeralda also, half a ton of ore, after being pulverized and roasted, three hundred pounds of water, and one hundred pounds of wrought iron, in little fragments, are put into a barrel, which revolves on a perpendicular axis. At the end of two hours the mass has taken the consistence of thick cream, when five hundred pounds of quicksilver are put in, and after the barrel has revolved four hours more, the amalgamation is complete. More water is now poured in; the barrel revolves very slowly to let the amalgam all settle to the bottom, the mud runs off through a cock four inches above the bottom, and the mercury and amalgam are then drawn off through a little hole in the bottom of the barrel.

§ 215. *Quicksilver Mining.*—The ore from which quicksilver is obtained is a sulphuret. The sulphur is driven off by heat, and the metal, which rises in fumes from the ore, is collected by condensation. The miners are Cornishmen and Mexicans. The ore is in large masses underground, not in a connected vein of regular thickness; and after one mass is exhausted, much labor is often vainly spent in search of another. There are, however, usually little seams of ore running from one large deposit to another, and it is the business of the mining captains to observe these veins closely, and trace them up when a "fault" occurs. There are no scientific rules for finding the ore; and the business of searching for the large deposits is never intrusted to educated mining engineers, but always to mining captains, who have themselves been laborers, and have learned by experience where to seek. The New Almaden mine produces two hundred and twenty thousand pounds of metal in a month. The *hacienda*, or reducing establishment of the mining company, has fourteen brick furnaces, each fifty feet long, twelve feet high, and twelve feet wide. At one end of each furnace is the fire chamber, which may be nine cubic feet inside; next that is the ore chamber of about the same size; and beyond that is the condensing chamber in which there are a number of partitions alternately running up from the bottom and down from the top, with a space for the fumes to pass, their course being up and down, and up and down again, and so on, for a distance of thirty feet to the chimney, which is forty feet high. In the bottom of the condensing chamber is water. The walls between the fire chamber and the ore chamber, and between the latter and the condensing chamber, are built with open spaces, so that the heat, smoke, and fumes can pass through. The ore is placed in the ore chamber in such a manner as to leave many open spaces. The heat drives off the sulphur and mercury of the ore in fumes, which in passing through the condensing chambers, deposit the mercury, and the smoke and sulphur escape through the chimney. In the Enriqueta and Guadalupe mines the quicksilver is condensed in a close iron retort, and the sulphur is absorbed by quicklime.

Copper ore is dug from several mines in California, but it is all exported to be smelted elsewhere.

§ 216. *Platinum.*—Platinum, iridium, and osmium, three white metals of about the same specific gravity with gold, are found with the latter metal in the placers in the basin of the Klamath and Trinity Rivers. Their particles are usually fine scales, very rarely reaching a quarter of an ounce in weight, and the largest piece of either ever found was less than an ounce and a half. They cannot be separated from the gold by washing, but they do not unite with quicksilver, and therefore they are separated from the more precious metal by amalgamation. They have no regular market in the state; miners never make them the chief object of search, and they have not been studied, so it is not known to what extent they might be obtained.

§ 217. *Del Norte and Klamath.*—Del Norte county in the northwestern corner of the state, is about forty miles long from east to west by thirty from north to south. The mining population in it is small. Most of the mining is done along the banks of the Klamath River, which runs about twenty miles through the southeastern portion of the county. There are some miners on the head-waters of Althouse Creek, which runs northward into Oregon. The county assessor, in his report for 1860, does not mention the existence of any quartz-mill or mining-ditch in the county. The mining districts are very mountainous and difficult of access. They obtain most of their supplies from Crescent City. The mining is chiefly in shallow placers, in deep and narrow ravines, and on bars of the Klamath River.

Klamath county lies immediately south of Del Norte, and is about the same size. It is almost exclusively a mining county, and has a population of about eighteen hundred. The diggings are placers in the bars and banks of the Klamath River and its tributaries, the Trinity and Salmon Rivers, and many small creeks. The principal mining places are Orleans Bar, Gullion's Bar, Negro Flat, Cecilville, Weitspeck, and Red Cap. The whole county is very rugged and mountainous, and much of it is covered with heavy timber. The diggings are so difficult of access, and are so protected by mountains against ditches, that they will last for many years. There is probably no part of the state where the single miner, without capital, has a better chance to dig gold with a profit. Nearly the whole beach of the county is auriferous.

§ 218. *Siskiyou.*—Siskiyou county lies east of Del Norte and Klamath, is forty miles wide from north to south, one hundred miles long from east to west, and reaches to the eastern boundary of the state. It has a population of 7,629, the large majority of whom are engaged in mining. The mining district is all in the western end of the county, along the banks of the Klamath River and its tributaries the Scott and Shasta Rivers. The Klamath runs through a deep cañon; the Scott and Shasta Rivers have pleasant open valleys, but the diggings along their banks are chiefly among the cañons near the Klamath. Hydraulic and tunnel claims are rare. There are six quartz-mills in the county and fifteen mining-ditches, of which last the principal is the Yreka canal, forty miles long, bringing water from the head of Shasta River to the town of Yreka. In 1859 there were four quartz-mills in the county, one of which was at Mugginsville, one in Scott's valley and two in Quartz valley. I have no information about the situation of the two built since that time. The principal mining towns are Yreka, Scott's Bar, Hawkinsville, Johnson's Bar, Deadwood, and Cottonwood.

§ 219. *Trinity and Shasta.*—South of the western part of Siskiyou and the eastern part of Klamath, lies Trinity county, ninety miles long from north to south, and about twenty miles wide on an average. The northern part of the county is the basin of the Trinity River, and is auriferous. From the county Assessor's report for 1860, it is to be inferred that there is not a quartz-mill or a mining-ditch in the state. The county is very mountainous, and most of the mining is done in rugged cañons along the Trinity River. The chief mining towns are Weaverville, Cox's Bar, Big Bar, Arkansas Flat, Mooney's Flat, and Trinity Centre.

South of Siskiyou and east of Trinity lies Shasta county, which is on an average forty miles wide from north to south, and one hundred miles long, reaching to the eastern border of the state. There is a rich auriferous district about twenty miles square, in the vicinity of the town of Shasta, in the southwestern part of the county. The diggings are mostly in the basins of Clear Creek, Cottonwood Creek, Rock Creek, and Salt Creek, all of which enter into the Sacramento. There are four quartz-mills in the county, one at French Gulch, one at Middle Creek, one at Muletown, and one at Old Diggings. The county has twenty-seven mining ditches, with a joint length of one hundred and forty-one miles, an average of five miles each. The chief mining towns are Shasta, Horsetown, French Gulch, Muletown, Briggsville, Whiskey, and Middletown.

§ 220. *Plumas and Sierra.*—South of the eastern part of Shasta county lies Plumas, which is about seventy miles square. About one-third of the county, in the southwestern part of it, comprising that portion drained by the head-waters of Feather River, is auriferous. It lies high above the level of the sea, and the work of mining is interrupted during a considerable portion of the winter, by cold, snow, and ice. Hydraulic and tunnel claims in deep hills, furnish a large portion of the gold yield of the county. There are five quartz-mills, one at Elizabethtown, one at Eureka Lake, and three at Jamison Creek. The principal mining towns are Quincy, Jamison City, Indian Bar, Nelson's Point, and Poorman's Creek.

South of Plumas is Sierra county, which is fifty miles long from east to west and twenty miles wide from north to south. The North Fork of the Yuba River runs through its centre, and the Middle Fork is its southern boundary. Though small, it is one of the richest mining counties of the state, and in proportion to the extent of its mining ground, is much richer than any other county. All its territory is four thousand feet above the sea-level, at the lowest. Most of the mining is done in hydraulic and tunnel claims in deep hills. Near the centre of the county is a mountain called the Downieville Butte, or the Yuba Butte, eight thousand eight hundred and forty-six feet high, on the sides of which are found some rich quartz leads. In 1859 there were eleven quartz-mills in Sierra county, of which seven are at the Butte, two at Downieville, one at the Mountain House, and one at Sierra City. The principal mining towns are Downieville, Monte Cristo, Pine Grove, St. Louis, La Porte, Poker Flat, Eureka City, Forest City, Alleghany Town, and Cox's Bar. One of the most remarkable features of the placers of the state is the blue lead, which was first discovered in Sierra county, and has been more thoroughly examined there than elsewhere. The "blue lead" is a stratum of blue clay very rich in gold. It is found deep under other strata. The general opinion is, that the blue lead occupies the bed of a large antediluvian river, which ran parallel with the Sacramento and about sixty miles eastward of it. It has been traced twenty miles or more, passing near Monte Cristo, Alleghany Town, Forest City, Chip's Flat, and Zion Hill. Mr. C. S. Capp wrote thus to the San Francisco *Bulletin:*

"This is not one of the many petty leads, an inch or two in breadth and thickness, which, after being traced a few hundred feet, end as suddenly and mysteriously as they commence; but it is, evidently, the bed of some ancient river. It is often hundreds of feet in width, and extends for miles and miles, a thousand feet below the summits of high mountains, and entirely through them. Now it crops out where the deep channels of some of the rivers and ravines of the present day have cut it asunder; and then, hidden beneath the rocks and strata above it, it only emerges again miles and miles away. Wherever its continuity has been destroyed, the river or gulch which has washed a portion of it away, was found to be immensely rich for some distance below, and the materials of which the lead is composed are found with the gold in the bed of the stream. It is evidently the bed of some ancient stream, because it is walled in by steep banks of hard bed-rock, precisely like the banks of rivers and ravines in which water now runs, and because it is composed of clay which is evidently a sedimentary deposit, and of pebbles of black and white quartz, which could only be rounded and polished as they are by the long-continued action of swiftly running water. The bed-rock in the bottom of this lead is worn into long smooth channels, and also has its roughnesses and crevices like other river-beds.

The lighter and poorer qualities of gold are found nearest to its edges, while the heavier and finer portions have found their way to the deeper places, near the centre. Trees and pieces of wood, more or less petrified and changed in their nature, which once floated in its waters, are also everywhere encountered throughout this stratum.

"The clay and fine gravel in which these pebbles and boulders are found to be tightly packed, is of a light-blue color, which gives the name to the lead. Much of this clay is remarkably fine and free from coarse particles, and is smooth and unctuous to the touch. It is said to be strongly impregnated with arsenic, as was shown by chemical analysis, and contains large quantities of iron and sulphur in solution, for pyrites and sulphurets of iron are deposited in shining metallic crystals in every vacant crevice. Fine gold is found among this clay, and the heavier particles beneath it, upon the bed-rock. This stratum varies in thickness from eighteen inches to eight or ten feet, while the whole lead varies in width from a hundred and fifty to five hundred feet.

"The same lead has been found at Sebastopol, four miles above Monte Cristo, and also higher up among the mountains. It appears at Monte Cristo, which is four miles above the high-lying Downieville, and over three thousand feet above it, and at Chapparal Hill on the side of a deep ravine; then at the City of Six, which is also on very high land, about four miles from Downieville, across the North Yuba. It is next found at Forest City, on both sides of a creek, and is there traced directly through the mountain to Alleghany Town and Smith's Flat, on the opposite side. There it is again cut in twain by a deep ravine. It crops out on the other side at Chip's Flat, where it has been followed by tunnels passing completely through the mountain to Centreville and Minnesota on the other side. Here it is obliterated by the Middle Fork of the Yuba, but it is believed to be again found at Snow Point, on the opposite side of the river; and again at Zion Hill, several miles beyond. There is no reason for doubting that after thus reaching over twenty miles, it still extends further. Hundreds of tunnels have been run in search of it. Where the line it follows was adhered to, they have always found it, and have been well rewarded for their labor. Millions of dollars have been taken from this lead, and its richness, even in portions longest worked, is yet undiminished. These tunnels have cost from $20,000 to $100,000 each, and interests in the claims they enter sell readily at from $1,000 to $20,000, in proportion to the amount of ground within them remaining untouched, and the facilities which exist for working it. Many of these claims will yet afford from five to ten or more years' profitable labor to their owners, before the lead itself within them is exhausted. As in some of them quartz veins and poorer paying gravel have been found, many of them may be valuable to work from the top down as hydraulic claims."

This idea that the blue lead occupies the bed of an antediluvian river is however not universally accepted. Mr. B. P. Avery, who has written numerous newspaper articles upon the mineral deposits, asserts that the "blue lead," as it is called, is not a "lead" but an extensive stratum which is many miles wide, and is found all the way from the foot hills to the summit of the Sierra Nevada. In reply to this, it is said that while a bluish stratum of clay similar to that of the blue lead is found over a wide district, that it is evidently different in origin from the blue lead itself, which is confined to a narrow bed, and marked by the signs found in all the other ancient river-beds of the state.

The Sierra Butte Quartz Mining Company has some of the best auriferous quartz lodes in the state. One lode called the Cliff Ledge, is twenty-five feet wide; and another called the Aërial Ledge, is about three feet wide. In the Cliff Ledge, the paying rock averages about six feet in thickness next the foot-wall. The average yield is eighteen dollars per ton. The quartz is bluish-white in color, and very hard when first taken from the lode, but on exposure to the air it slowly crumbles into sand.

§ 221. *Yuba and Butte.*—West of Sierra county, and drained by the same streams, is Yuba, which reaches to the Sacramento River, lying half in the mountains and half in the plain, the mining district being in the former half. The principal mining towns are Camptonville, Timbuctoo, Foster's Bar, Texas Bar, and Long's Bar. In 1859 there were nine quartz-mills in the county, three at Brown's valley, and one each at Camptonville, Dobbin's Ranch, Dry Creek, Honcut, Indiana Creek, and Robbin's Creek. The assessor in 1860 reported only two quartz-mills in the county. There are twenty-two ditches in the county, with an aggregate length of nine hundred and fifty-two miles, an average of forty-three miles each. The most important ditch, called "Bovyer's," supplies Timbuctoo with five thousand inches of water in the winter, less in the summer. The diggings at Timbuctoo are in a deep hill, which is washed away by the hydraulic process.

West of Yuba and Plumas counties lies Butte, which is drained by the Feather River. The principal mining towns are Oroville, Bidwell's Bar, Forbestown, Natchez, and Whiterock. In 1859 there were seventeen quartz-mills in the county, of which four were at Oregon Gulch, at Columbiaville and Hansonville three each, two at Yankee Hill, and at Evansville, Gold Run, Long Bar, Nesbitt's Flat, and Spring valley, one each. The assessor reports for 1860, twenty-nine quartz-mills, worth fifty thousand dollars, and crushing in the aggregate one hundred and sixty-two and a half tons per day. There are sixty-four mining-ditches, with an aggregate length of five hundred and eighty-three miles. The bars and beds of Feather River were once very rich, and some of the most extensive enterprises of river mining in the state have been undertaken within the limits of Butte county. The greatest flume ever built in California was that of the Cape Claim Company, near Oroville, in 1857. It was three-quarters of a mile long and twenty feet wide, and furnished employment for two hundred and fifty men from May till November. The expenditures during that period were $176,985, and the receipts $251,426, showing a clear profit of $74,441. The next year, after the water had fallen, the company commenced its labors again; spent $160,000 and received $115,000, and thus lost $45,000. North of Oroville is a "table-mountain" with a top of basalt, covering a rich deposit of auriferous clay.

§ 222. *Nevada and Placer.*—South of Yuba and Butte is Nevada, the richest mining county of the state. Within its limits the tom, sluice, under-current sluice, and crinoline hose were invented, and the ditch and hydraulic power were first applied to placer-mining; and quartz-mining was first undertaken extensively. In 1859 there were thirty-two quartz-mills in the county, and twenty-eight mining-ditches, with an aggregate length of three hundred and ninety-four miles. No part of the mineral region of the state is better supplied with water than Nevada county. The richest quartz district is in the vicinity of Nevada City, which has fifteen mills, and Grass Valley, five miles distant, has seventeen. The great Allison mine, which has the richest lode in the state, is in Grass Valley.

The quartz mines here are much troubled with water, and during the winter of 1860–'61 many of the mills were compelled to stop for weeks until the shafts could be drained by steam-engines, after having been filled by a long and heavy rain. The annual gold yield of Grass Valley has been estimated at four millions of dollars. North San Juan has the finest hydraulic claims, and Sweetland the largest tail-sluices. The Eureka Lake Ditch Company has more ditching and water than any other company in the state. Their main ditch is seventy-five miles long, and there are one hundred and ninety miles of branches, making a total of two hundred and sixty-five miles, which have cost nine hundred thousand dollars. The daily sale of water is six thousand inches, with a weekly income of six thousand dollars. The principal mining towns are Nevada, Grass Valley, North San Juan, Rough and Ready, Orleans Flat, Moore's Flat, and Humbug City.

Placer county lies next to Nevada going southward. The

North Fork of the American River runs through the middle of the county, and the Middle Fork forms its southern boundary. The principal mining towns are Auburn, Yankee Jim's, Gold Hill, Dutch Flat, Todd's valley, Michigan City, Iowa Hill, Bath and Wisconsin Hill. Two-thirds of the present gold yield of the county is derived from hydraulic claims.

Some of the placer diggings near Bath are found in a peculiar formation. The principal deposit is in a deep hill, at the bottom of which is a stratum of pay-dirt, consisting of a fine sandy sediment, with pebbles and pieces of quartz. The gold is round and coarse. Above that is a stratum of blue gravel, which varies from twenty to one hundred and fifty feet in thickness. This blue gravel is "spoted;" that is, in places it pays well; in other places it does not pay at all. Above this stratum is another layer of pay-dirt composed of a reddish gravel, which is about three feet thick on an average, and contains little scales of gold. The top of the ridge is composed of a whitish cement or tough clay, which, where exposed to the air, is reddish in color, and resembles the red gravelly clay found in most of the hills in the rich mining districts.

§ 223. *El Dorado and Amador.*—El Dorado county adjoins Placer on the south, and is drained by the South Fork of the American River, which runs through its centre, the Middle Fork, which is its northern boundary, and the Cosumnes its southern boundary. It is the oldest placer-mining county of the Sacramento basin, Marshall having made his discovery within its limits; and ten years ago it was called the "Empire County," because it cast the largest vote in the state, but it has now lost much of its population and fallen behind several others. The principal mining towns are Placerville, Coloma, Georgetown, Diamond Springs, El Dorado, Spanish Bar, and Indian Diggings. In 1859 there were forty quartz-mills in the county, of which six are at El Dorado, three at Steeley's Fork, at Placerville, Nashville, Grizzly Flat, Loafer Hollow, and Logtown two each, and the others are scattered about. The county has fifty-one ditches, twelve hundred and fifty miles in aggregate length, the average length being twenty-four miles.

"Indian Diggings," says Mr. Capp, "is a mining village twenty-five miles southeastward from Placerville, on the bank of Indian Creek. In this district, a belt of limestone, or blue and white marble, rises in ridges through the slate bed-rock, and is in places cut by the water into long and deep channels, some of which serve as natural tail-races for the miners, but it oftener renders large amounts of blasting necessary. The claims in the bed of the creek formerly paid well, as the tailings washed down from the hydraulic claims above continually enriched them. In some of the creek claims, in the middle of the channel, deep holes were found, filled with a kind of dirt different from that above it. This was sometimes extremely rich, and in one claim a single panful paid three dollars. Another singular feature connected with these deep places, is that they seem to have subterranean outlets, for in one instance a hundred inches of water poured in for three days, with all the dirt it washed down, failed to have any perceptible effect in filling it up. It was finally stopped with bushes and gravel, and the water turned off. A mile or more above this, in another claim, a similar hole was discovered, and forty inches of water poured in for several hours produced no visible progress toward filling it. Here the miner was in doubt whether there was a rich deposit of gold awaiting him down there, or whether the bottom of his claim had fallen out altogether."

Amador county adjoins El Dorado, with the Cosumnes for its northern boundary and the Mokelumne for the southern. The principal mining towns are Jackson, Volcano, Butte City, Dryton, Fiddletown, Sutter Creek, and Lancha Plana. Much of the bed-rock is marble. The county is rich in auriferous quartz, and has thirty-two mills, of which six are at Sutter Creek, five at Amador City, four at Dry Creek, at Volcano, Clinton, Contreras, and North Fork of the Mokelumne, two each, and at Big Bar Bridge, Butte City, Drytown, Grass Valley, Gales' Ranch, Herbertsville, Oneida, and Rancheria

Creek one each. The county has eight mining-ditches, with an aggregate length of one hundred and sixty-seven miles.

§ 224. *Calaveras and Tuolumne.*—Next to Amador is Calaveras county, bounded on the north by the Mokelumne River, on the south by the Stanislaus River, with the Calaveras River running between them. The principal mining towns are Mokelumne Hill, San Andreas, Murphy's, Angel's, Vallecito, West Point, Campo Secho, Douglass' Flat, Carson, Jesus Maria, and Esperanza. The county has thirty-three quartz-mills, of which twelve are at Angel's, at Carson and the South Fork of the Mokelumne River four each, three at the North Fork of the Mokelumne, at West Point, Rich Gulch, Murray's Creek and the Middle Fork of the Mokelumne, two each, and at Bear Creek and McKinney's Humbug one each. Mr. Capp says, "The main wealth of the district about West Point consists in its quartz leads, which are so numerous that several of the residents informed me, that starting three miles north of West Point, and proceeding south for a distance of nine miles to the junction of the forks of the Mokelumne, a person would cross a quartz vein in every hundred yards. About one hundred of these veins have been prospected upon the surface, and scarcely any have been found that did not prove to contain gold. As a proof of the richness of the veins of this district, it would be sufficient to state that large numbers of Mexicans and other Spaniards are now working them successfully, although they pay from one dollar to one dollar and fifty cents per cargo of three hundred pounds, to have the rock ground in arastras, to which freight from the leads to the mills along the river has also to be added. Mexicans who do their own work, cannot possibly afford to work rock that does not at least pay three dollars per cargo, or twenty dollars per ton, and in fact they seldom do work rock that pays less than six dollars per cargo and forty dollars per ton.

"There is very little slate in this district, and nearly all the quartz veins are encased in granite, which is usually much decomposed. Occasionally, the granite appears to 'pinch' the quartz leads until they become very thin, but by tracing them on further, or downwards, they again swell out to their original size, and sometimes bulge out beyond it. In such places, and at the intersection of small veins, very rich deposits of gold are frequently found, which, from their narrowness and the depth to which they extend, the Spaniards call *clavos* or nails. The principal portion of the rock about Angel's is of a greenish and gray color, and contains large quantities of the sulphuret of iron. Mixed with this are streaks and veins of white quartz or limestone. The sulphurets are found, either in irregular bright crystalline masses, or small threads and veins. Some of these veins are as much as eight inches in thickness. In other portions of the green rock, the sulphurets are scattered all through it, as separate and minute square crystals. The whole formation will probably become one solid vein when any considerable depth is reached; but near the surface it is cut up into separate veins by streaks and wedges of slate, which do not appear to contain any gold. These streaks of slate are from a few inches to several feet in thickness. The poorer portions of the rock contain from twelve to sixteen per cent. of the sulphurets, while the richest are nearly pure crystals, among which the gold is seen shining in small particles and scales."

In the southeastern corner of Calaveras county, thirty-five miles from Stockton, are mines of carbonate and sulphuret of copper. During the first three months of 1861, four hundred tons of ore, containing about thirty-three per cent. of metal, were taken out, and one-half the amount was shipped to the Eastern states. The cost of getting the ore out and hauling it to Stockton has been eight dollars per ton.

Tuolumne adjoins Calaveras county, and is bounded by the Stanislaus River on the north, and the divide between the Tuolumne and Merced Rivers on the south, the former stream being within the limits of the county. The principal mining towns are Sonora, Columbia, Springfield, Shaw's Flat, Jamestown, Chinese Camp, Big Oak Flat, Garrote, Don Pedro's

Bar, and Pine Log. The first four are within a circle with a radius of five miles, a district which has been extremely rich in placer diggings, and especially in large nuggets. In 1850 two lumps of twenty-three pounds, one each of eighteen, thirteen, ten, five, four, and three pounds, were found; in 1851 one each of twenty-eight, twenty-four, twenty-three, and five pounds; in 1852 one each of nine and five pounds; in 1853 one each of twenty, ten, nine, eight, seven, and six pounds; in 1854 one each of seventy-two, twenty-seven, sixteen, and seventeen pounds; in 1855 one of thirty pounds; and in 1858 one of thirty-three pounds. Two of the largest mining ditches in the state supply water to the miners in the vicinity of Sonora and Columbia. The number of mining ditches in the county is twenty-one, and their aggregate length two hundred and seventy miles. There are thirty quartz-mills, of which four are at Quartz Hill, four at Tuttletown, three on the banks of the Tuolumne River, and as many on Turnback Creek, at Columbia, Soulsby's Ranch, and Wood's Creek two each, and at Bald mountain, Big Oak Flat, Italian Camp, Jackson Flat, Moccasin Creek, Rawhide Ranch, Sonora, Whiskey Hill, Wood's Crossing, and Yankee Hill, one each. The table-mountain in Tuolumne county is the most remarkable elevation of its kind in the state. It has an average height of five hundred feet, an average width of four hundred yards, a length of thirty miles, and a surface almost perfectly flat, slightly descending toward the west. It was evidently formed by an immense stream of lava, which was once confined between banks higher than its own surface, which banks have since been washed away, leaving the stream of lava standing like a mountain above the level of the adjacent country. The sides near the top are perpendicular and of solid basalt; farther down they are sloping, and composed of dirt and fragments of basalt that have fallen from above. Under the basalt lie the gravel and sand of the ancient river-bed, enclosed at the sides by ridges of rock, which rise above the level of the adjacent plain. When therefore the miners first wished to reach the auriferous deposits under the mountain, it was necessary for them to cut tunnels through a rim of rock; but now that many of the tunnels have drained away all the water, they are started up on the hill-side above the rock, and cut in sloping downward, so that the hard bed-rock is avoided, and in many places the tunnels made on this plan may run all the way through soft dirt. It is established beyond all reasonable doubt, that the auriferous deposit under the basalt was once the bed of an ancient river. Every mark indicates it. The wide water-worn bed, the bends, the bars, the deposits of gravel in eddies, the collection of coarse gold in the centre, the position of the large flat stones, all pointing down stream, the remains of fresh-water mollusks, and the beds of little tributary streams—all these are conclusive proof that a large river once ran where this mountain now stands. The pay-dirt is a tough clay filled with large stones, and is from a foot to six feet deep. In one place a claim one hundred superficial feet square yielded seventy-five thousand dollars. The distance from the outside of the mountain to the pay-dirt, varies from six to twelve hundred feet. About ten miles east of Sonora is the Soulsby quartz lead, one of the richest in the state.

§ 225. *Mono and Mariposa.*—Eastward of Tuolumne, east of the summit of the Sierra Nevada, and within the limits of the Great Basin, lies the county of Mono, which contains the gold placers of Mono Lake and Walker River, and the silver lodes of Esmeralda. The placers of Walker and Mono are neither extensive nor rich; water is scarce; and the winters are so cold that mining is necessarily interrupted. The Walker diggings are seventy-five miles southward from Carson City, and the Mono placers are twenty-five miles further in the same direction. The Esmeralda mines are in a nest of mountains of the same name, most of the ridges of which run north and south, and are composed of eruptive rocks, such as trap and basalt, with occasional greenstone and porphyry. The argentiferous region lies in a rugged part of the mountains, about five thousand feet above the level of the sea. The ore is all a sulphuret of silver, in a gangue of gray quartz, running through trap and porphyritic greenstone. The lode called the Esmeralda, the most prominent and apparently the mother lode of the district, runs with the meridian, and contains very little or no gold, while other leads, running at right-angles to it, contain much free gold; that is, particles of metallic gold which have formed no chemical union with the silver ore. The Esmeralda mines were discovered in August, 1860, by J. M. Cory; not much of the ore has been taken out, no mills have been erected, little of the ore has been reduced, and therefore not much is known of the real wealth of the district, although little doubt is entertained that it will in time produce much silver. There are several small ditches at the Mono placers, but the county has no quartz-mills, and thus far no auriferous quartz has been discovered.

South of Tuolumne lies Mariposa county, which is drained by the Mariposa and Merced Rivers. The mines are shallow placers and quartz. There is, I believe, not a hydraulic or tunnel claim in the county, and the mining-ditches are few. The towns are small; the population in the placers unsteady and irregular in their mode of life; and the county, taken as a whole, is considered one of the most unpleasant parts of the state for the home of a family. In consequence of the scarcity of ditches there is no water in summer, and the placer miners therefore lie idle during a great part of the year, and either go to other counties or spend their time in dissipation. The county assessor reports five mining-ditches with a total length of forty-two miles. The quartz lodes are numerous and rich. There are thirty-four quartz-mills, of which four are on Fremont's ranch; four at Coulterville; at Gentry's Gulch and Whitlock Creek three each; at Agua Fria, Bean's Creek, Bear valley, Burns' Creek, Mariposa Town, Mariposa Creek and Stockton Creek two each; and at Bull Creek, Corbett's Creek, Guadalupe, Mount Ophir, North Fork, Quartzburg, and Saxton's Creek one each. The Fremont ranch, which contains forty-eight thousand acres, is the most valuable mineral estate in California, and includes Bear valley, one of the richest quartz districts in the world. There are two principal lodes, called the Pine-Tree and Josephine, which unite and form a vein thirty feet wide. The rock crushed at the mills of the estate averages fourteen dollars to the ton.

§ 226. *Fresno, etc.*—Fresno county, south of Mariposa, has the Chowchilla River for its northern boundary, and the Fresno and head-waters of the San Joaquin within its limits. These streams are all auriferous, but the diggings are not rich, and the gold is not fine in quality. There are no mining towns of note in this county.

Between the San Joaquin River and White River, a distance of one hundred miles, the Sierra Nevada is barren of gold. White River is in Buena Vista county, and is not rich, but has some small districts of placer mines and a little quartz.

Kern River in the same county has a small extent of placer ground and a good deal of auriferous quartz, most of which is crushed and amalgamated with the arastra.

In latitude 36° 20′, east of the Sierra Nevada and Owen's Mountains, lies the Coso argentiferous district, a region of which very little is known as yet.

Gold placers are found in the San Francisquito Pass, forty-five miles northward from Los Angeles, in the San Gabriel cañon, twenty miles northeastward from the Mission of San Gabriel in Bear valley, fifteen miles eastward of San Bernardino, on the banks of the Colorado, twenty miles north of Fort Yuma, and in the Sierra of Santa Lucia, near the Mission of San Antonio. It is said that veins of auriferous and argentiferous quartz are also found in Bear valley, and in Holcomb valley in the slopes of Mount San Bernardino, and that there is rich auriferous quartz at Amargosa in the valley of the Mohave, north of the sink of that river.

NOTE.—The statement of the number of quartz-mills and mining canals in the several counties, is taken from H. G. Langley's *State Register*.

CHAPTER IX.

OTHER BRANCHES OF INDUSTRY.

§ 227. *High Wages.*—There are two great difficulties in the way of the productive industry of California, the high prices of labor and the high rates of interest. These may be blessings to certain classes of individuals, and perhaps even to the population generally, but they render it impossible for California to compete with foreign manufacturers in many branches of employment. The great distance of our state from Europe and Atlantic ports—San Francisco is about nineteen thousand miles from New York, by the route followed by the sailing vessels ordinarily—and the high freight on merchandise, is a great protection to our home industry, and the Federal tariff gives us a further protection on many articles; but nevertheless, a large proportion of the manufactured goods consumed here are imported from abroad, and probably will be for many years to come. We have no secondary coal in the state, and cannot expect to smelt iron ore, or to make cutlery or fine articles of hardware. Some cotton is produced on the shores of China and is now shipped to England; perhaps we may at some future day be able to import it raw and manufacture it here, but of this there is no certainty. We cannot hope to obtain much cotton from this side of the Pacific, for the western slope of the North American continent is not suited to cotton growing. Now coal, iron, and cotton are the raw material for a large share of the most profitable manufactures of our age. We shall produce a large quantity of fine wool, and it will in time be spun and woven here with a profit. Lumber is so bulky that it cannot well be imported from abroad, and our forests are so extensive and convenient to the market, that we can make our lumber as cheap as that made elsewhere. We have extensive fisheries and they must be developed. We must build our own houses, and construct our own roads, railways, mining-ditches, and flumes. We now import all our hardware, glassware, porcelain, stationery, silks, cottons, fine woollen goods, and nearly all our clothing, boots, shoes, and fine furniture, and many of our agricultural implements, mechanical tools, wagons, and carriages.

There has been a gradual fall in the wages of labor since 1849. For instance, in that year the wages of good carpenters were sixteen dollars per day; in 1851, ten dollars; in 1853, seven dollars; in 1856, five dollars; and now four dollars; and there has been a similar decrease of wages in all those branches of labor much in demand. Tailors, shoemakers, and cabinet-makers have never received high wages, because little is done in their trades. Millers, calkers, and ship-wrights now get from four to six dollars per day; bricklayers, stone masons and plasterers from four to five dollars; boiler-makers, machinists, and pattern-makers four dollars; carpenters, blacksmiths, and carriage-makers from three to four dollars; house-painters, paper-hangers, and stevedores three dollars; hod-men and washerwomen two dollars; common laborers one dollar and seventy-five cents. Of such persons as are hired by the month and boarded, gardeners get thirty-five dollars; farmers, teamsters, waiters, sailors, chambermaids, and seamstresses twenty-five dollars. Clerks in stores get from thirty to sixty dollars with boarding; from fifty to one hundred dollars without boarding. The best miners, of the class called "drifters," who cut and blast tunnels and dig shafts, get four dollars per day; common miners get fifty dollars a month and boarding. The wages of labor in California are now higher than in any other part of the world.

§ 228. *Lumbering.*—Lumbering, or the preparation of forest timber for industrial purposes, is an important branch of the industry, and the sale of lumber is an important branch of the trade of the state. Our houses are built of lumber, our streets are planked with lumber, our fields are fenced with lumber, and our flumes and sluices are made of lumber. And some parts of the state are very rich in timber and can readily supply the whole demand. Lumber is of three kinds, sawn, hewn, and split; the last two kinds being very small in importance as compared with the first. About one hundred and seventy millions of feet of lumber are sawn annually in California, and at the average price of twenty dollars per one thousand feet at the mill, the total amount of the lumber trade may be estimated at three millions of dollars. The chief lumbering districts are in the Sierra Nevada, and very near the coast. Mendocino is the largest lumbering county of the state, and according to the assessor's report for 1860, produced thirty-five millions of feet in that year. The mills are at Timber Cove and the mouths of the Noyo, Albion, and Big Rivers; and the timber is nearly all redwood. Humboldt occupies the next place, sawing thirty million feet per annum; and Santa Cruz county next, with ten million feet annually. All the timber cut in Santa Cruz is redwood; in Humboldt there are about equal amounts of redwood, spruce, and fir, and a little fragrant cedar. Santa Cruz ships much lumber to the southern part of the state, and Mendocino and Humboldt supply most of the redwood lumber for the San Francisco market. The lumber cut in the mining counties is mostly used near home, large amounts being consumed for sluices, flumes, and in other mining enterprises.

The method of getting the logs to the mills in Humboldt county is peculiar. The mills are all on the shores of Humboldt Bay, which is surrounded by flat land six or eight miles wide. Through this flat land run tide-water sloughs or channels, into which brooks run from ravines in the hills. The land in this county has all been Federal property, and has been open to pre-emption; and most of the lumbermen have laid claim to the tracts where they work, or have bought them with state-school warrants, under which any of the Federal land in the state open to pre-emption, may be purchased. The lumbermen owning claims along the sloughs, drag their logs to the water and tumble them in. Those owning claims along the ravines, at the heads of the sloughs, have a wooden tramway made by laying down long poles, about six inches in diameter and four feet apart. On this tramway runs a wagon with four wheels, each wheel of solid wood eight inches wide, and from two to three feet in diameter, made of a transverse section of a tree. On this wagon one or two logs are placed at a time, and two mules easily haul the load down hill to the slough, and then haul the empty wagon back again. The slope in these little ravines is very gradual, so there is no difficulty either in hauling the load down, or the wagon up. The thickness of the logs varies from sixteen inches (nothing smaller is sawn) to nine feet. The average thickness is four feet and a half; seven feet is a common thickness in redwood. Of pine and spruce logs the largest are five feet through; the average thickness is three feet. The greater the thickness of a log, the shorter it is cut. The ordinary lengths of saw-logs are fourteen, sixteen, eighteen, twenty, twenty-four, and thirty-two feet. Redwood is rarely sawn more than twenty feet long; spruce and pine usually of that length or longer. A good lumberman will cut down a redwood tree three feet in diameter in an hour; a tree five feet in diameter in three hours and a half; a tree seven feet in diameter in six hours. Ordinarily two choppers work together, one on each side of the tree, and then, of course, they fell it in half the time that would be required for one man alone. They use the American axe and American axe-handle; the handle being about a foot longer than is used in common chopping. After the tree is down, it is cut into saw-logs with a cross-cut saw, managed by one man. It has been found that one man can make a longer stroke than two, and the length of the stroke is a matter of much importance to "clear the saw," or throw out the saw-dust; so the handle at one end of the common cross-cut saw is knocked off, and it is then held like the ordinary handsaw. The logs in the ravines are all cut within a hundred yards of the tram-way, to which they are dragged by oxen. The logs once thrown into the slough, are made into rafts from fifty to one hundred yards long, and from ten to forty feet wide. The outer logs of the rafts are fastened to each other at the ends, by a little chain with a tooth or dog at each end, and one dog is driven into each log. There are then ropes running across in several places to keep it from spreading out in the middle. This is a very easy and simple method of making a raft and the logs are not injured in the least. When the raft is all complete, the lumbermen get on it, and float with the tide down to the mills, which are on the shore of the bay near the mouth of the slough. The distance from the point where the rafts are made to the mills, varies from three to eight miles. When the tide turns, the raft is made fast to a tree or stump on the shore, and the loggers wait until the ebb commences again. Two tides will usually carry a raft to the mill. Every mill has a boom or enclosure for logs. This enclosure consists merely of large and long logs chained together and floating on the surface of the water, of a small slough or cove. When the raft arrives the boom is opened, the raft pulled in and surveyed. The logs are generally cut by companies of "loggers" who devote themselves to that business, and sell their logs to the mills. The survey is made by a surveyor, a public officer of the county, who is under bonds. He receives ten cents per one thousand feet of lumber, board measure, one-half to be paid by the loggers and one-half by the mill. The thickness of the log is taken at the small end, and one-fourth is thrown off for waste. The large logs will usually produce more lumber, and the small ones less than the amount indicated by this mode of measurement. In some places along the streams it is not convenient to make tram-ways, so the logs are cut in the summer and are piled up on the bank until the heavy rains of winter come, when there is enough water to carry the logs down, and then they are thrown in. The current carries them down to the slough, where the channel is wide and the current more moderate, and there they are caught by a boom, which is made by driving a heavy pile into the bank on each side, and chaining a large and long log to it. These two logs are chained together at their lower ends, and thus hold all the logs coming down from above. The boom logs may be three or four feet in diameter, and seventy-five feet in length. A little stream, a foot deep and three feet wide in the summer, will increase its volume of water a hundred-fold in winter, and carry down logs ten feet in diameter. Such a brook often supplies five million feet (board measure) of logs in a season.

There is no place in the world where the average thickness of the logs sawn in the saw-mills, is so large as in Humboldt county, and the mills are built with special reference to the size of the logs. The frames are made very large and strong, and the saws are of proportionate length. The saws used in the mills are of four kinds: the single-gate saw, the gang-saws in a gate, the muley-saw, and the circular saw. The single-gate saw is fastened in a frame or gate, which plays up and down. This saw is used for sawing small logs, and is not found in very large mills. The gang-saws are a set of saws fastened in a frame parallel to each other. In some gangs there are twenty-four saws side by side, and they cut a log into boards at one movement. The gang-saws move slowly and make smooth lumber. All the boards, plank, joists, rafters, and studding, are cut with gang-saws. The largest logs cut with gang-saws are not more than three feet and a half through. The muley-saw is an upright saw, which is fastened at the lower end to a shaft connecting with the steam-engine or water-power, and the upper end is loose, though it plays in a groove to keep it straight. The muley is used for cutting the largest logs into bolts, and taking off slabs, so as to reduce the logs to a size suitable for other saws to work upon. The muley-saw makes three hundred strokes in a minute, whereas the gate-saw makes only about one hundred. The gate of a gate-saw large enough to saw logs nine feet in diameter, would be too heavy for convenience or quick work. The circular saws are used for cutting all the thin siding, pickets, and laths. The largest circular saws are fifty-two inches in diameter, leaving about twenty-four inches on each side of the axle. When large logs are to be sawn with circular saws, one saw is put above the other, and one cuts into the log from above, and the other from below, so that they can saw a log four feet in diameter. The frame-work for the circular-saw mills is lighter than that necessary for the upright saws; and circular-saw mills are not unfrequently moved about on wagons from one place to another. The circular saw is used almost exclusively in the Sierra Nevada. In Mendocino county and at Santa Cruz, the muley and circular saws are used without the gang-saws. The cost of sawing lumber is about the same with the circular saw as with the upright saw. The circular saw is more dangerous to the sawyer, and requires more skill in mechanics, so the wages of a man to manage a circular saw are one hundred and fifty dollars per month, while the wages of a man for an upright saw are only sixty dollars. The bark of the redwood is always taken off before the log is put into the mill. A tree seven feet in diameter will make four thousand feet of lumber, on an average. In Mendocino county the logs are cut in the summer along the little streams, and in the winter they are carried down by the high water to the mills, where they are caught by large booms, but as the rivers are large with fierce currents, these booms are sometimes broken, when millions of feet of logs are swept out to sea and lost. The lumber after being sawn, is shipped to San Francisco in schooners, varying from one hundred and fifty to three hundred tons burden. The freight is eight or nine dollars per one thousand feet. Most of the trees convenient for the lumbermen near the mills in Mendocino county, have been cut away; but about Humboldt Bay the supply will last many years. In the Sierra Nevada, the circular-saw mills ordinarily move from place to place, so as to have the timber near at hand. In the valleys of Eel River there are hundreds of redwood trees twenty feet in diameter, and one measuring twenty-eight feet. On the trail from Humboldt Bay to Trinidad, there are very large trees, and indeed all along the coast, from Eel River to the Klamath, there are numerous trees of mammoth size.

The spruce cut at Humboldt Bay is light and is used for packing-boxes. The redwood is used for siding, furniture, railroad-sleepers, telegraph-poles, fence-posts, and all kinds of house-work, inside and out. The fragrant cedar has an odor suggestive of a mixture of turpentine and ottar of roses, and is used for cupboards, clothes-presses, and inside-work of houses. Fir is used for fence-boards, studding, rafters, joists, and plank, but the grain is too coarse for inside-work of houses. Most of the lumber cut in the Sierra Nevada is sugar-pine, a clear, good wood, which is the chief material for inside work and furniture in the mining districts.

The principal kinds of split lumber produced in California, are fence-posts, rails, pickets, and shingles. The redwood-tree splits very freely, smoothly, and straight; and furnishes nearly all the split lumber of the coast. It is a favorite tree for fence-posts, telegraph-poles, and railroad-sleepers, because of its great durability under ground, lasting three or four times as long as any other wood in common use. For split lumber, the tree is cut down and divided with a cross-cut saw, in the same manner as for saw-logs. The choice of the trees for splitting is important, as they differ greatly in the straightness of the grain, and the facility of splitting. Those trees which grow in places exposed to the wind are often twisted, and the wood is full of curls, and will not open without splintering. Where the wood twists, indications of the course of the grain will usually be found in the course of the seams in the bark. The best trees are those with a straight, perpendicular growth, preserving nearly the same thickness one hundred feet up, as at the surface of the ground, with no limbs for one hundred and fifty feet, with thin, smooth bark, seamed with perpendicular lines. It is a pleasure to strike an axe into such a tree.

The length of the log depends upon the kind of lumber wanted. Posts are seven feet long; fence-rails ten or twelve feet; pickets from five to seven feet; rafters, joists, and boards from twelve to twenty feet. The work of splitting usually commences at the end nearest the root of the tree. The lumberman draws a straight line with a pencil across the end of the log, and through the centre, avoiding any knot or curl apparent in the wood. He then sets his axe in this line and drives it with a stroke or two of the maul half an inch into the wood, takes it out, and moves a little further along till he has opened the line across the log. Then he sets the axe in the log near the centre, and drives it in with the maul several inches deep; then takes another axe and drives it in on the other side of the centre, and when the second axe has entered two inches, the log, if a good one, will split across the end. He introduces iron wedges in the end and drives them in, and takes out the axes; and then uses wooden wedges or gluts, first at the end, and then at the side of the log. In subdividing the log, he always tries to have as much wood on one side of the wedge as on the other; for if there be more than an equal share of wood on one side of the wedge, the split will not run straight with the grain, but will gradually approach the weaker side, so that the piece split off will be smaller at one end than at the other. The redwood when well managed splits so beautifully, that boards are frequently made twelve feet long, a foot wide and an inch thick, and almost as smooth as if sawn. One of the most important places in the state for the manufacture of split lumber is Lexington, on the Santa Cruz mountain, fifteen miles southwestward from San José. Fence-posts made four inches thick, six inches wide, and seven feet long; fence-rails ten or twelve feet long, two or three inches thick, and six inches wide; pickets six and a half feet long, three inches square; and shingles eighteen inches long. The posts and rails are sold at eight cents apiece; the pickets at twenty-five dollars per thousand; and the shingles at four dollars and fifty cents per thousand.

§ 229. *Fishing.*—The rivers of California and the waters of the ocean near its coast abound with fish. Trout are caught in the little streams, salmon in the Sacramento, and San Joaquin, and the rivers emptying into the ocean north of San Francisco Bay, and a great variety of fish are caught in the ocean.

Our fisheries are as yet so limited in extent that few fish are salted, nearly all going while fresh to supply the market of the towns on the coast. Salmon is the only fish salted for export. The species of salmon caught in our waters is called the Quinnat salmon. They are born in the rivers, go out to sea when three or four months old, stay out at sea some months, probably not less than fifteen months, and then return to the river in which they were born, there to spawn. The Quinnat salmon, as found in our waters, averages ten pounds in weight and sometimes grows to sixty pounds. It enters our rivers in November and remains about four months. Before our rivers were kept in a continual state of muddiness by the gold miners, the salmon ascended every brook in the Sierra Nevada, large enough for a fish to swim in, but now they do not leave the large rivers nor ascend them far. The salmon in clear water offer fine sport to the fisherman with the fly, but in California they are caught only as a matter of business, and always in the gill-net, which has meshes just large enough to let the fish get his head in, and then the twine catches him behind the gills and holds him. The net is not dragged, but is stretched across or partly across the river and is allowed to drift with the current down stream, a distance of some hundreds of yards, perhaps a quarter or even half a mile, the fishermen accompanying it in a boat. The net has lead sinkers at the bottom and cork floats at the top, so as to keep it upright, and it is not so deep as to catch on the bottom. The fish are swimming up the river, so they of course run into the net. A large number of salmon are taken in Eel River, Humboldt county, and great quantities might be caught in the Klamath and other streams along the northern coast. A few young salmon varying from three to six inches in length, are caught while on their way out to sea, with fine nets in the shallow waters of San Francisco Bay. The Quinnat salmon is fat when it enters the fresh waters from the ocean, but gradually grows lean, and the color, which is light yellowish-red, changes to a deeper shade as it ascends the rivers. The meat becomes leaner, poorer in flavor, and redder in color in proportion to the length of time that it remains in fresh water; but the little ones, which have never seen the salt water, have a more delicate meat than the larger ones fresh from the ocean. No attempt has yet been made to breed fish for our rivers, though it might evidently be done to a profit in many of the streams; but whether in the Sierra Nevada, where the mud abounds, is doubtful. Yet the probabilities of success are sufficient to justify the trial. Fifteen years ago the salmon regularly ascended all, or nearly all, the mountain streams to points above any of the present mining camps, where the waters are as clear now as they were in 1847. The rule is known to be general and supposed to be universal, that the salmon leave the ocean in the stream from which they entered it; and it is supposed further that they go to the very branch or brook in which they were born. It is well known that there is a salmon in the Klamath River never seen in Humboldt Bay, and various species in the Columbia never found in the waters of California, and salmon in the Quiniault River, Washington territory, not found yet in any other stream; and the Indians of Oregon say that certain tributaries of the Columbia have species never caught in any other place. If then a million of eggs were hatched at the head-waters of the Sacramento River, there would be reason to hope that they would return to spawn there.

The legislature has passed an act in regard to the salmon fishery. It provides that the run of salmon shall not be obstructed by any dam, weir, fence, or fixed stop-net; and that no person shall catch with a net in the San Joaquin River, or any tributary, in August, September, and the first half of October; or in any other salmon stream of the state in August, October, December, and January; nor shall any person, save

Indians, sell fresh salmon during the season when fishing with the net is prohibited. There is no legal prohibition against taking other fish, or shell-fish at any season of the year.

The halibut are not sufficiently abundant on the coast to make the fishery for them a distinct branch of business. They are caught with a hook at sea in water varying from thirty to fifty fathoms deep, on rocky bottoms. The line called a "trawl-line" is about six hundred yards along, with numerous short lines and hooks, and is left six or eight hours in a place, and when drawn up has halibut, flounders, rock-fish, turbot, cod, and nearly all the large biting fish that come to the market. The bait used is chiefly sardines and herrings.

The mackerel (*Scomber diego*), a good fish, but smaller than the Atlantic mackerel, is caught with a hook off the coast south of Point Conception. It is a surface fish, and bites greedily at a bit of white rag or shining fish-skin jerked through the water. It does not frequent bays, but is caught in the harbors of Catalina Island. A number of vessels are now employed in the mackerel fishery of California.

The little brown rock-fish (*Sebastes auriculatus*), is caught in San Francisco Bay about the wharves; but the other species are only found out in the open sea. They stay where the bottom is rocky, eat crabs and shell-fish, and bite freely at hooks. Most of them are caught near Punta Reyes and the Farallone Islands. The rock-fish are in the market and of equally good quality throughout the year.

The turbot is caught with the trawl-line throughout the year. Soles are caught with small mesh-nets in the shallow waters of the bay of San Francisco at all seasons of the year. There is no separate fishery for them; they are caught with numerous other species of small fishes, among which the smelts have an important place. The smelts are much more abundant than on the Atlantic coast, go in large shoals, and are caught at all seasons. A large business might be done in salting them, but they are caught only for the fresh market. The anchovies are very numerous in San Francisco Bay, where they try to keep in shoals by themselves, but do not succeed, and are caught with other small fishes in nets. They are fully equal to the European anchovy, and will soon become an important article of commerce. At present most of those taken are eaten fresh, and only a few are potted. They are caught at all seasons of the year. Sardines are also abundant, and of a flavor equal to those on the coast of France, but larger. They are found in all the bays along the coast from May to October. An attempt was made several years ago to pickle sardines for the market, but it failed. The herring is not abundant on the coast of California, or at least is not found here in such dense shoals as in the Atlantic, and our species is smaller. It is caught with a net in the shallow waters of the bays. There are no shad in the waters of California. Shrimps are caught in the shallow waters of the bay of San Francisco with small mesh-nets, but are becoming very scarce The sturgeon visits the rivers of the Atlantic states for only a couple of months in a year, but it is abundant in the Californian rivers at all seasons. It never bites, the mouth being a round hole, always open, surrounded with gristle. In the Eastern states the sturgeon is often harpooned, but here it is caught only with nets. The meat is coarse, and is sold at one-fourth or one-sixth the price demanded for the meat of other fishes. The sturgeon might be salted, but nothing has been done in that business yet. An attempt was made several years ago in San Francisco to establish the business of preparing caviare from the roe of the sturgeon, but it did not prove profitable and it was abandoned. Sea-bass, a fish of fine, delicate flavor, and highly prized by epicures, is caught with hand-lines outside the heads of San Francisco Bay, and in the bay near Saucelito with nets during the spring and summer. It is not abundant. The sheepshead, an excellent fish, is caught off Santa Barbara with hand-lines during the summer. It should be brought to the market alive in smacks, for it loses its delicacy of flavor soon after death. The jewfish is abundant south of Point Conception, and may easily be taken with a hook or harpoon. It spends most of its time at the bottom in both deep and shoal water, but frequently comes to the surface, and according to report sleeps there. It also goes into lagoons, and likes to be near the kelp. They grow very large, sometimes to weigh five hundred pounds, and as their flesh is very good, a profitable business might be made of fishing for them.

Sharks are taken by Chinamen for food and by Americans for their oil. The common sharks caught by the Chinamen, perhaps more properly called "dog-fish" (*Acanthea suckleyi* and *Triakis fasciatus*), are taken in nets during the summer months and are dried in the sun. They are from three to five feet long, and contain a large amount of meat, which is never eaten by white men, but seems to have favor among the Mongols. The fish is cut open by a dexterous and quick stroke of a large knife along the back-bone, and is then dried without the use of salt. The fins are considered a delicacy. In Humboldt Bay the true shark (*Notorhynchus maculatus*), from five to twelve feet in length, is taken with spears. Three men have a flat-bottomed boat twenty feet long and four feet wide, with which they go into the shallow waters of the bay, whither the sharks resort in pursuit of the sardines. The liver is taken from the shark and the remainder thrown away. Each liver yields from one to eight pounds of oil. The spears have a handle eight feet long, which is loose and comes out of the spear-head after the shark is struck. If the handle were fastened in the spear-head, it would be broken by the struggles of the fish. A rope attached to the spear-head suffices to hold the shark, and by its means he is drawn up to the side of the boat, where he is struck by an axe on the head, and thus dispatched. The shark season lasts only about two months, during July and August. The oil is used for lubricating the machinery of the saw-mills about the bay, and sells for one dollar per gallon; and so long as the season lasts the fishermen make from five to ten dollars per day.

§ 230. *Hunting.*—The principal game quadrupeds and birds of California are grizzly bear, elk, deer, antelope, hare, rabbit;

the gray Canada brant, the white goose; the canvas-back, mallard, sprigtail, spoonbill, and summer ducks; the widgeon, the teal, the English black-breasted, sand, and dowiches snipe; the curlew, the mountain partridge, the valley quail, and various kinds of grouse. Nobody makes a business of hunting the grizzly bear; to attack him is so dangerous and to kill him so difficult, that many hunters will not shoot at him even when he comes in their way. A large number of them, however, are killed every year, and their carcasses are seen in the meat markets of San Francisco at all seasons of the year. The meat resembles pork in its greasiness, but it is coarser in texture and rank in flavor. It nauseates some delicate stomachs. The Spanish-Californians sometimes lasso the bear. When four or five of them, well mounted and provided with good saddles and reatas, surprise a bear in an open plain, they all beset him at once, and while one throws the lasso over his head, another catches him by a hind-leg, and a third by a fore-leg, and then two horses in front, but at a little distance from each other, drag him along, and the third and perhaps a fourth horse follows him, each one keeping his lasso stretched, so that even if the bear should succeed in breaking one reata or slipping it off, he will still be held fast by several others. He is thus dragged to a pen, where he is kept for a bull-fight or some other amusement.

It is only a few years since the elk were abundant on the Sacramento and San Joaquin, but they have now disappeared in those places, and are found in small numbers along the northern coast, where they will soon be exterminated. The meat resembles that of the deer, but is a little coarser in grain. The elk are shy animals, have a very quick ear, and are more difficult to approach than any other game animal in the state, unless the mountain sheep be excepted. They ordinarily lie hidden in thickets during the middle of the day, and feed about sunrise and sunset, at which times the hunters seek them.

The black-tailed deer are good game for the hunter. They

may be approached with more ease than the Virginia deer, run with a steady gait, and when disturbed do not run so far. The deer east of the Mississippi go with a run and a jump; the Pacific deer move with a steady run. Their meat is not so sweet as that of their eastern congeners. The deer live near the timber, and are found along the coast and in the Sierra Nevada. They were at one time very abundant but are now rapidly decreasing. The best place for hunting them is in Mendocino county. There is no deer-hunting on horseback, nor by large parties. The hunters go out alone or in small parties. Occasionally a deer is caught with the lasso, but this requires an excellent horse, a first-rate vaquero, and a good piece of open ground where the horse can have a fair chase and the vaquero can swing his reata.

The antelope lives in the open plain and in the desert. The valley of the San Joaquin was once full of great herds of them, but they, like other large game, have become rare now. They are shy, but inquisitive also, and are easily enticed to approach the hunter, who hides himself behind a rock, and fastening a white handkerchief to his ramrod, waves it back and forth. The antelope, like the deer, is occasionally caught with the reata, but these occasions do not occur once in the year, and, when they do occur, they establish the fame of the horse and rider engaged in the exploit.

There is one pack of hounds in the state, and they are sometimes, but rarely, used for hunting coyotes and foxes, as well as deer.

The wild geese and ducks are very abundant in California from September to March. They spend the winter in the tules of San Francisco Bay and tributary waters, and in the spring they migrate to the north. While here, they afford profitable employment to a number of hunters, who are of two classes—the "boat-shooters" and the "ox-shooters." The boat-shooters go in parties of two or three, each party having a sloop of its own. The sloop goes to the slough where the game abounds, and there every man starts in his skiff, with three double-barrelled shot-guns. He usually shoots first at the ducks or geese while they are in the water, and afterward again and again as they rise and fly. Sometimes he goes ashore, to shoot them while feeding. The geese spend the night in the water—generally in a slough or pond—and rise about daybreak, to feed in the fields of grain, grass, or wild oats. They remain there during a considerable part of the morning, return to spend the middle of the day in the water, go back to the fields in the afternoon, and at sunset take to the water again for the night. The ducks get most of their food in the tules, and are not often shot on the land.

The ox-shooter stalks his game. He has a trained ox, which walks before him and hides him from the geese or ducks until within good shooting-distance. The boat-shooters average thirty ducks a day during the season; and a good ox-shooter will sometimes kill one hundred and fifty geese in a day.

Snipe, curlew, and quail, are the game for sportsmen who hunt for their amusement, and the modes of hunting them are the same as those in the Eastern states.

Hunting is an unimportant interest in California as compared with fishing, and must continually decrease in importance, while the fisheries will increase.

A game-law prohibits the killing of quail, partridge, mallard and summer duck, from the 1st of March to the 15th of September; and elk, deer, and antelope, during the first half of the year; and prohibits the selling of the game slain during the forbidden season.

§ 231. *House-building.*—In the building of houses, the Californians, like Americans generally, are expert and quick. It is not uncommon to see a wooden dwelling-house commenced and finished within a month. Brick houses are built so fast, that the mortar has scarcely time to dry and harden as the walls go up. Most of the houses are of wood, and of the kind called "Balloon" or "Chicago" frames, fastened together with nails, without tenons and mortices, and with no upright posts thicker than two by four inches. This kind of a frame, called

"Balloon" from its lightness, and "Chicago" because they first came extensively into use in that place about fifteen years ago, appears very strange to a carpenter familiar only with the old-fashioned frames held together by tenons and mortices; but weak as the balloon-frame appears, it is really the strongest kind of a wooden building; and it is not unfrequently made four or five stories high, whereas the heavier frame very rarely reaches three stories.

In the balloon-frame, the sills, instead of being eight, ten, or twelve inches square, are only two or three inches by six or eight; and they rest on numerous studs, which again rest on the ground. The sills are nailed together at the corners. The studs are not morticed into the sills, but nailed upon them. The lower joists stand upon the sills, and the upper ones rest upon an inch board "let into" the studs to which they are nailed. On the top of the studs is no heavy plate, but only a board. At the corners two studs are put side by side. Each stud is hoisted to its place separately, so there is no "raising." Wooden houses are all covered with shingles. White pine, imported from the Eastern states, is used to a considerable extent for the frames and casings of doors and windows, and for other inside-work; and nearly all the doors and window-sashes are imported ready made.

Three-fourths of the houses in the state are of wood; the other fourth are of brick and adobes. Stone houses are very rare. Brick buildings are numerous in the business-streets of the cities and towns. Every town of note has its fire-proof brick stores, with iron doors and window-shutters, and its roof of brick laid in mortar. The bricks are made in this state, and the lime is burned here. Brick buildings not constructed to be fire-proof, have shingled roofs. There are a few buildings with fronts of granite, which for one house was brought from China, and that for others from the Eastern states.

Stone houses are very rare in California; it would almost be possible to count all of them on the fingers. Nearly all the dwellings in the counties bordering on the coast, from Monterey southward, are made of adobes, or sun-dried bricks; but most of the houses built of late, and all the elegant structures, are of wood or brick.

When an adobe house is to be built, the adobes are made very near the place, for they are too heavy to be hauled far. The earth to make adobes should be a sandy or gravelly clay, which will not crack in drying. A pure loam or pure clay will not do. The material having been found, it is dug up, and mixed with water and some straw, until it makes a thick mud. Not far from the pit the earth is levelled off, as a yard where the adobes are to be dried. The adobe is from three to six inches thick, from ten to twenty inches wide, and from a foot to two feet long. The mould is made of the size which the adobes are to have, without top or bottom. It is placed on the ground; the mud is thrown in, until the mould is filled; the top is scraped off level with a board; and the mould is lifted up, and moved to another place. The mud must be so thick, that it will retain its shape after the mould is removed. In twenty-four hours the adobe is so hard, that it will stand on one side, and in three or four days it is dry enough to be used for building. Its sides are rough and its corners broken, but it serves to make a house. Adobe walls are often made three feet thick, rarely less than a foot and a half. The mortar used is of the same mud of which the adobes were made. The walls are usually protected by wide eaves, and sometimes the roof projects six or eight feet, so as to make a corridor running entirely round the house. If the adobes be exposed to the rain, they are soon washed down. The walls have a foundation of stone laid about eight inches high on the surface of the ground.

In many of the adobe houses there is no floor save the bare earth. These dwellings are very cool in summer and warm in winter; and in old times, when the work was all done by Indians, they were cheap, but now it costs more to build a neat house of adobes than of wood. Many of the old adobe houses are still covered with the tiles baked by the Mexicans; those built of late have nearly all shingle roofs. In Los Angeles county, the roofs are generally made of asphaltum, obtained from the bituminous springs, of which there are a number along the southern coast. The rafters are covered with boards or cane, upon which earth is thrown, and upon that the asphaltum is placed. Sometimes the asphaltum is poured upon the earth in a melted condition; sometimes it is thrown on in hard lumps during the summer; and the melting is left to the heat of the sun. The asphaltum cracks in the cold and melts in the sun; so the eaves are dripping in July and August, and in the winter, if a rain comes immediately after a severe cold, the roofs are certain to leak. The cracks are always filled up again when a hot sun shines on the roof and melts the asphaltum. Most of the dwellings are surrounded by verandas or corridors, which are the most pleasant parts of the houses in the summer-time. Some of the dwellings are built in the form of a hollow square, with a paved court inside, planted with trees, and with a corridor running round. This corridor is the favorite place for spending the day, and here visitors are generally received.

§ 232. *Furniture, etc.*—Nearly all the fine furniture of the state is imported. The cabinet-makers' shops are few and small. The costly articles of fine wood-work, made on a large scale in California, are the billiard-tables; and these are made of unsurpassable excellence and with unsurpassed elegance. Our agricultural implements, wagons, carriages, omnibuses, and coaches, are mostly imported; and when they are made here, imported wood is used. No hubs, spokes, or felloes, are made in the state. It may be that all our agricultural implements will soon be manufactured at home, for a contract has lately been made to employ one hundred of the state-prison convicts in this labor; but we shall probably continue to import our wagon-lumber, for very little of the Californian timber is strong enough for such uses. Our tubs are mostly made here, and are of good quality. We have many well-built wooden bridges in the state.

California has built many schooners, and one war-vessel of fifteen hundred tons, but will never construct many ships. Her evergreen-oak, laurel, and madroña, are good for ships' knees and frames; but the trees are rare and costly, and far from the shipyards: while Puget Sound is so near, and has such very great advantages over us, that we cannot compete with Washington territory.

There are a great many excellent roads in the mountainous districts of the state, but the general character of our highways is poor—muddy in winter and dusty in summer. Macadamized roads are few. Slowly we are getting railroads. There is one, twenty miles long, between Sacramento and Folsom; another, forty-five miles long, is being made from Folsom to Marysville; one has been commenced between Vallejo and Marysville; and another is to be soon begun between San Francisco and San José.

All our cotton and cotton goods are imported. There is not a cotton-carding machine, spinning-jenny, or loom, in the state. No linen is grown or manufactured here. All our fine woollen goods are imported. We have only two woollen-mills in the state, and they make only blankets and coarse cloths.

Large quantities of leather are tanned in California, but it is all heavy. We tan no fine calfskin, morocco, or kid. Nearly all our boots and shoes for men's wear, and all for women, are imported. There are small shops where men's boots and shoes are made by the single pair to order after measure taken, and the customers in most cases are desirous of having a very elegant fit; but there are no large factories to make boots by the thousand.

We make no gunpowder, fire-works, fire-crackers, pitch, tar, or turpentine. It is said that our pine-trees do not contain the resinous sap to make turpentine, pitch, and rosin, and therefore we shall always have to import these articles.

There are two paper-mills in California, one of which makes printing-paper and the other straw wrapping-paper. We still import much of our printing and wrapping paper, and all of our writing-paper. Much of our coarse soap is made here, all fine soaps are imported. Nearly all our candles are imported. We have no furnaces for smelting iron-ore, and no forges to make wrought iron or steel. All our cutlery, stoves, cast iron, firearms, cooking-utensils, nails, screws, locks, hinges, tin-ware, copper, and copper-ware, are imported. So also are all our lead and shot, wire, tin, zinc, solder, and sheathing-metal. All our bottles are imported. A bottle-factory was established in San Francisco, to make bottles out of broken glass, of which much could be obtained in the large towns. This factory, however, was shortly abandoned; but there is a report that it will soon be established again.

CHAPTER X.

COMMERCE.

§ 233. *General Advantages.*—California is now the most important commercial state on the shores of the North Pacific, and it will continue to hold its present relative position. Nearly all the foreign commerce will be done by San Francisco, which has advantages superior to those of any other place on the coast between the Isthmus of Panama and Puget Sound, or on the coast of Asia. Her only equal, in the waters of the Pacific, is Melbourne. Several ports of China and Japan may have as much shipping; but the people of those countries are only semi-civilized, and we shall take their trade, and make them tributary to us.

In estimating the commercial advantages of California, we must consider the yield of our gold, silver, and quicksilver mines; the produce of our grain-fields, vineyards, orchards, and saw-mills; the mineral resources of Western Mexico; the lumber and fisheries of Washington territory and Oregon; the gold of British Columbia; the whale-oil and whale-bone of the North Pacific; and the industrial products of China and Japan. California is in the position to be the common carrier for all the countries bordering on the North Pacific. Between Puget Sound and Valparaiso there is on this coast no port that can prove a dangerous rival to San Francisco. As the population, wealth, and industry of this coast increase, and as the foreign commerce of China and Japan becomes more extended, the trade of San Francisco must continue to grow in at least an equal ratio. The increase, in every event, is certain.

Our chief commercial city is forty days' sail from China, and fifteen from Honolulu. It has been proposed to establish a line of steamers between San Francisco and Shanghae, but I doubt the necessity of such a line. Whenever the trade will justify it, there will be a line of swift-sailing packets, which can make almost as good time as steamers. There is no sea where sailing-vessels can make such regular and swift passages throughout the year as on the North Pacific, particularly if they sail across it from east to west or from west to east. The trade-winds, which blow over it toward the south, are constant, and equally favorable for vessels bound in either direction. Side-wheel steamers would be entirely unsuited for the route, because the wind is always from the north, and one wheel would be out of water nearly all the time; but propellers, to be used when the wind is low, might be of service; though there is reason to doubt whether they would be of sufficient service to justify the additional expense. Six or seven years ago, a steam-propeller was employed between San Francisco and Honolulu, but it was soon driven off by the sailing-vessels, which made the trip ordinarily in about the same time, and on several occasions beat the steamer.

We cannot foresee clearly the manner and rapidity of the development of the foreign commerce of China and Japan, but that it will reach a high development is certain. Three hundred and fifty millions of industrious people are not to be shut out from intercourse with Christendom. They have wants which must be supplied, and which white men alone can supply; and as California is the nearest state containing a large white population, and having a large commerce, we must have much to do with the business of supplying them.

§ 234. *Tributary Population.*—The present population of Nevada territory is about six thousand men, and the yield of silver and gold is estimated to amount to about one million dollars per year. It is now the general and confident opinion of Californians that the population and silver production of that territory will rapidly increase, and of course the increase must have a relatively favorable influence upon the commerce of California, from which Nevada will obtain all her supplies.

Oregon has a coast line of two hundred and seventy-five miles, from latitude 42° to 46°, and extends nearly four hundred miles from east to west. All that portion lying east of the Cascade Mountains (which are a continuation of the Sierra Nevada), comprising about two-thirds of the state, is barren, or nearly so. It may contain good pasture-lands and valuable minerals; but, with the exception of a few fertile valleys and bottom-lands near the Columbia River, these have not as yet been discovered, or at least not occupied by white men. The western part of the state contains some rich placers, fine forests, and valuable land for farming, but the country is difficult of access. The only entrance to it from the sea is by the Columbia River, the mouth of which is dangerous to shipping. There is a land entrance to Oregon from California on the south, and from Washington territory on the north. Oregon has a population of some fifty thousand, and produces about one million dollars in gold-dust annually.

Washington territory has a sea-coast of two hundred miles, and extends six hundred miles eastward to the Rocky Mountains. The fertile strip of the territory is only about one hundred and twenty miles wide, along the Pacific; the eastern portion is for the most part barren, but it contains extensive placers of gold, which yield about three hundred thousand dollars annually, and, if reports be true, will soon yield much more. The population numbers about ten thousand. The fertile district west of the Cascade Mountains is penetrated to a depth of one hundred miles by Puget Sound and Hood's Canal, the finest bodies of land-locked tide-water, for the purposes of internal navigation, in the world. The timber of the territory is very valuable, abundant, and accessible, and is now shipped to all quarters of the globe. No country can furnish large spars in such great abundance, or at so cheap a rate; and this resource, if there were no other, would secure wealth to Washington. Yet, in addition to these, she has extensive fisheries, fine farming-lands, and great facilities for commerce. For these reasons, Washington will occupy an important place on the Pacific, will have a dense population, and will contribute much to build up the commerce and increase the wealth of California. The climate of the western district of Washington is very moist throughout the year—cool in summer, and not cold in winter. Ice seldom forms more than two inches in thickness, or remains more than two or three days; but the sun is hidden nearly every winter by fogs and clouds, for weeks at a time. The climate of Washington bears as close a resemblance to that of England as does the climate of California to that of Italy.

British Columbia has a population of six thousand, yields about one million dollars in gold-dust annually, and has fine forests of ship-timber along the shores of the Gulf of Georgia. The soil is not rich, either on the mainland or on Vancouver Island, and the land will not be extensively cultivated there until after the richer land of Washington territory has been occupied.

The western part of New Mexico and Arizona is a barren country, unfit for tillage, but rich in minerals, which may at some day attract a large population. At present, its white inhabitants, west of the Rocky Mountains, do not number more than one thousand. The present silver production of Arizona may amount to ten thousand dollars per month. Western Arizona and New Mexico obtain most of their imported goods from San Francisco.

The population of Utah is about forty-five thousand, and they do much of their trading with Placerville and Los Angeles.

The western coast of Mexico, from San Blas northward, has a population of about a million and a half, and these obtain many of their imported goods from San Francisco.

The Hawaiian Islands have a population of seventy thousand, including two thousand whites, and they obtain most of their imported goods from San Francisco.

Adding together the inhabitants of Nevada, Oregon, Washington, the western parts of New Mexico and Arizona, the northwestern part of Mexico, British Columbia, Vancouver Island, and the Hawaiian Islands, we have a total of about one million seven hundred thousand people; and mineral, commercial, and industrial resources and advantages that must attract many more inhabitants at no distant day.

§ 235. *Imports.*—The imports of California are about $55,000,000 per year; that is, we import about as much as we export; and our exports amount, with gold, silver, grain, wool, wine, and sundries, to $55,000,000 annually. We have, however, no exact table of our imports, for most of them come from New York, and, if of foreign production, have paid duty there, and no report of their class and value is made at the custom-house in San Francisco. Therefore, we cannot obtain such tables of the imports into our chief port as are made at New York, where all the foreign goods are received direct. In the absence of accurate statistics, I must make an estimate of the values of the several classes of imported articles, which, while confessedly inexact, may yet serve to convey information to those who know nothing of the subject:

The clothing and material for clothing imported in a year may cost us $15,000,000; provisions (among which butter is perhaps the most important), coffee, tea, and spices, $3,000,000; tobacco, $3,000,000; cutlery, hardware, and metallic articles, $4,000,000; articles of wood, such as wagons, agricultural implements, &c., $1,000,000; drugs, $1,000,000; boots and shoes, $3,000,000; jewelry, $1,000,000; coal, $1,000,000; liquors, $2,000,000; and sundries, $2,000,000: total, $36,000,000. For freight on imported goods we pay $4,000,000; for insurance to foreign companies, $1,000,000; for interest-money on foreign capital, $1,000,000; for passenger-fares, $2,000,000—making a total sum of $44,000,000, which is less by several millions than the amount of annual exports.

§ 236. *Exports.*—Gold has the first place, but the exact amount exported is not known, for considerable quantities not reported at the custom-house, are taken away by passengers; and in the amount of treasure reported by manifest for exportation, the gold is not distinguished from the silver. The amount of treasure exported in 1861 was $40,639,089 57; and probably not more than $1,000,000 of silver is included in the sum. Of this treasure, $32,628,010 were sent to New York, $4,054,-436 to England, $3,525,325 to China, $338,536 to Panama, $58,220 to Japan, $12,459 to Honolulu, $9,000 to Manilla, $6,000 to Singapore, and $7,100 to Mexico. As compared with previous years, the export of treasure in 1861 shows a falling off from 1860 of $1,664,256; of $7,025,908 from 1859; $6,908,936 from 1858; $8,337,608 from 1857; $8,247,454 from 1856; $4,543,542 from 1855; $10,689,564 from 1854; $16,691,935 from 1853; $5,947,055 from 1852; and $1,943,606 as compared with 1851. The total treasure export in 1850 was $27,-676,346, and $4,921,250 in 1849, as manifested at the custom-house. From 1848 to 1854, large sums were carried away by passengers.

After the gold, in importance as an export, we may place silver; which, however, is not all produced within the borders of California, and which is surpassed in value by wheat. But as one of the permanent exports of San Francisco, silver will undoubtedly have the next place to gold. The silver contained in the gold of California amounts to about $200,000 annually; and that produced by Washoe may have amounted in 1861 to $2,500,000. Hitherto very little silver has been exported; but it is accumulating so rapidly, and it is so inconvenient, in a country where so little small coin is used, that it must be sent away hereafter. It is already at a discount of two per cent.

For several years past, California has exported large quantities of wheat and flour. In 1861, we exported 1,350,783 sacks of wheat and 170,562 barrels of flour, worth together $3,500,000. In 1860, the value was about $2,500,000; in 1859, about $1,100,000; in 1858, about $120,000; in 1857, about $70,000. However, we cannot expect that this ratio of increase can be maintained; we have now reached the summit, and the statistics of 1862 will show a large decrease as compared with the previous year, and perhaps an entire stoppage of exportation.

The quicksilver exported in 1861 amounted to 35,995 flasks, containing 2,699,625 pounds of metal, worth $1,079,850. The export was 9,348 flasks in 1860, 3,399 in 1859, 24,132 in 1858, 27,262 in 1857, 23,740 in 1856, and 27,165 in 1855. The great falling off in 1859 and 1860 was caused by the closing of the New Almaden mine, under an injunction issued at the instance of the federal government.

Among the other exports of 1861 were 181,166 hides, worth $554,951; 3,721,998 pounds of wool, worth $507,297; 373,852 sacks of barley, valued at $355,224; silver-ore, valued at $188,815; copper-ore, worth $122,580; lumber, valued at $76,748; tallow, worth $65,982; and wine, worth $12,399.

The total value of the exports in 1861, other than treasure, was $6,988,375; of which we sent $2,744,537 to New York, $1,283,391 to England, $1,078,118 to Australia, $566,860 to China, and $453,953 to Mexico.

§ 237. *Shipping.*—In 1861, there were 1,981 arrivals of vessels, with an aggregate of 599,233 tons, at San Francisco. Of these, 1,538 vessels, with an aggregate measurement of 267,698 tons, were from ports of California, Oregon, and Washington (domestic Pacific ports); 106 vessels, with 121,342 tons, were from domestic Atlantic ports; 42 vessels (25 from Great Britain, 11 from France, 4 from Hamburg, and 2 from Malaga), with 30,573 tons, from Europe; 31 vessels, with 12,334 tons, from Australia; 27 vessels, with 28,286 tons, from China; 54 vessels, with 15,704 tons, from Mexico; 41 vessels from Vancouver Island, 17 from the Hawaiian Islands, 13 from the Society Islands, 7 from Japan, 17 from Chili, 11 from the Russian possessions in America, and 15 from whaling-voyages. The average size of the vessels which arrived from domestic Atlantic ports was 1,144 tons; of those from China, 1,047 tons; of those from Europe, 745 tons; and of those from domestic Pacific ports, 174 tons, most of the latter being little schooners, while those from New York, Boston, and China, are chiefly American clipper-ships, the finest class of sailing-vessels in the world.

§ 238. *Unsteadiness of Business.*—The mercantile business is very lively and fluctuating in California. There is no people in the world who, in proportion to their numbers, import so largely of foreign goods, or who pay such high prices for them. The amount consumed is not very great, the supplies are irregular, and the time required to obtain shipments round Cape Horn about four months, and therefore it is not a very difficult matter to forestall the market; and it is a frequent occurrence that a few wealthy men combine together and try to buy up all of a certain kind of merchandise, and then control the market and raise the price. In other countries it is impossible to get any accurate information about the supply of an article, which is stored in large quantities in fifty cities and owned by hundreds of merchants and producers; but in California the main stock of all imported goods is stored in San Francisco and is held by a few men. Our business of dealing in merchandising is therefore full of speculations, which, though dangerous, particularly to the stupid, are agreeable to the bold and enterprising, and contribute to render our trade peculiar and different from that of other states.

The business of California is conducted boldly. Men make money rapidly, spend it freely and hastily. Changes in occupation are frequent, and in wealth rapid. Hazardous speculation is the body of our commercial system. Most of our business men are young, and they still are under the influence of the feverish times of '49. Our business is unsteady. Hereditary wealth is unknown. Our rich men all came to California poor, and they are prominent advertisements of the victories that may be achieved by enterprise and bold speculation. We are speculators by our very position. The people who come to California are bold adventurers naturally. We were dissatisfied with life in Europe and the Eastern states, because it was too slow. We came here to enjoy an exciting life and to make money rapidly. Slow and sure plans are unsuited to us, and besides, from the unsettled nature of our trade, there is not the same sureness about slow modes of doing business as in older states. In any little random gathering of a dozen men in San Francisco, you will probably find some among them who have worked in the mines, and this amongst the wealthiest and most intelligent, as well as among the poorest. It is no uncommon thing to see men who have been wealthy on three or four different occasions and then poor again. "A fire," "an unfortunate speculation in merchandise," "a revulsion in real estate," "a crash among the banks," "an unlucky investment in a flume," these are phrases used every day to explain the fact, that this or that man of your familiar acquaintance, though once rich, is now poor. When men fail they do not despair—and least of all those who have been rich and have failed by some sudden turn in business—they hope to be rich again. A number of the prominent business men in San Francisco have been insolvent within the last five years.

§ 239. *Insolvencies.*—Insolvencies, legally declared and cancelled by the courts, are more frequent in San Francisco, in proportion to its population, than in any other part of the world. Our laws provide that any man who declares himself unable to pay his debts, and petitions to be released from them, may obtain a judicial discharge, unless he has been guilty of fraud; and as the fraud must be distinctly proved upon him before the discharge will be denied, the release is almost invariably obtained. The number of suits brought (with their debts and assets) in 1860, as compared with each of the five preceding years, has been as follows:

YEAR.	NO. SUITS.	DEBTS.	ASSETS.	DEFICIT.
1855	197	$8,377,827	$1,519,175	$6,858,652
1856	146	3,401,042	657,908	2,743,134
1857	125	2,375,899	812,417	1,563,482
1858	96	1,940,662	658,782	1,281,880
1859	60	706,219	208,044	498,175
1860	68	1,019,416	76,787	942,629
Total 6 years,	692	$17,821,065	$3,933,113	$13,887,952

The display looks very serious. There was a regular decrease from 1855, when the great failures began, down to 1859, and then the increase began again. The amount of assets is proportionably smaller for 1860 than at any previous time, but in fact the assets are almost invariably nominal, consisting of bad debts that never can be collected, and property estimated at cost, but worthless in the market. It is rarely that a man declares himself insolvent so long as he has property which he can turn into money. Our insolvent law is very liberal to debtors, and no doubt that contributes, with the very speculative temper of our population, the facility for getting credit, and the unsteady course of our trade, to make our insolvent lists so large.

§ 240. *Interest of Money.*—Again, in the matter of the interest of money, California occupies a peculiar position. The current rates are higher here than in any other Christian land. The common rate on long terms and the best security is one and a half per cent. per month, for interest is always calculated by the month. On short loans, or with security in the least doubtful, the interest is from two to two and a half per cent. per month. The laws place no more restriction upon contracts for interest than for any other commercial contracts. The man who promises to pay interest must do so at his own risk; for the law will not assist in any plans to violate his bargain. The high interest of money is owing in part to the unsettled character of the people, many of whom have no permanent abode, and wish to have their money so that they can get it at any time. The fluctuations of the market present numerous speculations to merchants. The titles to real estate are, in many cases, questionable, and the capitalist may not like to buy, and will demand a high rate of interest if he must take a mortgage on a doubtful title. Capital always makes an extra charge for running a risk. It is a general rule that interest will be high where wages are high. There are then many people who can afford to pay interest, and many people who will wish to obtain those comforts and luxuries only to be purchased by the investment of considerable sums of money. The wages of common laborers and mechanics are very high in California, and a large portion of the people are in hopes to attain wealth, and many of them make a practice of engaging in speculations, to do which they must borrow money. The high rate of wages, the unsettled habits of the people, the questionable character of land-titles generally, and the fluctuating nature of our commerce, all contribute to keep the rate of interest at a high figure. In San Francisco a peculiar custom prevails of loaning money from "steamer-day" to "steamer-day." Steamer-day is the business-day which precedes the sailing of the steamer for Panama, which steamer always carries away a shipment of about a million dollars. As the steamer starts early in the morning, all the business in arranging the shipments must be done on the previous day, and then importers must send their money to the Eastern houses from which they obtain their supplies, and they must then dun their customers, the jobbers; and the jobbers must dun their customers, the retailers; and the retailers must dun every body. So steamer-day is a great day for the payment of money, and as every body expects to get his money on steamer-day, so he borrows promising to pay then. There are men who make it a business to lend money from steamer-day to steamer-day, a period of ten or fourteen days, the rate for that period being from one to two per cent., almost invariably with "collateral security" of merchandise, which is on deposit in a warehouse, and is transferred by warehouse receipt to the money lender. Nearly every body borrows and lends money in California. It is no disgrace to lend money for high interest, nor is it ever considered bad management for a merchant to borrow and pay high rates; that is, if the amount be small or the term brief. But as a general rule, the merchants of California borrow altogether too much money. There are not less than fifty mercantile houses in San Francisco, that have paid more than ten thousand dollars each of interest on money borrowed during the last ten years, and of one house which lately failed it was asserted that they had paid one hundred and eighty thousand dollars of interest. It is a general practice among farmers to borrow money, and, if before harvest, give a mortgage on the crops; if after harvest, store the grain in the lender's warehouse as security. One of the most profitable branches of business in the state is "grain commission." The commission merchant has a quantity of ready money which he loans to the needy farmer, who stores his grain with the lender, and the latter charges about two per cent. a month for the money, a dollar a ton per month for storage, and two per cent. commission for selling.

§ 241. *Speculation in Land.*—Speculation in land is one of the most important branches of business in the United States, and the increase in the value of land has been one of the main sources of the wealth of the country. The American farmer going into a new district, expects to purchase his land at a low price and see it gradually rise in value, until it makes him, if not wealthy, at least comfortable. Lands in Illinois which, twelve years ago, were to be obtained from the government at one dollar and twenty-five cents per acre, are now worth thirty and forty dollars per acre; and a similar remark will apply to most of the land in the Western states, though the average increase has not been so great. This increase has been one of the greatest inducements to the rapid settlement of the Mississippi Valley. Unfortunately, California has not been permitted to derive much benefit from this source of wealth. The fertile valleys near the coast had been granted away in large ranchos by the Mexican government, and the American government refuses to sell the land in the mineral districts.

It appears to me a matter of the utmost importance to the welfare of California, that the land in the mineral districts should be sold. Not more than one acre in forty within the limits of the mineral region is now occupied for mining purposes, and not more than one acre in ten will ever be worked.

§ 242. *No Paper Money.*—There is no paper money in Cali-

15

fornia. The Constitution of the state prohibits "banking," and "the creation of paper to circulate as money." No bank-notes have ever been current in this state or on this coast; nor are bank-notes used on any part of the coast between Acapulco and Sitka, and we are so far from the countries in which paper money is current, that no attempt is made to introduce it here. In our banks there are great piles of double eagles, but no bank-notes are visible. Wherever you go, or whatever you buy, you see only gold and silver; people do not think of paper. All large sums are paid in double eagles, and three-fourths, if not nine-tenths, of our coin is of that size, which is far more convenient than the smaller coins common in other countries. A large proportion of our shipment of treasure abroad is in double eagles, and nine-tenths of the gold coined at the San Francisco Mint is in pieces of that size. In 1860, $11,178,000 of gold were coined, and $10,899,000 were in double eagles. The general character of our coin is large. No copper money is current, nor can any thing be bought with a cent. The smallest coin commonly used is a dime; half-dimes are rarely seen, and when used two of them are ordinarily put together to make a dime. The general sentiment among the people is opposed to the use of any coin less than a ten-cent piece; they like high wages and high prices, and think that the introduction of half-dimes and cents would have a tendency to make us feel poor and to introduce low wages. Many retail dealers, even in the sale of candies and fruits, will therefore not take a half-dime, and not a few persons would be ashamed to offer to purchase half a dime's worth of any thing. A half-dime is looked upon with more contempt and is far more rare in California than a cent in New York. During the last three months, for instance, though I purchase little articles every day, I have not seen a half-dime. That coin is not made in our mint, nor is there any demand for it. Three-cent pieces, coppers, and nickels are never seen here except as curiosities, and are of no value to make purchases.

§ 243. *Opportunities for Investment.*—The opportunities

for investing money securely in California are few compared with other countries, and the chief causes of the difference are to be found in the defects of our land-titles, and the unsteadiness of business. There is no place where capital needs the constant attention of a prudent manager more than in this state. There are so many revolutions in business, that a brief neglect may cause the loss of a fortune. It is a very unsafe country for money left in the charge of agents. Their commission is high, and they will usually give much less attention to a bailor's property than to their own. The interest of money is too high for investment in state stocks. The bonds of the state of California are all sold in New York, and the interest is made payable there, and nearly all are held there or in Europe. Most of these bonds bear only seven per cent. interest annually, not half the current rates here. So of the bonds of our towns and counties, for nearly every town and county has its debt, the bonds are mostly held abroad, and the interest is too low for the demands of capital in California. There are then no state stocks that can be owned here, and thus we are cut off from one of the best and most extensive fields of investment. Labor is so high that we have few factories. The titles of farming-land are as a general thing insecure. Farming, as a business conducted on a large scale, is extremely uncertain. Our population is so small that the market is easily overstocked, and freight to Europe and the Atlantic states is so expensive, that a surplus can only be sent away with difficulty; and we are so remote that any shipment is accompanied by many risks, for between the time when the vessel sails from San Francisco and the time when she arrives at Liverpool, there is an interval of three or four months, during which period a ruinous depreciation may occur. The fluctuations which occur in the grain market, have their parallels in all other kinds of agricultural products: cattle, sheep, and wine. There is no title to mineral land save occupation; and mining on a large scale is more dangerous than farming. Whenever any branch of business becomes so established as to be permanent and to yield a certain profit, capitalists abroad will wish to take possession of it, and they can afford to pay more for it than we can, because our money is dearer than theirs. A man in New York can afford to pay more for stocks than we can in San Francisco. He gets his money for six per cent. per annum while we pay eighteen. Whenever therefore any large railroad or any extensive improvement, that will certainly be permanently profitable, is established in California, foreign capitalists will buy it up. We cannot compete with them until our money becomes cheaper. Our ocean steamships are owned in New York. Our river steamboats are mostly owned by a large association, called the California Steam Navigation Company, but the stock is very fluctuating and dangerous as an investment. There is no purchasable bank-stock in California. Most of the capitalists of San Francisco either invest their money in houses and lots, or let it out at interest under bond and mortgage. The taxes are very high in California, in no county less than one per cent. a year, and in some places four per cent. Indeed, when streets are repaired in the large towns, the taxes sometimes amount to ten per cent. on the value of the property. There is a considerable amount of French and Swiss capital invested in San Francisco, most of it loaned on mortgage, and under the charge of French and Swiss bankers.

In no part of the United States is there so small an investment of capital, and so small an amount of real and personal property held in fee simple, by individuals and local corporations, in proportion to the area, population and amount of business done, as in the gold mining districts of California. The custom-house manifests show, that during the last fourteen years, we have exported $550,000,000 of gold, and no person at all familiar with the business and history of the state, will estimate the amount of exportation, not manifested, at less than $150,000,000 ; and yet what is there to show in the mining districts for all that immense wealth? There are many fine mountain roads, and yet they are few, and bad, as compared with the number that there should be; and even those which we have, are mostly private property, and are made the means of levying severe taxes upon all travellers and merchandise passing over them. The counties have no fine roads, no large or durable bridges. There are great ditches—with a total length of six thousand miles—and wonderful flumes, but they are made as cheaply as possible for use, during a few years—not more than five or ten—and in too many cases, cheapness has not been found to agree with durability. Immense labors of many kinds have been undertaken, but simply with a view to present profit, and without conferring any permanent benefit, or beautifying the country. There is scarcely an elegant county building in the state. The mining towns contain many fire-proof brick buildings and elegant residences, and yet any mining town set down in New York or Ohio, would be considered a very shabby village. Wherever we turn, in the mining districts, we see no investment of capital at all commensurate to the wants of the state.

The whole policy of the federal and state governments, in regard to the mineral districts of California, has been not to enrich the state, but to benefit individuals who did not and do not intend to be permanent residents, whose interests are adverse to those of the state, and to whom no inducements are offered to become permanent residents. Rather the inducements are on the other side, and obstructions are thrown in the way of establishing homes. The improvements which should have been made in previous years are wanted, but the capital is not so abundant now. The main body of wealth has been transferred to the agricultural counties and commercial towns. The mining districts, instead of growing richer, from their immense treasures, are growing poorer every year, all of them comparatively—many of them absolutely.

Miners come to capitalists residing out of the mineral districts, and invite them to invest money in gold-mining enterprises. The investments would furnish employment to laborers, increase the production of gold, and benefit the state in many ways, but the capitalists refuse because of alleged insecurity. The miners must abandon their schemes, and leave the wealth of the mines undeveloped, or obtain money in their own neighborhood, at rates of interest two or three times higher than those prevalent in larger cities, where undoubted security is given. Hence, serious obstacles to the development of all branches of industry, and to the establishment of that well-ordered society, made up of permanent residents, dwelling in homesteads fixed for life, with a proper proportion of women and children. The least of all the evils is, that the mining counties contribute far less than they should to the treasury of the state.

The main object of a government should always be to build up a permanent, dense, intelligent, industrious and moral population, skilled in the mechanic arts, and so employed that their capital increases rapidly and regularly. Such is the condition of all the most prosperous countries, and such would have been the condition of California now, if she had been properly governed.

If application be made to a San Francisco capitalist to buy a gold mine, or loan money on the security of a gold mine, or invest money in any gold mining operation, an offer of sale will be received with the same refusal. A promise of ten per cent. a month will not induce him to examine the proposition. He will reply, that all gold mining is insecure. He treats quartz-mining, ditching, hydraulic claims and sluice, alike; different as they are from one another, he makes no distinction between them; they are alike, and equally dangerous. If he be asked how he knows that they are insecure, he will say that he lost an investment some years ago, or, that this and that friend of his lost investments in gold mines, and everybody says they are insecure, and that is enough for him. He cannot afford to travel through the mines and investigate the matter. He can find what he considers safe investments in San Francisco, in agricultural lands, from Sonoma to San Diego, and in Washoe silver mines. There is now more San

Francisco capital invested in Santa Clara county alone than in all the gold mines of the state. There is scarcely a county near the coast, between latitude 40° and 33°, in which there is not a large amount of capital invested by non-residents, to the evident benefit of the county. Although gold mining has been the chief industry of the state for fourteen years, and although Washoe is of late discovery, and has less population than single counties of California, yet more San Francisco capital is invested in Washoe silver leads than in all the gold mines of the state.

There is a great difference of opinion among business men in regard to the causes of this alleged insecurity. One will say that the cause is in the want of fee-simple titles; that the ownership of mining claims depends upon possession, and is a kind of personal property with a still weaker title, for a man may lay his purse down, or let his horse go, and both will belong to him in law for years; but when he shoulders his pick and blankets, and leaves his claim for three days in the working season, he has not the least interest in it. Another consideration, however, will show that this cause alone, although it may contribute much to the general result, is not sufficient in itself to render gold-mining investments insecure, even in any one case of note. Most of the valuable mines in the world are held by mere possession, and abandonment or cessation of work would lead to a forfeiture of title. This is the rule throughout the rich mineral districts of Spanish America, in most of the countries of Europe, and even in Washoe; and yet the value of these mines, whose title depends only on undisputed possession, is almost as good as if held in fee simple, for it is plain that so long as the mine is profitable there will be no cessation of work.

Another theory to explain the aversion entertained by capital for gold-mining enterprises is, that the business is too uncertain as to its profits; that prudent business men avoid great risks, and will not venture their principal; and that they cannot invest in gold mines without the greatest perils, because it is impossible for them to know beforehand whether a mine contains much rich ore or dirt. But if we look a little further at this explanation, we shall find that it has no good foundation. It is not true that the prudent business man always avoids great risks; in other words, prudent business men often make investments with a probability that the entire investment will be lost, but in clear view of a possibility that an immense profit may be made.

It is a mark of sound judgment and prudence in a business man to avoid risks that may much endanger the loss of his whole fortune, but he will not necessarily refuse to venture a tenth part of his fortune, if he see a possibility—as of one chance in five—of making a profit of a hundred-fold. It is under the guidance of principles like these—stating the risk in its least favorable aspect—that mining for gold and silver is conducted throughout the world, to the great general profit of those engaged in the business, and it is in view of principles like these that mines of all kinds have a certain market value. They may be hidden hundreds of feet under the earth, but there is a probability that a metalliferous vein, if tolerably rich, and similar to other veins not far distant, and found to be rich through a considerable extent, will also be rich in like manner in other parts out of reach. There is such a chance for profit that a high value attaches to mines where there is not even a regular vein or layer of mineral to guide the miner. For instance, cinnabar is not found in veins, but in masses, usually connected with each other by little seams. At New Almaden, sometimes a mass of ore is found ten feet cubic, and when it is worked out, only a little seam indicates where other masses may lie. Whether they do lie there is a matter no way certain. The Enriqueta mine, which promised about two years since to rival the Almaden, has fallen to the level of the Guadalupe. Some rich masses were found; and more may be found, but nobody knows whether they will be, and so with the New Almaden; and though these facts are perfectly understood, the latter mine is worth millions in the market.

There is such a presumption of the richness of its unexplored portions, that a man worth one hundred thousand dollars can be justified in venturing an investment of fifty thousand dollars of it in New Almaden stock at the market price. Any piece of property having a steady market price, promises security to justify the investment of money. This theory, therefore, is not satisfactory; it explains nothing.

A third theory adduced in explanation is the natural dishonesty of men, and the peculiar facilities with which dishonesty may be practised in gold mining. It is said that if a mine is found to be valuable, the resident shareholders keep the knowledge to themselves, levy heavy assessments upon remote partners or stockholders, and take care that no dividends shall be declared, by which policy all who do not know the mine are at last driven to sell, and usually at a sacrifice. The managers of the mine are, of course, either directly or indirectly the purchasers. This system of management is neither imaginary nor rare; it is familiarly known as "freezing out," and is not confined to gold mining, but extends to silver mines, and commercial corporations. In mining for gold, however, there are greater opportunities and more speedy rewards for this kind of fraud than in any other kind of business; but whether the frauds are more numerous or greater may well be doubted. Theft is another form in which dishonesty is dangerous to the owners of valuable mines. In all those mining enterprises where considerable amounts of capital are invested, numerous laborers must be hired; and in California most of these men are strangers to their employers, without family, permanent home, or any tie that can give security for their good conduct. The great value of gold as compared with the space it occupies, gives the thief fine opportunities for seizing and hiding it. Neither placer nor quartz mines are exempt from this danger. The laborer employed in deep hydraulic claims, or far from the daylight in shafts or drifts, cannot fail to see the large lumps or the rich portions of the auriferous rock. He has abundant means of hiding enough every day to equal his wages, and as no one else saw the piece previous to the time when he took it, it is almost impossible to prove theft, even if found upon him. But it is not customary to search laborers when they leave their work, for the honest men would probably not submit to the humiliation. In quartz mines the term "knocking down" is given to the system of theft of rich bits of the rocks by the laborers, and it is said that some mines where the rock is very rich, are plundered so extensively that they are far less profitable than those wherein the metal is equally diffused through a wide vein, so that the workmen can find no small pieces worth stealing. To what extent the influence of such dishonesty—the existence of which, as we have said, is recognized in the popular speech—may go, we have no information or estimate; but we are satisfied that though they may be serious, they are in themselves far from sufficient to account for the utter distrust with which capitalists look upon enterprises for the developing of our auriferous wealth.

I have thus mentioned three theories which, in my view, whether considered each separately or all jointly, are unsatisfactory and insufficient to account for the evil. If there be any other theory I have not heard of it, or it does not occur to me now. I have already spoken of the greatness of the evil; it strikes at the very basis of our prosperity as a state, destroys the field of labor, drives our capital away, renders trade unsteady, and contributes to deter immigration and to make our mining population a set of wanderers.

There is something radically wrong in the financial and industrial condition of that country where the capital is not invested in the main branch of industry. So it is in California. Many important improvements are needed in the mineral regions, but the money cannot be obtained to make them. We send away $40,000,000 annually, because people abroad are not satisfied with the security which we offer, though we would, and could with advantage, pay three times as much interest as it can earn in the Atlantic states or Europe.

Great numbers of poor men seek employment, but the capital is not here to pay them. The invested capital does not bear a proper proportion to the population and labor. The lack of regular employment makes business unsteady, population unsettled, exposes the state to desertion at every cry of rich diggings at Cariboo or Salmon River, and prevents immigration, which we should by this time understand, is one of the greatest sources of wealth and prosperity. It is, or at least, previous to the great rebellion, was a common assertion, that California was no place for a poor man; that it was far inferior to the younger states east of the Rocky Mountains, in the probabilities of steady employment, and the opportunities of gradually acquiring a capital of five or ten thousand dollars, such as a large proportion of those who were poor men in the states in the Mississippi Basin, a score of years ago, have in that interval collected together.

Is this complaint about the insecurity of the gold mines, as a sphere for the investment of capital, true? If so, why and how? Is it equally true about the quartz-mining, the placer-mining and ditching? Certainly it cannot be true in the same manner, for the three branches of industry are very different from one another. If it be true, is there a remedy? These are questions which private enterprise has not solved, and for the solution of which a good government should make some provision. If we can but show how capital can be invested securely here, it will be invested, and population, industry, and the prosperity of residents and the state will swiftly follow.

§ 244. *Assurance.*—The assurance of property against fire on land and against wreck at sea, has come to be an extensive branch of modern business. It is one of the chief channels through which capital runs to the great centres of trade. It is a legitimate, safe, and profitable investment. It is managed on careful principles. All its operations are based on statistics, collected in the course of many years. The premiums are always set at such a figure that according to the ordinary course of human events, the company must realize a profit in the long run. The occasional losses are more than covered by the regular collection of the premiums. The amount of the premium is determined by the risk, and the risk is known from long observation to within a very small percentage.

We have only two small assurance companies in California, and nearly all our assurance business is done by companies of the Eastern states and Europe, represented by agents here. These agents do a very large and profitable business, a business which ought to be done by companies organized in our own state. Not less than $600,000 are annually sent from this state for assurance premiums, and the total losses do not exceed $200,000, leaving a clear profit of $400,000, or sixty-six per cent. to the assurers. These figures are not extravagant estimates, made without a knowledge of the subject; if there be any error, they represent the profits less than they really are. Nearly all the policies, for which these $600,000 of premium are obtained, are issued on property in San Francisco, and the losses by fire in 1860 were $162,000, an unusually large amount, the average during the previous five years having been $105,000, or thereabouts.

The average premium paid is about one and one-half per cent., the lowest being three-quarters of one per cent., and the highest three, or even four per cent., whereas one-quarter of one per cent. on an average would pay the losses.

CHAPTER XI.

CONSTITUTION AND LAWS.

§ 245. *Outlines of Constitution.*—California is a state, and a member of the United States of America, with rights equal to those of the other states. The sovereignty of government is divided between the federation and the state, the former taking precedence. Every officer of the state, before entering upon the duties of his position, must take an oath to "support the constitution of the United States, and the constitution of the state of California." A double allegiance is thus imposed, that due to the state authorities being secondary to that due to the authorities of the nation at large. The Union has exclusive power to regulate commerce, naturalize foreigners, coin money, make treaties, declare war, make peace, and maintain an army and navy. Beyond these powers, the state is sovereign. The state government is republican. All the officers of the government are chosen by the people. The legislature is composed of two houses, which sit separately, and the consent of both is necessary to the passage of any bill. The longest term of office for any important executive or legislative position is two years; the members of the assembly and many of the county officers being elected annually. The voters thus have an immediate control over the government. The judges of the Supreme Court and of the District Courts are elected by the people, and the term of office is six years. Suffrage is universal; that is, every sane, adult, white male citizen, not a felon, may vote at every election. No ownership of property or payment of taxes is required as a qualification for voting or holding office.

§ 246. *Inferiority of Colored Persons.*—All white male citizens are equal before the law of California; but negroes, Indians, and Chinamen are not permitted to vote or to testify in the courts against white men. In a criminal case, one-eighth negro blood and one-half of Indian blood, in civil cases one-half of either, disqualifies a witness for testifying against a white man. Slavery is forbidden by the constitution.

§ 247. *Laws Favorable to Debtors.*—The laws of California relating to the collection of debts are very favorable to the debtor. His homestead, the property owned by his wife previous to marriage, that given to her afterward, his household furniture to the value of two hundred dollars, his tools, if a mechanic, his horse and wagon, if a mechanic, and his library, if a lawyer, are exempt from execution. A married man, a widow, or widower with children, or any head of a family, is entitled to a homestead worth five thousand dollars, secure against creditors. An unmarried person may have a homestead worth one thousand dollars. Such laws may prevent much oppression of poor people, but they also protect and encourage much rascality. A man may own a homestead worth five thousand dollars, and that may include a very elegant dwelling. His household furniture, worth as much more, may have been presented by some friend to his wife after marriage. She may have a separate estate of one hundred thousand dollars, and may derive an annual income of ten or twenty thousand dollars from it, and both may live in an extravagant style, and yet creditors have no hold upon him whatever. There is no imprisonment for debt except in cases of fraud, and that the laws are so drawn that it is almost impossible to prove. One important class of testimony, admissible to prove a man a thief or murderer, is not admissible if he be accused of fraud in contracting a debt. In trials for obtaining money by false pretenses—the most common kind of fraud in contracting debts—the oral testimony of witnesses, as to the representations made by the accused, is not sufficient to convict; there must be some writing or token furnished by the defendant.

In many ways the debtor is fenced about, so that the laws seem to have been devised by men who had had experience in swindling creditors, and wished to secure themselves against trouble in the future. Every precaution is taken against the creditor, as though he were a public enemy; while the men who do not pay their debts are treated as though they were the soul of the state, and as though their mode of doing business should be encouraged at all costs. When a man gets in debt, he can get out again without difficulty. Our insolvent law provides that he has only to petition a court for release, set forth his debts and assets, swear that he cannot pay his debts and give up his assets. Unless the creditors can then prove that he has committed fraud, either in contracting the debts or in concealing his property from his creditors, he is entitled to be released. He may go through this process repeatedly; and many cases have occurred of two discharges in insolvency granted to one person within two years. The favor shown by our laws to debtors, is, I think, a grave error, and contributes much to establish a dishonest tone in general society, and to encourage dishonest actions. On every side can be seen men who have swindled creditors out of large amounts of money, and are themselves now living in extravagant, or at least luxurious style. Such laws encourage habits of rash speculation, with the expectation that riches will come with success, and no discredit or loss to any one save creditors, with failure. Honest and prudent men suffer, business is thrown into disorder, and reckless adventurers and knaves get positions and influence which they never should have.

The property owned by either the husband or wife before marriage, and by gift, bequest, or inheritance after marriage, belongs to each separately; and the property acquired after marriage, by other means than gift, bequest, or inheritance, is common property belonging in equal shares to both. The husband, however, has sole control of it. The wife has no right of dower, and the husband has sole control of the common property, and may sell, without the consent of the wife, any of it except the homestead; a deed or mortgage for which without her signature and seal, is absolutely void. The husband cannot convey his interest unless she conveys her interest at the same time. The husband has the management of the separate property of the wife too, but, if she desire it, she may apply to a court and have the property placed in the charge of any trustee whom she may select. For a valid sale of the wife's separate property, husband and wife must join.

The laws of California, like the customs and trade, do not favor the perpetuation of wealth in families. There is no right of primogeniture. All children inherit equally. The eldest son gets no more than the youngest. Public opinion runs with the law. The rich man who should express an intention to give all his property to his eldest son, merely because of his seniority, would be hated. Entails are forbidden. A man cannot tie up lands in the county for more than ten years, or town lots for more than twenty. How different is all this from the state of affairs in Europe! There, at least in some parts of the continent, all the property goes to the eldest son; property is entailed in the family for many generations; the debtor is subject to imprisonment; there is no release for insolvents; the property of the woman is by marriage vested absolutely in the husband, and does not revert by inheritance to her blood relatives by her death; the limitations for commencing law-suits are very long, and sales, if not made at the market price, or contracts, if made so that one party appears to have obtained an advantage of the other, may be rescinded. The habits and opinions of the people give strength to their laws; and wealth once in a family is almost as certain to be transmitted through many generations by inheritance in Europe, as its loss in the second or third generation is certain in the new states of America.

§ 248. *Tenure of Land.*—Four-fifths of the land in California is owned by the federal government, which acquired it from Mexico by treaty. This federal land lies in the mineral regions, and in all the unsettled districts of the state. Most of it has been surveyed, and, with the exception of the land in the mineral districts, is offered for sale at one dollar and twenty-five cents cash per acre, in lots of forty acres, or tracts of any size of which forty is a multiple. There is no limit to the amount which one man may buy. Any man may occupy any of the federal land. It is open to the use of the public. The tule land, of which there are about eight hundred square miles, belongs to the state, and that is offered for sale at one dollar and twenty-five cents per acre, but the payment may be made in instalments at long intervals. One-eighteenth of the land has been given to the state for school purposes, and this is also for sale at one dollar and twenty-five cents per acre in instalments.

Most of the land held in private ownership in the state, is under grants made by Mexico previous to 1846. Of these grants there are eight hundred and thirteen, covering a total of 9,828,181 acres. Of these claims about one hundred and fifty, covering about 3,000,000 acres, have been finally rejected, and a number are as yet undecided. The grants were for large tracts called ranchos, intended to be used chiefly or exclusively for pasturage, and the average size was about 12,000 acres, or three square leagues. The grants were made, not by the acre or by the mile, but by the square league, containing 4,438 acres and a fraction, or, to be precise, 4,438.683 acres. Every ranch had its name, for it was a kind of principality, and these names have in many cases been transferred to towns and townships under the American dominion.

The common tenure of land in California is fee-simple. Such conditional tenures as are common in Europe are very rare here, and many of them are prohibited by our laws. We have no life estates, nor is any lease or limited conveyance of land good for a longer period than ten years, unless it be a town lot, and then the limit is twenty years. All conveyances of real estate are placed upon record in a government office, and without such record they are not valid as against persons not parties to the conveyance.

§ 249. *Ownership of Minerals.*—It has been a much disputed question whether the owner of land, under a Mexican title, owns the minerals in the soil. Under the Mexican law he did not; but the latest decision of the highest court of California is, that the owner of the land owns all the minerals in it. The Mexican and Spanish law is unsuited to the American system. By law the mines are open to everybody. Any man or woman, citizen or alien, white, black, or yellow, may lay claim to an unoccupied tract of auriferous quartz or dirt, and if his claim be not larger than is customary, he has a good legal title so long as he may work it, provided that, if an alien, he shall pay a license of four dollars per month. However, although he may have a good legal title, the white miners in some districts will not permit Chinamen to come among them, and their legal right is of no value against the omnipotent mob. There is no express law throwing the mines open to all the world; but the intent of the government is plainly to be inferred from the debates of Congress, and the fact that no restraint has been placed upon mining. All the federal land in the state, save that in the mining districts, is thrown open to pre-emption for agricultural purposes; but the mines are reserved, and evidently for the miners, from whom the federal government has never demanded any tax or share of the gold obtained. The state legislature has imposed a tax of four dollars per month on all alien miners, and has promised that all who pay that tax shall be permitted to mine as freely as citizens; but that promise is not kept.

§ 250. *Titles of Mining Claims.*—The miners do not own the land on which they work; they only have a right of possession so long as they occupy it; and their right is called a "claim." The size of the claims varies in different districts, and depends entirely upon the regulations or custom of the district. Every district has a written code of regulations, determining, among other points, the size of mining claims; and these regulations are recognized by the state as valid, so far as they are not in conflict with the statutes or constitution.

Quartz claims are usually two hundred feet long, following the course of the lode. In some districts the miner holds one hundred feet of ground on each side of his lode, so that he cannot be disturbed by other persons in his vicinity; in others he holds only the width of his lode, and if another lode, or placer diggings, be found within a few feet of his place of working, other claimants may come so near as to interfere greatly with his convenience. Placer claims vary in size, according to the nature of the ground. On bars, a claim usually has a front of fifty, a hundred, or two hundred feet on the river, and runs back across the whole width of the bar. Tunnel claims have a front of fifty, a hundred, or two hundred feet on a hill-side, and run transversely to the middle of the hill, or entirely across it, parallel with the direction in which the tunnel is commenced. Ravine claims include the bed of a ravine for a distance of not more than three hundred feet. Placer claims not on bars or ravines, or in hills that must be tunnelled, are usually rectangular, containing as much as one hundred or two hundred feet square. Persons who discover new placers or auriferous quartz veins are entitled to two claims.

The method of getting a claim is very simple. The miner finds a piece of unoccupied ground that suits him; drives stakes at the corners; fastens on one of these stakes a piece of paper, containing a written notice similar to the following:

"I hereby lay claim, for mining purposes, to a piece of ground a hundred feet square, of which this is the northeastern corner, the other corners being marked by stakes.

"JOHN SMITH."

He then goes to the mining recorder of the district, requests him to record the claim in his book, and pays a fee of half a dollar. In some districts the recorder must go out and look at the claim, to see that it is not on ground previously claimed. In other districts one of the stakes must have a piece of tin on it, marked with the number which the claim has on the recorder's books. In other districts it is necessary to mark the limits of claims with a trench. But the common custom is to require only a notice of the simplest intelligible form and a record; and by these the miner acquires a perfect title as against all other persons, so long as he continues to work the claim. But if he does not work the claim, he forfeits his title. The time within which work must be done to preserve the title good, varies greatly in different districts—from three days to a month. The period is usually brief in proportion to the shallowness of the diggings and the ease of working them. In most districts no amount of work will secure a claim against forfeiture. A miner may work his claim every day for years, and then, if he deserts it for a day beyond the period allowed in the mining regulations of the district, it is legally forfeited, and may be taken by the first comer. There are certain classes of diggings which can only be worked during half of the year, and in these, claims are subject to forfeiture only during the workable season. It is not expected that a miner will work a dry ravine claim in the summer, or a bar claim in the winter; and the mining regulations, being based on reason, and well designed for the convenience of the miners and the development of the mining interest, impose no penalty upon such actions as would be approved by industrious, prudent men. There are deep diggings where all the claims cannot be worked at once: those in front, or nearest the place where there is an outlet for the water, must be worked out before the miner can begin on those behind. In such cases, the claims in the "background" will not be forfeited until there is a possibility of working them, even though years elapse. The mining regulations impose no penalty for a neglect to do impossible things. Among the claims which may be left unworked for a long time without danger of forfeiture, are claims for mining purposes to land occupied by other parties for tailings. A miner may hold a tailing claim merely for the purpose of preserving his tailings, and with no intention to wash the natural dirt; and another miner may lay claim to the same place for mining purposes, but he must wait until the prior occupant has removed his tailings, and until such removal the title is not subject to forfeiture. And the prior occupant is under no obligation to be in haste; he can take the same time to remove the tailings that he would have taken if no claim for mining purposes had been made to the land. It is not necessary that the owner of a claim should work upon it in person; the labor of a hired man is as good to maintain occupation and secure title as any other labor. In the Esmeralda silver district, work to the value of seventy-five dollars secures the title forever, according to the mining regulations.

A company of miners may take up as much ground as the aggregate amount which they could take up separately; and when a company takes a claim, every member has an equal and undivided interest in every part of it. When a claim is to be taken for a company, it is sufficient that one member of the company should appear at the recorder's office, demand that the claim be recorded, and offer to pay the fees. The recorder has no right to make any preliminary inquiries as to who the people are, or where they live, or whether they live at all. One man, finding a good tract of rich ground, may write down the names of a dozen persons living at a distance, and, without consulting them, have their names recorded as members of a company owning a large and valuable claim, and by such record their title becomes good.

In most districts a miner may hold only one claim by his own original location, but an unlimited number by purchase. In districts where there are both dry ravine and bar-claims, he may at the same time hold one bar-claim and one ravine-claim by original location, the former being workable in summer and the latter in the winter. After having exhausted or abandoned one claim, the miner may always take up another.

The manner of taking up claims in the argentiferous and quicksilver districts, is the same as in the gold districts.

The water in the mineral districts is also subject to claim. The streams may be diverted from their courses and used for washing dirt, driving saw-mills, grist-mills, and quartz-mills, or irrigating tilled fields. The first claimant has the prior right, no matter for what industrial purpose he wishes to use the water. Miners have no monopoly of the water in the mineral regions, nor even a prior right. There is no limit to the amount which a man may claim. He may take the largest river in the mountains; he may take a dozen of them and hold them all. He may not only take all their water, but he may take all the land necessary to use it. He may make reservoirs covering hundreds of acres. He may make ditches a hundred miles long. All that is necessary to give him a possessory title to the water and land, is that he should drive stakes along the route of the ditch, post up notices of his intention, and commence work in building the dam and cutting the ditch. However, while the custom of the country would permit a man to become the owner of such extensive works, the practical result is, that the large ditches are almost invariably made by companies.

§ 251. *Marriage.*—Marriage, by the law of California, is a civil contract, which is complete with the consent of the man over twenty-one years of age, and of the woman over eighteen. No ceremonial form, license, publication of bans, consent of parents, blessing of priest, seal of magistrate, or presence of witness, is necessary to give validity to the contract. If either party be under the age mentioned, then the consent of the parent or guardian is necessary. Although the law does not require a ceremony, yet custom does, and the priests and preachers are usually called in to perform it. Divorce may be granted for adultery, habitual intemperance, extreme cruelty, desertion for two years, conviction of a felony, and impotence. There has been much complaint that the law renders divorce too easy, but the general opinion of California is favorable to the law as it is.

CHAPTER XII.

SOCIETY.

§ 252. *Population.*—The total population of California is reported by the census of 1860 to be 380,015; of whom 333,530 are white, and 46,485 are colored. This census, however, was not taken with proper care, and I am confident that the total population is not less than 400,000, of whom about 340,000 are white, and 60,000 colored. Of the colored, 50,000 are Chinamen, 3,000 are negroes and mulattoes, and 7,000 are Indians. Most of the Indians live far from white men, and in a state of complete savageness; the negroes are mostly in the large towns; the Chinamen are mostly in the mines and in San Francisco. The details of the census have not yet been published, so I must guess at them. I suppose that the white population is made up of 155,000 men, 95,000 women, and 90,000 children or minors. Of the men, there may be 130,000 citizens and 25,000 aliens. The aliens may be composed of 12,000 Frenchmen, 3,000 Englishmen and Scotchmen, 2,000 Irishmen, 3,000 Italians, 2,000 Spanish Americans and 3,000 Germans, &c. Putting white and colored adults together, there are about 110,000 women and 200,000 men in the state. The men are about equally divided between the mining and farming districts of the state, but the proportion of women is much larger in the latter than in the former. The minors are about equally divided between boys and girls; and the children, under ten years of age, have a large majority over those between ten and twenty-one. Of the nativity of voters I make the following estimate: 50,000 Americans from the free states, 30,000 Americans from the slave states, 20,000 Irishmen, 15,000 Germans, and 15,000 miscellaneous, including Canadians, Britons, Hungarians, Spaniards, Danes, &c. The most populous counties, according to the census report, have the numbers of inhabitants, and, at the presidential election of 1860, cast the number of votes, indicated in the following table:

Counties.	*Population.*	*Votes.*
San Francisco	56,805	14,415
Sacramento	24,145	7,542
El Dorado	20,562	7,049
Nevada	16,447	6,965
Tuolumne	16,229	5,547
Calaveras	16,302	4,815
Yuba	13,671	5,058
Placer	13,270	5,824
Butte	12,107	4,438
	173,538	61,653

The following is a statement of the population of the state by counties, as reported in the census of 1860:—

Counties.	*White.*	*Col'd.*	*Total.*	*Counties.*	*White.*	*Col'd.*	*Total.*
Alameda	8,581	346	8,927	San Diego	3,785	541	4,326
Amador*	8,392	2,541	10,933	San Francisco	52,955	3,850	56,805
Butte*	9,718	2,389	12,107	San Joaquin	9,171	263	9,434
Calaveras*	12,579	3,723	16,302	S. Luis Obispo	1,560	222	1,782
Contra Costa	5,186	142	5,328	San Mateo	3,138	76	3,214
Colusa	2,157	117	2,274	Santa Barbara	3,100	443	3,543
Del Norte	1,374	618	1,992	Santa Clara	11,683	229	11,912
El Dorado*	14,695	5,867	20,562	Santa Cruz	4,327	618	4,945
Fresno*	4,025	580	4,605	Shasta*	3,922	438	4,360
Humboldt	2,502	192	2,694	Sierra*	9,574	1,815	11,389
Klamath*	1,315	588	1,803	Siskiyou*	7,056	573	7,629
Los Angeles	9,919	1,417	11,336	Solane	7,132	38	7,170
Marin	3,042	292	3,334	Sonoma	11,672	195	11,867
Mariposa*	4,215	2,028	6,243	Stanislaus	2,022	203	2,245
Mendocino	2,881	1,086	3,967	Sutter	3,358	42	3,390
Merced	1,114	27	1,141	Tehama	3,288	756	4,044
Monterey	4,147	592	4,739	Tulare and Buena Vista	4,059	579	4,638
Napa	5,454	61	5,515	Trinity*	3,318	1,807	5,125
Nevada*	14,236	2,211	16,447	Tuolumne*	14,277	1,952	16,229
Placer*	11,104	2,166	13,270	Yolo	4,679	37	4,716
Plumas*	3,730	633	4,363	Yuba	11,557	2,114	13,671
Sacramento	22,761	1,384	24,145				
S. Bernardino	4,860	694	5,554		333,530	46,485	380,015

In the counties marked with asterisks the population is chiefly occupied with mining; and there is some mining in Sacramento, Yuba, Del Norte, Los Angeles, Buena Vista, and San Bernardino counties. Since the census was taken, the eastern part of Calaveras has been organized into Mono county, and the northern part of Napa into Lake county.

§ 253. *Nativities.*—In January, 1848, thirteen years ago, the total white population of California did not exceed 15,000 in number, of whom two-thirds were Spanish American. Of the 340,000 present white inhabitants, therefore, 325,000 are immigrants or miners. Of the adult population not more than one in twenty is a native of the state. The migration has not been for a short distance. It was from one side of the world to the other—a long and for most an eventful journey. Of the children, about half were born in California. The average time that the immigrants have spent in California may be six years. During the last twelve years not less than 250,000 persons who had made their homes in California for a year or more, have left the state, and they are now scattered all over the civilized world. The Americans in California represent every state in the Union, and representatives from every state may be found in many of the counties taken separately. In the public schools of San Francisco during 1860, there were 1,454 children born in the state of New York, 1,176 born in California, 667 from Massachusetts, 245 from Louisiana, 194 from Pennsylvania, 149 from Maine, 72 from Missouri, 70 from Ohio, 61 from Maryland, 49 from Illinois, 47 from New Jersey, 46 from Rhode Island, 44 from Connecticut, 41 from New Hampshire, 36 from Kentucky, and a few from Vermont, Delaware, Virginia, North Carolina, South Carolina, Georgia, Alabama, Mississippi, Florida, Texas, Arkansas, Tennessee, Michigan, Indiana, Iowa, Wisconsin, Oregon, the District of Columbia, New Mexico, and Utah. Of foreign countries there were 214 from Germany, 150 from Australia, 121 from England, 77 from China, 54 from France, 51 from Ireland, 29 from Canada, 27 from Scotland, 25 from Chili, 16 from New Zealand, 13 born at sea, and others from Poland, Tahiti, Nova Scotia, New Brunswick, Cuba, Spain, West Indies, Russia, Holland, Hawaiian Islands, Norway, Italy, Tasmania, Belgium, Panama, Sweden, Switzerland, and Peru. The children from Ireland, France, and Spanish America are few, because most of them are sent to the Catholic schools. Otherwise the table presents a pretty fair indication of the mixed character of the population, and the brief period of its residence in this state. The proportions are, however, not the same in all the counties. The proportion of people from the maritime states of New York and New England is larger in San Francisco than in any other county; whereas Missouri is the leading state in Sonoma, Napa, Yolo and Santa Clara counties, which are agricultural. Texas has probably furnished more citizens to Tulare county than any other state of the Union; and Illinois and Iowa have furnished a considerable proportion of the residents of Siskiyou and Shasta counties.

§ 254. *Liberal Tone of Society.*—As a body, the people of California are intelligent. It is very rare to find a white man who cannot read. They have travelled far to reach this coast, and the journey required from most of them a considerable expenditure of money and an enterprising spirit. Many of them had seen much of the world before coming to California, and have travelled considerably since their arrival. They are industrious, energetic, brave, quick in speaking, ready to avenge an insult, accustomed to carry pistols and knives, quick to use them in quarrel, affable, unceremonious, and unreserved in their intercourse with acquaintances or strangers. These remarks will not apply to all individuals, but to the people generally, as compared with the people of other states and countries. The state has not yet had time to build up a national character among natives, but the people are different from all other people. They have much of the general traits of Americanism, but the traits are more striking because so many of the population are men, and men in the most vigorous period of manhood, and bold, adventurous men, who have seen strange ups and downs of life, and seen much of foreign lands and strange peoples. In their own state they meet and are thrown into familiar intercourse with intelligent representatives of all climes and continents. The consequence is, that there is no population more cosmopolitan than the Californians. They have no provincial stiffness; there is nothing of the little village about them.

In no place is society more free and cordial, and ready to give a friendly reception to a stranger than in California. The new-comer is looked upon with favor; nobody cares whether he belongs to a distinguished family, has moved in a fashionable circle, or possesses wealthy or influential friends or relatives. The great question is, "Is he or she well educated, polished, and entertaining?" Of course Californians are not entirely above such considerations as govern society elsewhere, but they are influenced by them far less than people in other states. The course of business is such that no profession has all the wealth. There are rich men of all occupations, and some of the mechanical trades are now as profitable, on the average to those engaged in them, as are the learned professions. Those who were rich in the older states, and received a thorough education and a polished training, may here be poor, while those who came hither poor and ignorant may now be rich. Besides, the changes are so rapid that our neighbor who is poor to-day may be rich to-morrow, and the neighbor who is rich to-day may be poor to-morrow. Again, California is pre-eminently a country of business. People came here to make money, and everybody tries to make it; and in a state where wages are high, and profits large, a man's business depends to a considerable extent on the multitude of his friends, so everybody wishes to make a friend of everybody else. The millionaire in Europe may treat his tenant as an inferior; in California the wealthiest land-owner is expected to treat his tenant as an equal. All these things have their influence in preventing the separation of our society into those classes which prevail elsewhere.

In no part of the world is the individual more free from restraint. Men, women, and children are permitted to do nearly as they please. High wages, migratory habits, and bachelor life, are not favorable to the maintenance of stiff social rules among men, and the tone of society among women must partake, to a considerable extent, of that among men, especially in a country where the women are in a small minority, and therefore are much courted. Public opinion, which as a guardian of public morals is more powerful than the forms of law, loses much of its power in a community where the inhabitants are not permanent residents. A large portion of the men in California live alone, either in cabins or in hotels, remote from women relatives, and therefore uninfluenced by the powers of a "home." It is not uncommon for married women to go to parties and balls in company with young bachelor friends. The girls commence going into "society" about fifteen, and then receive company alone, and go out alone with young men to dances and other places of amusement. In this there is a great error; too much liberty is allowed to the girls in the states on the Atlantic slope, and still greater liberty is given here, where, as they ripen earlier, they should be more guarded.

The absence of restraint in society, the exciting character of business, the mildness of the climate, the interesting associations of life, in a state where a man lives more and sees more in a year than he would see in a lustrum in older countries, have given the people who have resided here an attachment for California. It has been observed that a large proportion of those who have left the state, intending to spend the remainder of their lives in their native places, have returned, declaring that they could not accommodate themselves to the slow, quiet, dull ways of more antiquated states.

§ 255. *Publicity of Life.*—Life in California is very public. Many of the people live in hotels and at large boarding-houses. Travellers are numerous; theatres and balls are abundant and well attended; celebrations and festivals are frequent; the population is excitable; all take the newspapers, and all are interested in the events of the day; and the history of the country is full of eventful incidents, which always present fruitful topics for discussion. Money is abundant, and is easily earned, and of course it is spent freely; and the favorite method of spending is in public festivities and attending places of amusement. In no part of the United States is so much of life public, and so little of it private.

§ 256. *Amusements.*—San Francisco is a city of public shows and processions. Dancing is almost universal. The children of every public school in our chief city must have picnics and dances in May and at Christmas. Some of the Sunday schools do without the dance; others have as many as the common schools. One has four dances every year. The regular dances, picnics, and festivals of various schools and associations in San Francisco, will average several for every week in the year. Theatres and operas are most liberally patronized, in proportion to the population. But perhaps the amusement which has found the most favor in California is billiard-playing. Billiard-tables are found everywhere. In many little villages where there is but one inn a fine billiard-table will be found. In San Francisco there are numerous large billiard-saloons, containing each from eight to twelve of the largest and most elegant billiard-tables, at which men are constantly playing. The climate along the coast is peculiarly favorable to dancing, for, as the evenings and nights are always cold, it is as pleasant to dance in the summer as in the winter. Among the other regular amusements of Californians is that of "going east." About ten thousand of them go to the Eastern states every year, to make visits and see their relatives. Among the fashionable places of resort in the state are the Yosemite Falls, the Mammoth Groves of Calaveras and Mariposa counties, the Alabaster Cave in El Dorado county, the Geysers in Sonoma county, the Sulphur Springs in Napa county, the Warm Springs in Alameda county; and in September and October, when the grapes are ripe, the towns of Sonoma and Los Angeles. Gambling was in 1854 commonly and openly practised in all the towns of the state; now it is prohibited as a crime, and is practised only in secret, and is a rare offence save in the mining towns. Horse-racing is common, and the state has some of the best thorough-bred horses in the world. Gambling was from 1849 to 1854 a common public amusement in all the towns and mining camps. In San Francisco a dozen of the largest and finest halls on the first floor, with doors always open to the street, and in several cases extending across whole blocks and opening upon two opposite streets, were occupied by gamblers, and were filled by crowds of people every day—no exception for Sunday—from dark until long after midnight. Fine bands of music and voluptuous pictures were among the common attractions of these places, and some of the tables had handsome women to deal the cards or gather in the money. Gambling is now prohibited by law, but is practised openly in the little mining towns.

Some years ago bull-fights and fights of bears against bulls were not uncommon, but they have gone out of favor, and nothing is heard of such exhibitions now.

§ 257. *Luxurious Living.*—The mode of living among Californians is luxurious. They all try to make the most of life. Everybody wants a neat house, elegant tableware, fine mahogany furniture, Brussels carpets, and a good table. In the towns especially, the people live well, even the poorer mechanics and common laborers. Every man dresses in broadcloth, and nearly every woman in silk. The exceptions are so few as scarcely to be worthy of notice. Gold watches are worn by draymen and washerwomen. In every occupation men get rich, and stingy men and misers are rare.

§ 258. *Health.*—Of the Americans in California, it may be remarked that they generally have the same marks as the Americans in the Eastern states. Their eyes are deep set, their foreheads high, their features regular and finely cut, their faces expressive and free from grimace, their lips thin, their mouths grim, their bodies tall, slim, and slightly bent in the shoulders, thin in the chest, the voice loud, the enunciation slow and clear, with little modulation. These general characteristics of the nation, as compared with Europeans, are not wanting in California. However, the Californians of the third or fourth generation will be different from those of the present day; they will be a heavier, and healthier, and probably a taller race. The tendencies in this direction are, I think, already evident, in the forms and growth of the children; and such influences might be inferred from the vigorous development, in this state, of the forms of animal and vegetable life generally.

Most parts of the state, especially those near the coast, are very healthy. Indeed I do not think that in any part of the world is nature more favorable to long life than in California from Sonoma to Santa Barbara, within thirty miles of the ocean. The regularity of the temperature, and the entire absence of both extreme heat and extreme cold, with a clear sky, a dry atmosphere, and a constant breeze, are the conditions most favorable to health, and they are nowhere more happily united than here. In the low land of the Sacramento and Colorado basins, where the summers are very hot, and in the high land of the Sierra Nevada, where the winters are very cold, the health of the inhabitants is not so good. In many of the mining towns, where much water escapes from the ditches, and keeps the earth constantly moist, and where new ground is thrown up every day, fever and ague are common. That disease prevails also in the moist lands along the Sacramento and San Joaquin Rivers, and about the Four Creeks, in Tulare county. Rheumatism is very common in the mines, and neuralgia throughout the state. Whether this latter disease is more common here than in the Eastern states, is a matter of dispute, but certainly I never heard so much of it elsewhere as I have heard in California. Diseases of the eyes are common here, caused probably by the dust, the dryness of the atmosphere, the glare of the sun, and the evaporation of mercury in open pans, by miners. The dense fogs which visit the coast render the climate unfavorable to persons afflicted with consumption; while I presume a residence in the mountains, high above the sea, would be very likely to cure that disease, at least in its early stages.

The American women of California are not healthy, as a class. They are trained up in the dark and in idleness, as though sunshine and work would ruin them. Pastry, pickles, and sweetmeats form a considerable portion of their food, and they are taught to abhor coarse strength and robustness as worse than sins. Of course they cannot, as women, be healthy. The girls are beautiful—beautiful as angels, but at twenty-five the women begin to wither, and at thirty-five they lose nearly every trace of physical beauty. The diseases peculiar to women are very common, and whether from this reason or some other cause, it is very frequent to see women of thirty or thirty-five, who have one or two children ten or twelve years old, and none younger. The native Californian and Irish women have large families; the American women have but few, and yet there is no country where children are, on an average, so large at birth (if the opinion of some experienced physicians of my acquaintance be sufficient authority to establish the fact), or where so many twins are born. It has been remarked that a multitude of instances have occurred of couples who, after having lived together for ten, fifteen, or twenty years in other countries, before coming to California, in a year after their arrival here have had children. Travelling and a change of climate will no doubt always have a favorable influence in this respect, but perhaps the extraordinary productiveness of California may be perceptible here too.

§ 259. *Proportion of the Sexes.*—There are not enough women for the men in California. The relation between the sexes is unsound. Unfortunate women are numerous, and separations and divorces between married couples frequent. No civilized country can equal us in the proportionate number of divorces. Our laws are not so lax as those of several of the states east of the Mississippi, but the circumstances of life are more favorable to separation. The small proportion of women makes a demand for the sex, and so when a woman is oppressed by her husband, she can generally find somebody else who will not oppress her, and she will apply for a divorce; whereas, in another state she would submit to much harsher treatment and not demand a separation. The abundance of money is here felt also. To prosecute a divorce suit costs money, and many cannot pay the expense in poorer countries. During 1860 eighty-five divorce suits were commenced in San Francisco, and in sixty-one of these, or three-fourths of the cases, the wives were the plaintiffs. During the six years from 1855 to 1860, inclusive, the number of divorce suits commenced in San Francisco was four hundred and forty-seven, and in more than three-fourths of the suits divorces were granted, and divorces were denied in very few. Some of the suits were discontinued or abandoned. The proportion is probably about the same in other parts of the state.

§ 260. *Education.*—The state has made a liberal provision for education. Common schools, free to all white children, are maintained by the public treasury, and the large fund provided for their support is declared in the constitution to be inviolable. The common schools in San Francisco are as good as any common schools in the world. Those in the country districts are not so good, and yet will compare favorably with most country schools. Boys and girls are taught together in these schools—an arrangement which is thought, by many parents, to be bad for girls over twelve; and therefore private schools for girls have many pupils. The Catholics in San Francisco have their own schools, and support them with their own money. They dislike the common schools, because pupils are not required to study the Catholic catechism there. The state constitution provides that there shall be a state university, but it has not yet been organized, nor is it likely to be for some years to come. There are a number of high schools called "colleges" and "universities," mostly maintained by religious sects, but they have not yet become so large or strong as to deserve special mention.

§ 261. *Vigilance Committees.*—In the last chapter I spoke of the constitution and laws of the state, here I must speak of the assumption of unconstitutional and illegal powers by the Californians, in a manner unexampled in other parts of the Union. The vigilance committees of San Francisco are famous the world over. There were two of these organizations, one in 1851, the other in 1856, the latter being a revival of the former. The laws for the punishment of crime in California are so loose and so favorable to criminals, and the officers of justice have been so corrupt, that the people have at several periods felt that the only safe method of having justice executed, was to take the matter into their own hands. The vigilance committee of 1851 was in existence but a short time, and did nothing save execute a few thieves. That of 1856 was the more important organization by far, and a brief statement of its actions may be necessary to a proper understanding of the character of the people. Under the American system of universal suffrage, the control of the government must fall into the hands of a political party, which again must be managed by a few individuals, and these, especially in large cities, are often base men who make a business of politics. It is so in New York city; and it was so in San Francisco in 1855 and the early part of 1856. The party then in power did not hesitate to elect to high office men entirely unfit to be seen in decent society—men of notoriously bad character, guilty of numerous crimes. Many of the officers of elections were scoundrels, who, after the polls were closed, threw away the genuine ballots and substituted others in their place, and thus declared their own men elected to office. Such conduct was not the exception but the rule; it was practised in many districts and for year after year, and the men guilty of it were patronized and protected by men high in office. Among the ballot-box stuffers, one of the basest and most prominent was James Casey, a man who had been convicted in New York of grand larceny, and had been punished by a term of imprisonment in the state prison. He had a little education, some experience in the management of political meetings, much tact, and a ready tongue. In San

Francisco he became a politician, and obtained the position of inspector of elections in one of the wards; and there he had complete control of the ballot-box, and used it to keep himself in office and elect his friends or those who paid him. In the fall of 1855 Casey managed to be elected member of the board of supervisors (which board has the same powers in San Francisco as the common council has in most cities), and this election was conducted in such a manner that there was no doubt in the mind of any reasonable man that the whole affair was fraudulent. This man Casey shot a San Francisco editor who had denounced him as a convict, a ballot-box stuffer, and a scoundrel. The shooting took place in the street in open day, and the wound was mortal. The editor had made himself prominent and popular by exposing various abuses, and no sooner was it announced that he was shot, than a great excitement arose, men collected by thousands in the street, and acted with a passion little short of raving. The vigilance committee, which had been dissolved for six years, was reorganized. Out of twelve thousand white male citizens nine thousand enrolled themselves as members of the committee. They formed themselves into military companies, obtained arms, chose officers, and established an armory and a fort. The governor ordered them to disband, and threatened to use the military power of the state against them, but they set him at defiance, invested the jail, took from it Casey and another man, who had committed homicide, imprisoned them in their fort, and subsequently hanged them, after trying them secretly. The committee arrested a large number of persons on charges of various crimes, executed two others for murder, banished about a dozen, and maintained their fort and their military organization for eight months, during which time they were really masters of the city. They inflicted no punishment without trial, held all their trials with much deliberation, and were in no haste to execute their sentences. The governor tried to call out the militia of the state, but the people generally sympathized with the movement, so neither the militia nor the legislature would do any thing. A few troops were collected in San Francisco, but the vigilance committee took them prisoners and deprived them of their arms. The governor applied to the federal government for assistance, but the authorities at Washington decided that they would not interfere, except at the joint request of the governor and legislature. The governor was thus left powerless, and the committee maintained their fort and their military organization, and really had control of the city for eight months. No judicial process was of any service to release a man whom they had imprisoned. They hanged a few scoundrels, drove others from the country, frightened still more, destroyed the influence of bad men, rendered life and property secure, gave good men an influence in the city government, purified the elections, gave a better tone to society, commenced a new era of decent management of the municipal finances, and gave respectability to the city. It is perhaps not more than proper that I should add here, that I personally was neither a member of the vigilance committee nor an advocate of its policy of setting the laws at defiance. This vigilance committee transacted all its business secretly, and no record of its transactions has yet been published. It is universally understood that the whole control of the committee was vested in a secret executive committee of thirty-three, who had been chosen at one of the first meetings after the reorganization began, when there were but few members present. After the time when they were chosen, their names were never submitted for approval to the great majority of the vigilance committee, who joined subsequently, nor were their names ever announced to the citizens by authority, nor did they ever appear in such a manner that either the general public or the members of the vigilance committee could know the names of this executive committee, which was thus vested with an irresponsible and absolute power. The meetings of the executive committee were secret, and the greater part of their proceedings was never reported even to their constituents. When any military or other action was to be taken, nothing of the main purpose was communicated, save to a few officers. The executive committee had really an absolute power. Their orders were implicitly obeyed. Whom they ordered into arrest, was arrested; whom they ordered into banishment, was banished. Their trials were secret; a counsellor was granted to the accused, but a counsellor who was himself a member of the executive committee, and of course bound to sustain his associates. The executive committee exercised these great powers with great moderation and wisdom, and without pay; and all the men whom they executed or banished, undoubtedly deserved all the punishment inflicted on them. After the vigilance committee was disbanded, the executive committee lost all its powers, and returned to the position of simple citizens. Such is a brief statement of the character of the vigilance committee of San Francisco, an organization without its like in American history; a secret society, ruled by a secret executive committee, whose names were unknown to the public, and even to the men pledged to obey them, whose meetings were secret, and of whose proceedings no report was published either at that time or since. The maintenance of the organization in defiance of the law, was expensive and troublesome, and it is not expected that it will ever be revived; yet the temper of the people is such that, rather than submit to the sway of ruffians, who had power in San Francisco in 1855, they would certainly re-establish the vigilance committee.

§ 262. *Lynch Executions.*—No association similar to the vigilance committee has ever been formed in California outside of San Francisco. There have been many executions by lynch law, but there was no permanent organization, deliberate trial of the accused, and delay in the execution of the sentence, which marked the proceedings at San Francisco. In the interior, the lynching is always done by a mob; they may act without much noise, and give the accused several hours' respite before swinging him up, but it is only a mob after all. The people come together in a state of excitement, and dispatch the offender while their blood is still hot. These scenes have been comparatively rare of late, but they happen at least several times in the course of a year. In the year 1855, no less than forty-seven men were executed by mobs in California, twenty-four for theft, nineteen for murder, one for arson, one for rape, and two Indians, for being spies to watch the movements of some white men who were making war on their tribe in Northern California. In nearly every instance when a man is executed by mob law in California, he gets his deserts. There may be exceptions, of course, but all the probabilities are against the victim. Though the mob are excited, they are by no means unreasonable; they frequently give a man a formal trial, hear the testimony against him, and do not execute him until a jury has rendered a verdict against him. Sometimes the execution is hasty, but in such cases either the crime is great and indubitable or there is danger that the offender will be rescued by the officers of the law. The method of procedure at a lynching is simple. The people collect about the place where the prisoner is kept; if he is in jail and the jailor refuses to give him up, they break the door open with crowbars, and take the prisoner out to a tree. If they have leisure, a jury is organized and put under oath or promise to give the accused a fair trial. Some one, no matter whether a lawyer or not, is appointed to examine the witnesses for the prosecution, and another for the defence; and after a brief hearing of the testimony a verdict is rendered, and in nineteen cases out of twenty, the accused must swing. Hanging is always used as the mode of execution. The main excuse for these lynchings is, that the law is so badly administered that there is no security that a criminal will be punished; but as the population becomes more permanent, there is less foundation for this plea, and the lynch executions become less frequent. They are confined altogether to the remote places in the state, and in a few years they will entirely cease.

§ 263. *Squatter League.*—There is now, and has for years been, a squatter league secret society, with hundreds of members, organized for the purpose of defeating Spanish land claims on which squatters have settled, and for the purpose of resisting the ejectment of squatters when possible. This secret society has the support of thousands of squatters who are not members. In May, 1861, these squatters refused to allow some trespassers to be ejected under legal process from the Chaboya rancho near San José. Fifteen hundred armed men collected to set the law at defiance, and when the governor proposed to send troops, the squatters of Sacramento, Sonoma and Marin counties promised to send two armed men for every soldier. At one time there was serious danger of bloodshed, but the affair was settled by granting some squatter demands not at all consistent with the dignity of the government. This secret organization, and the seditious feeling among a multitude of persons not members, still exist, and may yet cause serious trouble.

§ 264. *Anti-Chinese Mob.*—The white miners have a great dislike to Chinamen, who are frequently driven away from their claims, and expelled from districts by mobs. In such cases the officers of the law do not ordinarily interfere, and no matter how much the unfortunate yellow men may be beaten or despoiled, the law does not attempt to restore them to their rights or avenge their wrongs.

§ 265. *Deeds of Blood.*—California has obtained a sad notoriety for its deeds of blood, and although the number of these has very much decreased, still fatal affrays are common. During 1855 a list was kept of all the homicides in the state, and no list has been kept since that year, and therefore I refer to a time so remote for statistics. It appears, then, from the records of that year, that five hundred and thirty-eight persons died by violence. Of these three hundred and fifty-seven were whites, one hundred and thirty-three Indians, thirty-two Chinamen, and three negroes. The number of Indians killed by the whites was no doubt much greater, for the two races were at war in various parts of the state, and the skirmishes, which almost invariably proved disastrous to the red men, were in many cases not reported in the papers. Exclusive of the Indians slain there were four hundred and five homicides. In many instances no particulars of the killing were given, but of those cases where the mode of death was well ascertained, there were forty-seven executed by mobs, nine were executed according to law, ten were criminals killed by sheriffs or policemen in attempting to arrest or detain them, six were foreigners killed by the collectors of the foreign miners' license, thirty-two were killed by Indians, seventeen were killed in justifiable homicide by men who were compelled to defend themselves, twelve were killed in fights about mining claims, eight were killed in fights at gambling tables, sixteen were murdered for purposes of robbery, and forty-six were found murdered under circumstances that did not indicate whether the motive was deliberate murder, sudden anger, robbery, or self-defence. These homicides occurred in nearly equal proportions in the four seasons of the year, and in the different populated districts of the state. The homicides of each year between 1850 and 1855 were at least as numerous as those of the last-named year, but since then there has been a great falling off, and now I think the number does not exceed seventy-five in a year, exclusive of the Indians, of whom the yearly average slain since 1856 has not been less than one hundred. It may appear a very singular and rapid change, that the annual number of homicides in a state should decrease eighty per cent. within a period of six years, but it must be remembered that California is in a condition of swift transition, and that our society is far more stable now than it was in 1855.

Why is it that there are so many bloody affrays in California? The first and great cause is the high temper of the people. The Americans are an arrogant race. Every man thinks himself as good as his neighbor, if not better. They are a people who will not be insulted. They consider harsh words insulting. They are fond of using harsh words to one another. Custom authorizes them to apply the word "liar" frequently, and custom justifies a man called "liar" in repaying the epithet with a blow; and the same custom justifies a man who is attacked with the fist, in defending himself with a pistol or knife. In the mining districts, many of the men are constantly armed, so that the weapons of death are rarely out of reach. The Californians are reckless of life; and the men, living alone, free from the influence of family and relatives, are dissipated, and when heated with liquor and excited by gambling, are easily provoked to settle every little quarrel with blood. The wild condition of affairs in the early times has been impressed upon our society, and we have not yet been able to reform it altogether. The multitude of people from the cotton states have introduced among us their views and practices in regard to the use of deadly weapons, and the leaven has attached itself to the mass. The majority may have different opinions, but one quarrelsome man can compel a dozen to carry arms for self-defence. Adultery is held by public opinion to be a justification of homicide, and this is a fruitful source of deadly quarrels. And then it is to be remembered, too, that many of the residents of California are men who have committed crimes in the Eastern states, in Australia and Europe, and have fled hither to escape punishment. Such persons, although they may form a small proportion of the population, contribute to give trouble. As compared with the total number of homicides, the assassinations for revenge, and the murders for money, are few. Previous to 1856, street fights were among the institutions of San Francisco. It would frequently be announced by conversation, or even by the newspapers in the morning, that a street fight might be expected that day between two men whose names were mentioned; and the curious would collect on Montgomery street, the main business street, to see the fun. The belligerents, armed each with a concealed revolver and a bowie-knife, would walk along the street, and on coming near each other would draw their revolvers, and, with or without speaking, would commence firing. The fight would be strictly one of self-defence on both sides. In the use of deadly weapons, California resembles the Gulf states far more than the North. The wild condition of affairs in the early times has been impressed upon our society, and we have not yet been able to reform it altogether; and in the matter of carrying deadly weapons and in street fights, we have imitated the example of the cotton states. So, too, in the matter of duels, of which there have been many in California, and some of them of a character so remarkable as to attract attention all over the civilized world. Duelling is forbidden by the law of the state, and is made punishable as a felony by imprisonment for not more than two years, and if either party be killed the survivor may be imprisoned not less than one nor more than seven years in the state prison; and the constitution of the state provides that no person who has been a principal or a second in a duel shall be allowed to vote or hold office in the state; but a hundred duels have been fought in the state, and about one-third of them have proved fatal to one of the principals, and yet no man has been legally punished for duelling, nor has any one been prevented from voting or holding office for that reason; on the contrary, many of the duellists have until within a late date held offices among the most honorable and profitable in the state.

§ 266. *Religion.*—We have no accurate statistics of the churches and church memberships in California. Of the 155,000 white men in the state, I presume that 10,000 are communicants of some Protestant church; 10,000 of the Catholic church; 2,000 are Jews, and 1,000 are Mormons, making 23,000 in all, or 14 per cent. Of the other 86 per cent., two-thirds never go near a church, and scarcely claim to be Christians; while the remaining third go to church occasionally, some of them regularly. The principal Protestant churches are the Methodist Church (North), which has about 3,000 communicants, 60 churches, and 65 preachers; the Methodist Church (South) has about 1,000 communicants, 20 churches, and 40 preachers; the Old School Presbyterian Church has 1,000 communicants, 15 churches, and 17 clergymen; the New School Presbyterian Church has 500 communicants, 11 churches, and 13 clergymen; the Congregationalists have 600 communicants, 11 churches, and 12 preachers; the Cumberland Presbyterians have 800 communicants and 28 preachers; the Baptists have 1,000 communicants, 42 churches, and 30 preachers; the Episcopalians have 600 communicants, 20 churches, and 18 clergymen; the Unitarians have 100 communicants, with one church, and one preacher. The total number of communicants, as mentioned in this list, including women, who are a majority, is less than 10,000; but there are many persons staying temporarily in the state, and claiming membership in congregations in the Eastern states.

The Catholic priests claim to have 80,000 communicants in their church; and they have 70 churches, and 75 priests. The Catholics are more attentive to the forms of their faith than are the Protestants.

There are many Mormons in California, but they are scattered about, and have no meetings or church organization. Almost the entire population of the town of San Bernardino was once made up of them, but they abandoned the place, at the order of Brigham Young, and went to Salt Lake. Such Mormons as there may be in the state, are generally considered to be industrious and good citizens.

There are 10,000 Jews in the state, and they have three synagogues, and two rabbies. The Jews of California are divided into two main classes, the German Jews and the Polish Jews; and in San Francisco each class has its own synagogue, cemetery, and benevolent society. The German Jews are infected by the progressive spirit of the age, and have adopted a reformed ritual; whereas the Polish Jews adhere to the old forms and ceremonies. The Chinamen are, with the exception of three or four score Christians, all Buddhists. Buddhism, which has been called "the Christianity of the east," and in its purer forms is not unworthy of the name, is with them a gross superstition, accompanied by idol worship and meaningless ceremonies, with no accompanying conception of high spiritual truths. The Chinamen have several places of worship in California, where, at least once a year, public ceremonies are held. One of these places in San Francisco, a chapel about eighteen feet wide, by thirty long, and twelve high, has at one end an idol of life-size in a squatting position, painted with Venetian red. Around this idol are placed banners and curtains of silk, and boards with inscriptions painted and carved in Chinese characters. In front of the idol is a table, on which are lighted candles, sticks of incense burning without a blaze, and filling the place with smoke, and various fantastically shaped dishes of white metal and brass. On another table are placed flowers, vegetables, and various articles of food, such as pieces of roast pig and chicken, fish, and other articles, the names of which are unknown to Americans. The ceremonies are performed by men dressed in long robes and acting as priests, who march round the room, make genuflexions and bows before the idol, and occasionally kneel and recite forms of prayer. Many of the Chinamen have in their dwellings little images of Buddha, before which they keep tapers burning. Once a year they go out in procession with music to the public cemetery, where their dead are buried, and set a table there as a feast for the spirits of the deceased.

The wild Indians have no clearly defined religious ideas; and the tame ones, or those who live among the whites, are so ignorant that their faith is little better than a gross superstition.

§ 267. *Californianisms.*—The Californians have introduced certain words into the English language, or at least have adopted them in common use in the state, and a list of them, with their pronunciation and definition, may not be out of place here:

Aparejo (a par ay′ ho), a Mexican pack-saddle.

Adobe (a do′ ba), a large sun-dried, unburned brick, sometimes two feet long, a foot wide, and four inches thick.

Arroyo (ar ro′ yo), a brook, or the dry bed of a brook or small river.

Arastra (a ras′ tra), a primitive mill for crushing quartz.

Alforja (al for′ hah), a bag, usually made of raw cowhide, used for holding the articles to be carried by a pack-horse.

Bar.—A low bank of sand or gravel, at the side of a river, deposited by the stream.

Bummer.—An idle, worthless fellow, who does no work and has no visible means of support. The word "loafer," like "lounger," does not designate the general conduct or permanent character of a man, but only a temporary idleness. A respectable, industrious man may become a "loafer" by making idle, impertinent visits in business places during business hours; but the word "bummer" implies a low, lazy character. It is probably derived from the vulgar German words *Bummeln* and *Bummeler*, which are about equivalent to "loafer" and "loaf." Its origin has been attributed to *Boehmen*, the German name of Bohemia, a nation celebrated for the number of its sharpers and adventurers. The Gipsies are called *Bohemiens* in France, because of their roving lives and worthless character. "Bummer" is generally supposed here to be a Californianism.

Bumming, acting the bummer, used in such phrases as "he is bumming around."

Caballada (ca bâl yah′ da), a herd of broken horses.

Cañada (can yah′ da), a small cañon, a deep ravine, a narrow valley with steep sides.

Cañon (can′ yon), originally a tube, and hence applied to mean a deep gorge with high, steep walls. Comparatively few cañons and cañadas are to be found in that portion of the United States east of the Mississippi, but they are abundant in California. The Spaniards place the accent on the last syllable of cañon (can yone′), but in ordinary American usage the accent is on the first syllable. It is frequently spelt "canyon," and "kanyon."

Corral (cor ral′) a pen into which a herd of cattle or horses is driven, when one is to be caught.

To corral, to drive into a corral; to drive a person into a position from which he cannot escape.

To coyote, a mining term, to dig a hole resembling the burrow of the *coyote*, or small California wolf.

Claim, the tract of land claimed for mining purposes by a man or party. There are various kinds of claims, such as bank, bar, hill, tunnel, flat, &c.

Color, a visible quantity of gold found in prospecting. If the prospector finds one or more particles of gold in his search, he says he has found the *color*.

To dry up, a slang phrase, meaning to stop, fail, disappear, become silent. It is very expressive to Californians accustomed to see the whole face of the country dry up in the summer season.

Diggings, a general name for placer gold mines. *Wet diggings* are in the banks and bars of creeks or rivers; *dry diggings* are in flats or the beds of gullies, which are dry the greater portion of the year.

Espediente, the original papers relating to some government business, filed in a public office.

Embarcadero (em bar ca day′ ro), a landing-place.

To freeze out, a miner's phrase, used to express the policy whereby stockholders or partners in mines are driven to sell out. For instance, if some rich men, owning part of a mine, discover that it is very valuable, they may conceal that fact, and at the same time levy heavy assessments for works which can bring no speedy return; and thus the poorer shareholders will be burdened and discouraged, and induced to sell out at a low price.

Fuste (foos′ te), a strong saddle-tree, made of wood and covered with raw cowhide, used for lassoing.

Gulch, a gully.

Habilitation, from the Spanish *habilitacion*, a certificate, or stamp on paper, which authorized it to be used for certain purposes. To *habilitate* paper, to place the mark of habilitation upon it.

To hydraulic, a mining term, to wash dirt by throwing a stream of water upon it through a hose and pipe.

Jaquima (hack′ ee ma), a head-stall used in breaking wild horses.

To knock down, a miner's phrase, meaning to steal rich pieces of auriferous quartz from the lode.

Manada (ma nah′ da), a herd of breeding mares under the lead of a stallion.

Mecate (may cah′ te), a rope of hair, used for tying horses.

Mochilas (mo chee′ las), large leathern flaps for covering a fuste.

Plaza, a public square in a town.

Playa, a beach.

Pozo, a spring or well.

Pueblo, a town.

To pipe, to wash dirt by the hydraulic process.

Pay-Dirt, auriferous dirt rich enough to pay the miner.

Placer, from the Spanish, a place where gold is found in dirt near the surface of the ground.

To prospect, to hunt for gold diggings; to examine ground or rock for the purpose of finding whether it contains gold, and how much.

Prospect, the discovery made by prospecting.

Rodeo (ro day′ o), a collection of wild or half-wild cattle, made for the purpose of separating or marking them.

Recojida (ray co hee′ da), a similar collection of horses.

Rancho (ran′ tsho), before the Americans took California, meant a tract of land used almost entirely for pasturage, rarely less than four square miles in extent, sometimes as much as ninety-nine square miles, and in most cases not less than thirty square miles. Since the conquest, *rancho*, and its American derivation "ranch," are often applied to small farms, and sometimes, in the way of slang, to single houses, tents, and liquor shops. "Ranch" is sometimes used as a verb; thus a man who opens a farm, according to common parlance, "has gone to ranching."

Ranchero (ran tsha′ ro), a man who owns and lives upon a rancho. It is usually understood to mean a Spanish Californian.

Rancheria, an Indian hut or a village.

Reata (ray ah' ta), a rawhide rope, used for lassoing.

Rubric, a flourish, which Mexicans and native Californians append to their signatures, and which, in fact, they consider as an important part of their signatures, and the most difficult to imitate or counterfeit. They often use their "rubrics" alone as signatures. To *rubricate*, to sign with a rubric.

Sluice, a wooden trough about fourteen inches wide, and ten deep, and not less than thirty feet long, used for washing pay-dirt.

Ground-Sluice, a trough cut in the ground for washing pay-dirt.

Tail-Sluice, a sluice put in below a number of other sluices, and depending on them for its supply of dirt and water.

Sluice-Fork, a fork similar to a manure fork, but with blunt prongs, as wide at the point as at the heel. The fork is used for throwing stones out of the sluices.

Sluice-Head, the quantity of water used in a sluice; a constant stream of water running through an aperture, usually two inches high, and from five to fifteen inches long, under a pressure of seven inches.

Slum, slimy mud.

To strip, to throw off worthless dirt from the top of pay-dirt.

Sierra (see er' ra), originally a saw, a chain of mountains.

Square Meal, a good meal at a table, as distinguished from such meals as men make when they are short of provisions, a condition not uncommon among men who make adventurous trips into the mountains.

Tailings, the waste of a sluice, tom, rocker, or quartz-mill.

Tom, a wooden trough, from ten to fifteen feet long, for washing pay-dirt.

Tom-Stream, or *Tom-Head*, the amount of water used in a tom.

Rocker, or *Cradle*, a machine resembling a domestic cradle, for washing pay-dirt.

Wing-Dam, a dam in a creek or river, running partly across its bed and then down stream so as to shut out the water from part of the bed.

Vaquero (va kay′ ro—vulgarly pronounced buc ca′ ry), a herdsman.

Zanja (zan′ ha), a ditch for irrigating agricultural land.

Zanjero (zan hay′ ro) a person whose duty it is to keep irrigating ditches in order. Among the officers of the city of Los Angeles is a Zanjero.

§ 268. *Germans and French.*—The Germans of California have generally become citizens and permanent residents of the state, and most of them have learned the English language. They have gymnastic clubs and singing societies in all the larger towns. Most of the brewers and professional musicians in the state are Germans; and the Germans occupy a large place among the cultivators of the vine. The town of Anaheim, in Los Angeles county, was laid out by Germans, and is almost exclusively populated by them.

The French in California, as a body, are not citizens, or permanent residents, and have not learned English. Some of the largest bankers in San Francisco are Frenchmen, and the city owes to Mr. Pioche, a French capitalist, some of its most valuable public enterprises, such as the Mission Railroad, the Spring Valley Water Works, and the pleasure resorts of Hayes' Park and the Willows.

§ 269. *Spanish Californians.*—The people of Spanish blood in the state are mostly natives of California, Mexico, and Chili. As a class, they are poor and ignorant. The Mexicans and Spaniards who came to California while Spain held dominion of the country, brought few women with them, but took Indian women for wives; and the descendants of these women form a large majority of the Spanish Californians. Among the wealthier families, the Indian cast of countenance has almost disappeared. Although the features are thick, the expression of the face is mild and pleasant. The complexion is dark, and grows darker with age; the hair is black and straight, the eyes black, the cheeks ruddy. Many of the men are handsome, tall, broad-shouldered, large-boned, big-bellied, strong, healthy, and long-lived. They grow fleshy as they grow old; and the same remark applies to the women. They are a good-natured race, very kind and obliging to their friends, but out of place among Americans, who are too sharp for them in trading. Instead of increasing in wealth with the development of the country, the Spanish Californians have been rapidly growing poorer, and now they own scarcely one-tenth of the landed property which they had in 1848. Then they owned nearly every thing; now there is not a leading merchant or millionaire among them. They regret the conquest of their country. They lived in a very simple manner under the Mexican dominion, but they were secure in their property, and were the political masters. Now they form a small and powerless minority, among a people far superior to them in agricultural and mechanical skill and business knowledge—a people who are absorbing all their wealth, and who look upon them and treat them as inferiors. Although some of the Spanish Californians are content with the change of dominion, yet many hate the Americans, and hate them so bitterly that they would resort to civil war if there were any hope of success. Indeed, the condition of affairs, in some of the counties where the Spanish population is most numerous, was near civil war at various periods between 1851 and 1854. Most of the Spanish Californians live in the country; their chief wealth is in land and cattle, and the chief occupation of the poorer classes is herding cattle. Their dwellings are adobe houses, usually of one story, often with no floor save the bare earth, and without chairs.

§ 270. *Chinamen.*—The Chinamen in California are nearly all very ignorant and very poor. Their number is about fifty thousand, of whom more than half have been six or seven years in the state. Most of them are engaged in mining; and the remainder are merchants, fishermen, washermen, and a few are employed as cooks in hotels, and as farm laborers on farms owned by white men. Most of them come from Southern

China, and nearly all of them are members of five great companies, called the Yung-Wo, the Sze-yap, the Sam-yap, the Yan-wo, and Ning-yeung companies. These companies have each a large building in San Francisco, where they lodge and feed all the members of their company when they arrive from China, or when they come on a visit from the interior. The companies are benevolent associations, and take care of their indigent and sick. There are no Chinese beggars in the streets, and no Chinese patients in the public hospitals. The common laborers are brought to the state under contract to work for several years at a low rate of wages (from four to eight dollars) per month; and they usually keep these contracts faithfully. The employers in these cases are either the companies or associations of Chinese capitalists. The Chinamen generally are very industrious; indeed they are the most industrious class of our population, and also the most humble, quiet, and peaceful. The merchants are considered to be very faithful to their promises, and in San Francisco they can get credit among their acquaintances quite as readily as other men in similar branches of business. In the mines, the Chinamen work in the poorest class of diggings. They own no ditches, large flumes, hydraulic claims, or tunnel claims. The white miners have a violent antipathy to them, will not permit them to work in many districts, and will often drive them from their best claims in the districts where they are permitted to work. Sometimes the celestials venture to dam a stream, but not often. They use the rocker more than any other class of miners.

In San Francisco, the merchants are usually in partnerships, with not less than three nor more than ten partners; all of whom live in the store, and deal chiefly in Chinese silks, teas, rice, and dried fish. The two latter articles form a large portion of the food of the Chinamen in the state. They have not learned to use bread instead of rice. Those who can afford it, eat pork, chickens, and ducks. Beef, and most of our garden vegetables, do not find much favor with them, even among the wealthiest. The washermen are usually in companies of two or three, and they have numerous little shops in the streets of San Francisco, and in the smaller towns. They sprinkle their clothes previous to ironing, by filling the mouth with water and then blowing it over them. For ironing, instead of a flat-iron, they use an iron pan with a smooth bottom, and kept full of burning charcoal. There are not more than one thousand Chinese women in the state, and nine-tenths of them are prostitutes of the lowest class. The Chinese children are few.

The Chinese men, women, and children learn English very slowly; most of those who have been five or six years in the state cannot understand the most common English words. All the Chinamen in California adhere to their national costume, with some slight variations. They wear their hair long, use no white muslin or linen next the skin, and never put on a dress coat or stove-pipe hat. In the cities, they ordinarily use wooden-soled shoes, with thin cotton uppers. Instead of a coat, they have a short blouse, generally of dark-blue cotton, fitting close up to the neck. The wealthy have this blouse made of silk or fur. In cold weather, if of silk or cotton, it is wadded. The legs and lower part of the body are enclosed in breeches of cotton or silk, tight from the thigh down, and loose above. Some of the poorer men find trousers of the European pattern more convenient, and wear them. The miners generally wear coarse boots or shoes.

§ 271. *Indians.*—The Indians are a miserable race, destined to speedy extinction. Fifteen years ago they numbered fifty thousand or more; now there may be seven thousand of them. They were driven from their hunting-grounds and fishing-places by the whites, and they stole cattle for food; and to punish and prevent their stealing, the whites made war on them and slew them. Such has been the origin of most of the Indian wars which have raged in various parts of the state almost continuously during the last twelve years. Scarcely a month has passed since 1849, without some hostile encounters between the red men and the whites in some parts of the state.

At this time the American residents of Humboldt county are at war with the Indians there. The poor Indian, afoot, and armed only with the bow and arrow, is no match for the rich American, armed with rifle and revolver, and mounted on a horse, which saves him from fatigue, takes him swiftly to the best point of attack, or carries him still more swiftly from danger. For every white man that has been killed, fifty Indians have fallen.

In 1848, nearly every little valley had its tribe, and there were dozens of tribes in the Sacramento basin, but now most of these tribes have been entirely destroyed. Syphilitic diseases and brandy have co-operated with the bullet and the knife to make room for the white men. The Indians are fond of strong liquor, and when they can get it, frequently become habitual drunkards. The squaws drink as much as the "bucks." Among a tribe of drunken men and women, matrimonial constancy is not to be expected; nor is it found among the Indian women in California. The infectious disease which threatens to utterly destroy all barbarous and semi-barbarous nations, has slain many of the red men in this state, as well as in other parts of the continent.

The Indians of California, with the exception of the Mojaves, are supposed to belong to the general division of the Shoshonees, which includes also the Indians of Nevada Territory, and a majority of those in Utah. They are physically and intellectually inferior to their relatives in Nevada Territory, and far inferior to the Indians who dwelt during the last century east of the Mississippi River. The red men of this state have but a small share of the courage, military spirit, and intellectual activity of the Shawnees, Miamis, Delawares, and the other tribes who contended so stoutly for the possession of the valley of the Ohio. The majority of the Californian Indians never learned to use fire-arms, and never dared to meet the white men in battle. A few in the northern part of the state have obtained fire-arms, use them well, and fight stubbornly, but they are a small proportion.

The Californian Indian men are about five feet and a half high on an average, and the women four feet and ten inches. They are very thick in the chest, and have voices of wonderful strength. The children are clumsy, and heavy set. The women are very wide in the shoulders and hips, and strongly built. Men and women are large in the body, and slim in the legs and arms, as compared with Caucasians. When not affected by hereditary diseases, caught from the white men, the Californian Indians have healthy constitutions, and formerly they lived to a great age. During the last ten years, a number have died with the reputation of being more than one hundred and twenty years old. It is a common assertion, and one that I have never heard contradicted, though I have conversed on the subject with men who have seen much of them, that the wild Indians never take cold. During the winter of 1849–'50, I lived near a tribe in the mines, in what is now Shasta county, and I saw that the men never wore any clothing save a deerskin thrown over the shoulders; that men, women, and children went barefooted through a winter when snow lay on the ground for a week at a time, and that their huts were only about six feet wide, were open on all sides, and on two sides had holes large enough for men to get in and out; and I never saw one troubled with a cold or cough. In the tribes living far from the whites, the men usually go naked, and the women wear a petticoat made by fastening flags or strips of bark, about eighteen inches long, to a girdle. They are very filthy in their habits, and their houses are always filled with lice. Their form of government is very simple. They have hereditary chiefs who have very little power. The tribes are small, and have no wealth, and no laws. Occasionally a member of a tribe gives offence, and some of the leading ones of a tribe agree to kill him, and the sentence is carried into effect by waylaying him and shooting him with arrows. Their rule is blood for blood. They rarely keep men prisoners, but kill them immediately after capturing them. Women and children are held frequently as prisoners; and one of the most common causes of war is the capture of women. They have no hereditary slavery. They have no marriage ceremony, and the duration of the marriage relation depends entirely upon the pleasure of the husband. Polygamy is permitted by many of the tribes. The women are not prolific, or at least the children are few, and mostly boys. The girls are neglected, or intentionally killed soon after birth, and this policy will, of course, if continued, soon cause an extinction of the race in California. In certain tribes on the northern coast, if a mother, having an infant child, dies, the child is buried with her. Most of the tribes burn their dead, commencing the cremation in the evening, and keeping up the fire all night, while the friends watch, and the women relatives utter plaintive cries until daylight. They have no religious ceremonies; or no ceremonies to which they attach ideas clearly religious. Every year, usually in the spring, they have a dance, as it is called. They assemble, build a large fire, and the men surround it, and keeping their knees, elbows, and backs bent, they beat time with their feet to a monotonous song, which they sing with the assistance of the squaws, who sit off on one side. In some tribes, several of the men have pipes, from which they elicit a few notes as an accompaniment for the song.

The squaws are treated like slaves. They are required to do all the work, and to attend to every want of their husbands. They must collect vegetable food, prepare it, and carry all the movable property in times of migration. They are beaten on the slightest provocation. The men never consult them about the management of public or private affairs. They are bought as merchandise from the parent, and treated as slaves after the purchase.

Most of the wild Indians have no permanent place of residence. Each tribe has a territory which it considers its own, and within which its members move about. Each family has a hut, and a cluster of these huts is called a *rancheria*. The rancherias are usually established on the banks of streams, in the vicinity of oak-trees, horsechestnut bushes, and patches of wild clover. Such places are generally on fertile soil with picturesque scenery. The huts are made differently in different places. In the Sacramento valley, the most common plan was to dig a hole three or four feet deep and ten feet across; erect an upright post in the centre about six feet high; lay poles from the edge of the hole to rest on this post; and cover the poles with grass and then with dirt. In some districts the hut is made by taking large pieces of pine bark and laying them against a frame-work of poles fastened together in a conical shape. In the San Joaquin valley it was more convenient to make a frame-work of poles and cover it with rushes or tules. These huts may be deserted for a time, but are considered the property of the builders, who move, according to the seasons, to those places where they can obtain food most conveniently. In one month they go to the thickets; in another to the open plain; in another to the streams.

Their food is composed chiefly of acorns, clover-grass, grass-seeds, grasshoppers, horsechestnuts, fish, game, pine-nuts, edible roots, and berries. The acorns of California are large, abundant, and some of them are not unpleasant to the taste, but they do not contain much nutriment as compared with an equal bulk of those articles commonly used for food by the Caucasian race. The acorns are gathered by the squaws, and are preserved in various methods. The most common plan is to build a basket with twigs and rushes in an oak-tree, and keep the acorns there. The acorns are prepared for eating by grinding them and boiling them with water into a thick paste, or by baking them in bread. The oven is a hole in the ground about eighteen inches cubic. Redhot stones are placed at the bottom of the hole, a little dry sand or loam is thrown over them, and next comes a layer of dry leaves. The dough or paste is poured into the hole until it is two inches or three inches deep. Then comes another layer of leaves, more sand, redhot stones, and finally dirt. At the end of five or six hours the oven has cooled down, and the bread is taken out, an irregular mass nearly black in color, not at all handsome to the eye or agreeable to the palate, and mixed through with leaves and dirt. For grinding the acorns a stone mortar is used. This mortar is sometimes nearly flat, with a hollow not more than two inches deep; and occasionally one will be seen fifteen inches deep, and not more than three inches thick in any part of it. The pestle is of stone, round, ten inches long and three thick.

Horsechestnuts are usually made into a gruel or soup. After being ground in the mortar, they are mixed with water in a waterproof basket, into which redhot stones are thrown, and thus the soup is cooked. As the stones when taken from the fire have dirt and ashes adhering to them, the soup is not clean, and it often sets the teeth on edge.

Grass-seeds are ground in the mortar and roasted or made into soup.

Grasshoppers are roasted, and eaten without further preparation, or mashed up with berries.

Fish and meat are broiled on the coals. The intestines and blood are eaten as well as the muscle.

Clover and grass are eaten raw. The Indians go out into the clover patches, pull up the clover with their hands, and eat stalks, leaves, and flowers. They consider clover a great blessing, and get fat on it.

The Indians rarely have salt and never spices, and most of their food is such as a white man could not eat, unless reduced to near starvation. In eating they use no plates, cups, knives, or forks, nor do they use any utensils in preparing their food, save the mortar and waterproof basket. The pine-nuts, edible roots, and berries are eaten raw. Bugs, lizards, and snakes are all considered good for food. In those places where the tules grow, the roots of those rushes are eaten. Except one or two tules in the Colorado Desert, the wild Indians of California never tilled the soil.

They use very few tools. The bow was the only weapon for killing quadrupeds. It is made of a reddish wood said to be the western yew, and on the back the bow is strengthened with a covering of deer's sinews. The arrows are of reed, and have a head made of obsidian, which is a transparent, vitreous substance of volcanic origin, in appearance very similar to a coarse quality of glass. The arrow-heads are made two inches long, half an inch wide, an eighth of an inch thick, with a very sharp point and sharp edges. The head is fastened in a split of the shaft of the arrow by tying with deer sinews. Such an arrow-head can be used but once, for the obsidian is as brittle as glass and breaks at the first shock. Some tribes in the northern part of the state, poison their arrows by irritating a rattlesnake and then thrusting forward a fresh deer's liver, which it will bite. After it has bitten repeatedly, and thrown some of its poison at every bite into the liver, the latter is buried and allowed to putrefy. It is then dug up, the arrow-head is dipped in it and allowed to dry. An arrow thus poisoned will kill a man, a horse, or an ox in twenty-four hours, or less time; and it is said that the meat of an animal thus killed may be eaten with safety. I know that the Indians do eat the meat of animals killed with poisoned arrows, but I am not positive that the poison was prepared in this manner. The poison of the rattlesnake is not injurious when taken into a sound stomach; it is only when injected into the blood that its injurious influences are felt. The arrows, even when not poisoned, make very dangerous wounds, for the sinew used to fasten the head soon softens, and allows the head to remain when the shaft is pulled out.

The Indians are very familiar with the habits of wild animals. They know precisely the character of the brushwood and ravines in which the deer and bear hide during the day, and the places to which they go to feed in the morning and evening. In hunting deer and antelope, in places where there is grass eighteen inches or two feet high, the Indian will often hold the skull and horns of a buck deer before him, and thus crawl within bow-shot. The Pit River Indians dig pits about five feet cubic and cover them with brush and grass, and thus catch deer, hares, and so forth. For catching wild geese, various small and simple kinds of nets are used, and they are knocked down with clubs. Salmon are killed with stones and clubs in shallow water, and are caught with spears. Their most ingenious spear has a head of bone about one inch and a half long and sharp at both ends. To the middle is fastened a string, which is attached to the spear-shaft. One end of the head fits in a socket at the end of the spear-shaft. When the spear is thrown the head comes out of the socket and turns cross-ways in the fish, and then there is no danger that it will tear out. The Indians rarely hunt the grizzly bear. Along the ocean beach they get barnacles. Their method of catching grasshoppers is to dig a hole several feet deep, in a valley where this species of game abounds. A large number of the Indians then arm themselves with bushes, and commence at a distance to drive the grasshoppers from all sides toward the hole, into which the insects finally fall, and from which they cannot escape. The pine-nuts are sought at the tops of the pine-trees, which the "bucks" ascend by holding to the rough bark with their hands, and pressing out with their legs, so that they do not touch the body to the trunk of the tree in going up. It is more like walking then climbing.

The bow and arrow, the spear, the net, the obsidian knife, the mortar, and the basket, are the only tools made by the Indian. The obsidian knife is merely a piece of obsidian as large as a hand and sharp on one side. The baskets are all made of wire-grass, a grass with a round jointless stem, about a sixteenth of an inch thick and a foot long. The basket-work made with this wire-grass resembles the texture of a coarse Panama hat, and is waterproof. All the basket-work of the Californian Indians is made of this material. The most common shape for the basket is a perpendicular half of a cone, three feet long and eighteen inches wide, open at the top. The basket, carried on the back of the squaws, is used for carrying food, miscellaneous articles, and children. Neither the Californian Indians of the present, nor of any preceding century, made such mounds, circumvallations, arrow-heads, or spear-heads of flint, or pipes and battle-axes of stone, as are found in the state of Ohio. There is nothing to indicate that any of the inhabitants of the country, previous to the arrival of the Spaniards, were above a very low degree of savagism. They have no domestic animals save the dog, and that of a very low kind. They have so little skill in the preservation of food, that, like wild beasts, they grow grossly fat in the spring and poor in the winter. The Mojave Indians in the Colorado Desert, depend for their subsistence chiefly on cultivated food. They plant wheat, grass, pumpkins, and musk-melons. After the annual overflow of the bottom land, a small patch of ground is cleared off with the help of knives and fire; then small holes are made, the seeds are deposited, and the field is left to grow up as well as it may. The musk-melons are eaten fresh; the pumpkins are eaten fresh, or sliced and dried; the wheat and grass-seeds are ground, made into a paste with water and dried in cakes. The mezquit bean, next to the cultivated grains, pumpkins, and squashes, is the most important article of food with the Indians in the Colorado Desert. These beans are prepared for eating in the same manner with the wheat and grass-seed.

The preceding remarks relate to the wild Indians only, and are intended to illustrate the natural habits, character, and capacity of the race. During the last fifteen years, however, they have all been influenced so much by intercourse with the whites, that they have lost many of their wild habits and acquired new ones. In some districts they have fire-arms; in others they obtain much of their food and clothing from their Caucasian neighbors. In the counties along the southern coast, there are many civilized Indians, who live in adobe-houses, and support themselves by herding cattle, breaking horses, and working in the grain-fields, orchards and vineyards. They have lost much of the savage expression of countenance, and some of them have become very industrious and trustworthy laborers; but the majority are idle and dissipated in their habits. They have all learned a vulgar dialect of the Spanish, and a few speak a little English. The younger ones know nothing of any tongue save English and Spanish, but the elder Indians, when talking with one another, prefer to use the language of their fathers.

§ 272. *Cities and Towns.*—Below I have prepared a list of the principal towns of California, with the number of voters and inhabitants in each. The number of voters is ascertained from newspaper reports of the votes cast at the presidential election of 1860; and the taking the vote as a basis, I guess at the number of inhabitants. The proportion of voters to inhabitants is larger in the mineral than in the agricultural districts:

Towns.	Voters.	Population.
San Francisco	14,415	70,000
Sacramento	3,833	15,000
Marysville	1,871	7,000
Stockton	1,445	5,500
Nevada	1,423	4,500
San José	1,000	4,000
Grass Valley	1,292	3,500
Petaluma	822	3,500
Columbia	1,008	3,000
Placerville	964	3,500
Yreka	888	2,500
Los Angeles	795	3,500
Oroville	716	2,200
Folsom	629	2,400
Downieville	628	2,000
Sonora	598	1,800
Weaverville	571	1,600
Santa Clara	539	2,500
Dutch Flat	500	1,500
Big Oak Flat	491	1,450
La Porte	479	1,420
Mokelumne Hill	470	1,400
Vallejo	464	1,600
Forest Hill	457	1,300
Michigan Bluff	451	1,300
Georgetown	448	1,300

Towns.	Voters.	Population.
Santa Cruz	443	1,800
North San Juan	437	1,500
Shasta	374	1,100
Napa	379	1,500
Oakland	352	1,450
Diamond Springs	347	1,000
Murphy's	344	1,000
San Andres	328	1,000
Alleghany Town	319	960
Timbuctoo	309	1,000
Auburn	307	1,000
Shaw's Flat	301	950
Mariposa	230	700

This table is not offered as precisely accurate, but as a reasonable approximation, based upon the best materials within reach. The vote cast does not represent the number of voters in the town, but in the precinct, and this may extend four or five miles beyond the town limits. The total number of people residing in the towns is about 100,000, or one-fourth of the whole population.

§ 273. *San Francisco.*—San Francisco, according to the census of 1860, has a population of 56,805, of whom 52,955 are white and 3,850 colored; but according to the estimate of intelligent residents, the population is not less than 70,000; and I shall adopt the latter figure as correct. There may be 40,000 Americans, 12,000 Irishmen, 5,000 Germans, 4,000 Britons, 3,000 Frenchmen, 2,000 Chinamen, and 3,000 miscellaneous. The Federal census classifies the people according to their ages and color, as follows:

White Males.—Under 1 year, 1,730; 2 years, 777; 3 years, 730; 4 years, 627; 5 years, 645. Total under 5 years, 4,509; between 5 and 10 years, 1,842; 10 and 20 years, 2,915; 20 and 30 years, 10,184; 30 and 40 years, 9,390; 40 and 50 years, 2,581; 50 and 60 years, 842; 60 and 70 years, 162; 70 and 80 years, 36; 80 and 90 years, 2. Total white males, 32,463.

White Females.—Under 1 year, 1,563; 2 years, 739; 3 years, 677; 4 years, 600; 5 years, 551. Total under 5 years, 4,130;

between 5 and 10 years, 1,831; 10 and 20 years, 3,198; 20 and 30 years, 6,226; 30 and 40 years, 3,441; 40 and 50 years, 1,119; 50 and 60 years, 484; 60 and 70 years, 122; 70 and 80 years, 52; 80 and 90 years, 7. Total white females, 20,610.

Chinese.—Males of all ages, 2,168; females do., 448. Total Chinese, 2,616.

Colored.—Males of all ages, 711; females do., 435. Total colored, 1,146.

Recapitulation.—White males of all ages, 32,463; females do., 20,610. Total whites, 53,073; Chinese, male and female, 2,616; colored, 1,146.

San Francisco is the chief commercial and manufacturing city of this coast. The vessels which entered the port in 1860 numbered 1,686, and had an aggregate measurement of 500,000 tons. Of the 1,686 vessels, 325 were from foreign ports, 115 from American ports on the Atlantic, 1,231 from American ports on the Pacific, and 15 were from whaling voyages.

The first house was built in 1835, and the place was then called Yerba Buena, Spanish for "good herb," applied to a species of mint growing in the vicinity. In 1847 the name was changed to San Francisco. In 1846 the population was six hundred, and had grown to about one thousand in the spring of 1848, when the gold fever broke out. During July, August, and September, the town was deserted by many of its residents; but as the people became impressed with the richness and extent of the mines, and as adventurers began to arrive from abroad, the population of the town increased, and then suddenly it sprang from an obscure village to a world-famous city. In May and June, 1850, and in the same months the next year, great conflagrations swept away the wooden shanties with which the main part of the city was built up; and it was not until the latter half of 1851, that the citizens commenced to erect the numerous fine brick stores which now ornament the principal business streets. The sand ridges on the site of the city were cut down, and the hollows were filled in; and the shallow cove in front of the mainland was also filled in, and made the foundation for the busiest part of the town.

In 1861, according to statistics published in the *City Directory*, there were 11,265 buildings in the city: 9,308 of wood, 1,898 of brick, 47 of iron, 6 of stone, and 6 of adobe. These buildings are thus classified according to their height:

Wood—one story, 4,034; two stories, 5,090; three stories, 180; four stories, 4—9,308.

Brick—one story, 272; two stories, 1,125; three stories, 438; four stories, 59; five stories, 3—1,898.

Iron—one story, 8; two stories, 30; three stories, 6; four stories, 3—47.

Adobe—one story, 1; two stories, 3; three stories, 1; four stories, 1—6.

Stone—one story, 1; two stories, 3; three stories, 1; four stories, 1—6.

According to the same authority there were, in 1860, 800 grog-shops, 373 groceries, 288 lawyers, 276 tailor-shops and clothing-stores, 248 boarding-houses, 189 physicians, 179 brokers, 150 cabinet-makers, 150 butchers, 136 cigar-shops, 121 dry-goods stores, 120 carpenters, 95 barbers and hair-dressers, 85 dealers in coal and firewood, 84 restaurants, 84 watchmakers and jewellers, 78 fruit-stores, 66 bakeries, 65 house and sign painters, 64 stove and tinware stores, 33 lumber yards, 24 breweries, 20 auction-stores, 17 banks, and 8 assay-offices.

There are thirteen daily and twelve weekly newspapers, and five monthly magazines, published in San Francisco. Of the dailies, seven appear in the morning and six in the afternoon; seven are published in the English language, two in French, two in German, and two in Spanish. Of the weekly papers, one is an organ of the Methodists, another of the Methodist Church South, another of the Congregationalists, one of the Catholics, two of the Jews; one is commercial, one is filled with miscellaneous reading matter, another is agricultural, another devotes itself to topics interesting to firemen and soldiers, another to mining, and the last fills its columns with police reports. Of the monthly magazines, three devote themselves to general reading matter, one is medical, and the last Presbyterian. There have been Italian and Chinese newspapers in San Francisco, but there are none here now.

The Germans, French, Italians, Swiss, Scandinavians, Chinese, Illyrians, German Jews, Polish Jews, and Irish, have each a benevolent society, organized mainly for the purpose of rendering mutual assistance in case of illness. There is a Protestant and a Catholic Orphan Asylum, a House of Refuge, and an Asylum for the Deaf, Dumb, and Blind. The Masons, Odd Fellows, and Independent Order of Knights, are other benevolent associations established in our city.

§ 274. *Sacramento.*—Sacramento City, the political capital and second town of California, is situated near the centre of the Sacramento basin and of the state—is one hundred and twenty-five miles by the course of navigation, and seventy-five miles in a direct line, distant from the ocean, on the southeastern corner of land formed by the junction of the Sacramento and American Rivers, at an elevation of thirty feet above the sea, and in latitude 38° 33′ and longitude 121° 20′. The business part of the city is about twenty feet above low-water mark in the Sacramento River, which, in front of the town, during the dry season, rises and falls about a foot with the tide. (The figures in the preceding sentence were correct previous to the flood of 1861–'62; whether they will be after it has passed away is questionable. I write this while it still prevails; and it is said that, during the flood, gravel to the depth of twelve feet has been deposited in the river-bed in front of Sacramento.) The site is level, and in the midst of a wide plain, most of which is bare of trees, and much of which is not cultivated. The streets are wide and straight, run with the cardinal points of the compass, and are designated only by numbers and letters. Those parallel with the Sacramento are first, second, third, and so forth; those parallel with the American are A, B, C, and so on. The main business part of the city is near the

Sacramento, extending from first to sixth, and from H to L streets. The houses and stores here are mostly built of brick, one or two stories high. The streets are gravelled or planked; the side-walks are planked or paved with brick, and covered with awnings to give protection against the sun. In those parts of the town used for dwellings, the houses are chiefly of wood, neatly painted, and surrounded by gardens, and the streets are lined with shade-trees, such as cottonwood, willow, sycamore, elm, and locust. There are water-works and gas-works. The water is pumped up from the Sacramento River, which is so turbid, even at its lowest stage, that six inches of mud are deposited monthly in the reservoir. The gas is made from imported coal. A railroad twenty miles long connects Sacramento with Folsom, which is connected with Lincoln by another road of about the same length. A steamboat leaves Sacramento daily at two P. M. for San Francisco; thrice a week, starting in the morning, for Marysville; and at least thrice a week for Red Bluff. Stages run daily to Marysville, Auburn, Placerville, Coloma, Jackson, Stockton, and Fairfield.

The first settlement by white men in the city of Sacramento was made in 1839, by John A. Sutter, a Swiss by birth, who, after having served as a captain in the body-guard of Charles X. of France, came to the United States, where he was Americanized. He afterwards came to California and was admitted to Mexican citizenship. He obtained a grant of eleven square leagues of land on the eastern bank of the Sacramento River, and under that grant the title to the site of Sacramento City is now held. In 1841 he built some adobe buildings, which he dignified with the title of New Helvetia, while to the Americans it was generally known as "Sutter's Fort." It was, for a long time, the only place where white men had a permanent foothold in the Sacramento basin; and it was a place of importance, as the first point where the American trappers, travellers, and immigrants, entering the territory from the eastward, could obtain provisions, ammunition, and horses, and rest secure against Indians. Sutter treated all comers with the utmost generosity and liberality; no white man was turned away because of inability to pay for food or lodging. The first gold diggings were discovered about twenty-five miles eastward from the fort, which became the chief trading point between San Francisco and the mines. The adventurers ascended the Sacramento River to the mouth of the American, where they landed, and their goods were taken by ox-wagons to the fort, two miles distant, where they prepared themselves for the land journey. Before the first year of gold mining had come to an end, it was evident there must be a town on the bank of the Sacramento at the mouth of the American, so the present town site was laid off in October, 1848, and the New Year's day following the building of the first house (of logs), near the Sacramento River, was commenced. On the 8th January the lots were sold by auction, and were described as lying in the town of "Sacramento." The ford and its vicinity continued to be the chief place of business until April, '49, when the bank of the Sacramento was found to be much more convenient for purposes of business, and the merchants and traders moved. The town very soon became the most important centre of trade and population in the state, next to San Francisco, and it has continued to hold the same relative position, growing with the growth of the state, notwithstanding many severe disasters to which it has been exposed. In 1851 there was a serious riot about land-titles; on the 3d of November, 1852, the greater part of the town, including six hundred houses, was destroyed by fire, with a pecuniary loss estimated at the time at $5,000,000; and the city was flooded in January, 1850, in March, 1852, in January, 1853, and in December, 1861, and in January and February, 1862. In 1853 the business part of the town was raised about five feet, the streets being filled in with gravel to that depth, and a levee or embankment was built round the city, extending about a mile along the bank of the Sacramento and three or four miles along the bank of the American. The flood of 1861 and '62 proved extremely disastrous. It filled every part of the city, was three feet deep in every street; in some places fifteen feet deep. Gardens were destroyed, fences carried away, domestic animals drowned, furniture ruined, and many of the people driven to take refuge in San Francisco and other towns not afflicted by the general scourge. A long time will pass before the city will recover from the injuries inflicted upon it by the flood of 1861–'62.

The assessed value of the taxable property of the city is about $7,000,000; the public debt of the city is $1,800,000.

§ 275. *Stockton.*—Stockton is situated three miles eastward from the San Joaquin River, on the bank of a navigable tide-water slough, or creek (using the word in its British meaning), which is eighty feet wide and eight feet deep. The town site is in the midst of low, flat, tule land, which is intersected by numerous sloughs. The population is about six thousand. The town has a pleasant appearance. Many of the dwellings are neatly built and are surrounded by elegant gardens. Shade-trees are abundant. Fresh water is supplied to the city, for domestic purposes and for irrigating the gardens, by one hundred and fifty windmills, which pump it up through lead pipes, thrust down twenty feet deep into auger holes two inches wide. So abundant is the water in the soil at that depth, that there is no difficulty in obtaining it in this manner. Stockton is nicknamed "The City of Windmills," and indeed the name appears very appropriate to the traveller who approaches the town on a windy day, and at a distance sees little save a multitude of great arms revolving furiously above and among the trees and house-tops.

Stockton is the debarking point for the travellers and merchandise on their way from San Francisco to all parts of the basin of the San Joaquin River, including the important mining counties of Mariposa, Tuolumne, and Calaveras. The entire population of the eight counties in the basin of the San Joaquin River is, according to the census of 1860, 60,837, scattered over an area of about 16,000 square miles. During the winter, at least in wet times, the roads leading out of Stockton are very muddy, but when the ground is dry, immense wagons are used in great numbers, hauling goods out to the mining camps and to Visalia. Some of these wagons have bodies sixteen feet long and six feet high, and are drawn by teams of eight or ten mules. Occasionally a smaller wagon will be fastened on behind the larger one, and then at any steep hill one wagon is hauled up at a time. During high water, a steamer runs up the San Joaquin River from Stockton to Fresno City, a distance of one hundred and twenty miles by the river. A steamer runs every day from San Francisco to Stockton, and stages leave the latter place every morning for the principal towns of the southern mines.

The first settlement on the place was made in 1844 by Charles M. Weber and Mr. Gulnac, the latter of whom obtained a grant of the land from the Mexican government in that year. They had some trouble with the Indians, and Gulnac sold out to his partner, who would not give the rancho up, and afterwards, when the place became important for its commercial advantages, he became the founder and father of the town, where he still resides. The name was selected in honor of Commodore Stockton, who commanded the American naval forces on this coast during the war with Mexico, and contributed much to the conquest of California. The town, like Sacramento and Marysville, was overflowed during the great flood of 1862, the water having covered all the streets on the 11th of January, and stood for days more than a foot deep, in the highest of them.

§ 276. *Marysville.*—Marysville, the third town in the state, containing a population of seven thousand, is situated between the Feather and Yuba Rivers at their junction. The site, like that of Sacramento, is flat, and in the midst of the large valley, and has been raised artificially above its natural level, to protect the houses against floods. Marysville resembles Sacramento, though smaller, and its residents claim that it is the handsomest town in the state. The first settlement was made in 1841 by Theodore Cordua, a German, who built a couple of adobe houses and called the place New Mecklenburg. In 1849 a number of persons went there, and the place was called Yubaville. In January, 1850, the town was laid off and named after Mrs. Mary Covilland, the wife of the chief proprietor. On the 31st of August and the 10th of September, 1851, two large fires occurred, destroying almost the whole town. In the spring of 1852 the business part of the town was covered with water, and the next year it was raised twelve feet. The town was again flooded in December, 1861, and January, 1862. Marysville is at the head of navigation on the Feather River. The distance by water is about seventy miles from Sacramento; by the stage road it is forty-five miles. From Marysville the counties of Nevada, Sierra, Butte, Plumas, and Yuba, which contain an aggregate population of 67,977, obtain all their supplies of imported goods. Stages run to the main towns of these counties every day.

§ 277. *Nevada.*—Nevada, the fourth town of the state, cast 1,423 votes at the last Presidential election, and has a population of about 4,500. It is the largest town in the mining districts, but owes much of its importance to its trade. It is on an important route of travel across the Sierra, and does an extensive trade with the adjacent mining counties of Sierra and Plumas. The immediate vicinity of Nevada is the most productive quartz district in the state. The town of Nevada was founded in 1849. Grass Valley, which lies only five miles distant, is nearly as large as Nevada. The latter place is sixty-nine miles distant from Sacramento.

§ 278. *Los Angeles.*—The town of Los Angeles, formerly called *Pueblo de los Angeles*, or the Pueblo de la Reina de los Angeles—the town of the Queen of the Angels—is the largest town in the southern part of the state, and has a population of about 3,500. It was founded about 1780, and was built up to nearly its present size previous to the American conquest. The population was probably nearly as large under the Mexican dominion as at present, but the finest buildings in the place have been erected within the last twelve years. The town is situated on the western bank of the Los Angeles

River, where that stream breaks through the range of low hills, twenty-five miles north of the bay of San Pedro. The streets are mostly of good width, but they are not straight; they do not cross each other at right angles; they are not graded, nor are they paved. All the old houses are built of adobes, and most of them are of one story, with flat roofs of asphaltum. The new houses are of wood and brick. On the northwestern side of the town, and very near to the most busy part of it, is a hill about sixty feet high, whence an excellent view of the whole place may be obtained. The vineyards and gardens are beautiful. There are 2,500 or 3,000 acres of brilliant green—the largest body of land in vineyard, garden, and orchard within so small a space in the state. The fences fix the attention of the stranger. They are made of willow-trees, planted from nine inches to two feet apart, the spaces between the trunks being filled with poles and brush. After the fences, the stranger's notice is attracted by the *zanjas*, or irrigating ditches, which run through the town in every direction. These *zanjas* vary in size, but most of them have a body of water three feet wide, and a foot deep, running at a speed of five miles an hour. They carry the water from the river to the gardens, and are absolutely necessary to secure the growth of the fences, vines, and many of the fruit-trees, at least when young. One of the officers of the town is the *zanjero*, whose duty it is to take charge of the *zanjas*, see that they are kept in order, and that the water is equally distributed among those entitled to it. Entering the enclosures, we are among the vines, orange, lemon, lime, citron, pear, apple, peach, olive, fig, and walnut trees. Many of the vines are from ten to thirty years of age. The population of the place may be described as of nearly four equal classes, Americans, Europeans, Spanish Californians and Indians. The Americans own most of the houses and land in the town, the Europeans probably do most of its trade. The town is the seat of the county government, and the chief business place in this part of the state. Its merchants trade largely to Salt Lake. Los Angeles has mail communication with San Francisco twice a month by steamer, and twice a week by stage. It is also connected with the main towns of the state by a line of telegraph. The general impression upon my mind, after spending the last week in September in the place, is that it is one of the most pleasant places in the world, known to me, to visit. The luxuriant vegetation, with its semi-tropical character, is peculiarly agreeable to the sons of the North. The "clime of the sun," "the land of the cypress and myrtle," "where the citron blooms and the golden oranges glow amidst the dark-green leaves, have ever been with the poets of the colder lands the symbols of a terrestrial paradise, and some of the most brilliant verses of Goethe and Byron have been inspired by admiration of them. The song of Mignon came vividly before me as I walked through the gardens of the City of the Angels. Luscious fruits, of many species and unnumbered varieties, loaded the trees. Gentle breezes came through the bowers. The water rippled musically through the zanjas. Delicious odors came from all the most fragrant flowers of the temperate zone. Julius Froebel speaks thus of Los Angeles in his book *Aus Amerika:* "I could wish no better home for myself and my friends than such a one as noble, sensible men could here make for themselves. Nature has preserved here, in its workings and phenomena, that medium between too much and too little, which was one of the great conditions of high civilization in the classic regions of ancient times. Indeed, when we seek in other lands for places like Los Angeles and Southern California generally, we must turn our eyes to the Levant. In the United States there are no kindred spots." The town is situated on the banks of the Los Angeles River, twenty-five miles from the ocean. The population numbers about 3,500. The port of Los Angeles is San Pedro, twenty-five miles southward, where there are only a couple of houses.

§ 279. *Petaluma.*—Petaluma, situated on the flat bank of the navigable creek of the same name, ten miles from San

Pablo Bay, has a population of about 3,500, and owes its prosperity to the trade of the valleys of Petaluma, Santa Rosa, and Russian River, and the plain of Bodega, which communicate through this place with San Francisco. It is on the land mail route which connects Mendocino and Humboldt counties with the centre of the state. The houses are mostly of frame, and none are more than ten years old, the town having been laid off in 1851. Petaluma sends more butter and cheese to the market than any other town in the state.

§ 280. *San José.*—San José may have about 3,500 inhabitants. It is situated in the Santa Clara valley, ten miles south-eastward from the bay, and fifty miles, by the stage road, from the city of San Francisco. The town was laid out about the beginning of the century, and many of the houses are of adobe, and were built before the American conquest. The streets are lined with shade-trees, and gardens filled with beautiful fruit-trees and flowers are abundant. The place is dusty during the summer, but otherwise is very pleasant. There are eleven hundred acres of orchard in the vicinity of San José, more than in any other equally small district in the state. One of the boasts of San José is the "alameda," which properly means "a place of elms," but is here applied to a road three miles long, lined with willow and cottonwood trees. The trees stand close together, and are of large size, so that they form a dense shade. Unfortunately, the road under them is extremely dusty in summer, and muddy in winter. The port of San José is Alviso, seven miles to the north, on the banks of the Guadalupe, or Alviso slough.

§ 281. *Santa Clara.*—Santa Clara, three miles westward of San José, and connected with it by the alameda, has a population of about 2,000. It is a new town, and nearly all the houses are of wood. The principal building is the old mission church, erected in 1822. It is now used as part of a Jesuit college. The mission of Santa Clara was founded in 1777, and a church was built on the bank of the Guadalupe Creek, at a place called "Socoistika," the Indian name of the laurel-trees which grew there. Two years later this building was swept away by a flood, and in 1781 a new church was commenced, half a league distant from the river, in a grove of oak-trees, the Indian name of which, "Gerguensen," was given to the vicinity. This church was destroyed by an earthquake in 1818. The present church was consecrated in 1822.

§ 282. *Columbia.*—Columbia, the largest town of the southern mines, is sixty-five miles from Stockton. The place has a population of about three thousand, and includes many good brick buildings. The town was laid off in 1850.

§ 283. *Placerville.*—Placerville, though it cast only nine hundred and thirty-four votes in 1861, is really, after Stockton, the most important town in the state, and its population will soon be proportionate to its business. It is on the main road from Sacramento to Washoe and Esmeralda, and is the point where travellers going eastward prepare for their trip over the rugged portion of the mountain, and those coming westward rest from their toils. It is also on the route of the great overland mail and telegraph. It may be that there is less snow and an easier natural grade on the other roads across the Sierra Nevada, but the Placerville road has been more improved than any other. It is more convenient to the great body of trade and travel from San Francisco and Sacramento. It has more houses to accommodate and protect travellers, and, during the winter, has more travel, and receives more attention, and therefore it is, for general use, the best road.

§ 284. *Yreka.*—Yreka, the county seat of Siskiyou county, the largest town north of Marysville, has a population of about twenty-five hundred. It is situated at an elevation of fifteen hundred feet above the sea, in the valley of Shasta River, about twenty miles northwest from Mount Shasta. It is a mining town, being situated in a rich mining district, and founded on pay-dirt. The place is shut in by high mountains, the Siskiyou ridge on the north, the Sierra Nevada on the east, the Scott and Trinity ridges on the south, and the coast range on the west. The town is shut in by snows during part of every winter, and there is frost in every month of the year. Most of the merchandise sent out from this point, to mining camps in the vicinity, goes on pack-mules. The goods imported by Yreka are hauled one hundred miles, from Red Bluff, which is one hundred and seventy-five miles from Sacramento by the river. The town is on the main road between the Sacramento and Willamette valleys, and occupies a central position in the basin of the Klamath River, and will therefore probably maintain its importance.

§ 285. *Vallejo.*—Vallejo, a town of about eight hundred inhabitants, is situated at the mouth of Napa River, on the northwestern side of San Pablo Bay, from which it is separated by Mare Island. The town was laid out in 1850 by M. G. Vallejo, for the capital of the state. He supposed at the time, and so did the public, that he was a millionaire. He owned large tracts of land, then estimated to be worth several millions of dollars at least. Among his possessions was the Suscol rancho, and he was induced to believe that if he would lay off a town and make a liberal offer of land and money to the state, the capital would be established there, and increase the value of his land so much that he would profit largely by the affair. The suggestion appeared reasonable, and he adopted it, offering much land and three hundred and seventy thousand dollars in cash for the establishment of the capital at Vallejo—the three hundred and seventy thousand dollars to be spent in erecting public buildings. The offer was accepted, and the capital was located at Vallejo, but the legislature went thither at a time when there were no houses there, and they immediately went away. Señor Vallejo did not pay the money which he had offered, and finally the capital was established at Sacramento, where it is likely to remain. The business of Vallejo now depends chiefly upon the United States navy-yard and dry-dock, on Mare Island. The place is one which may have much importance in the future of California. It has the following resources:

First.—It is at the head of navigation of the waters tributary to San Francisco Bay for large sea-going vessels. It has a fine harbor, perfectly protected against all winds, with good holding-ground, and extent enough to accommodate all the commerce which will ever visit it. That portion sufficiently deep at low water for vessels drawing four fathoms, is three miles long by a quarter of a mile wide, and for smaller vessels the harbor is much larger. The best chart of the harbor is No. 61 in the U. S. Coast Survey Report for 1857, which report can be found in all public and many private libraries. The navigation to the harbor from the Golden Gate, a distance of twenty-six miles, is excellent, the channel being wide and deep, and the winds strong and regular.

Secondly.—The town is in the midst of a fertile district. The country north of San Pablo and Suisun Bays is the richest agricultural district of the state. The valleys of Petaluma, Santa Rosa, Russian River, Sonoma, Napa, Suscol, Suisun, and Vaca, form a collection which, for extent of fertile soil, abundance of water, and mildness of climate, have no equal of like extent in California. These lie to the northwest, north, and northeast of Vallejo, while south and southeast lie the San Pablo, San Ramon, Amador, and Diablo valleys.

Thirdly.—Vallejo has a magnificent site for a town. The present village is built on the slopes of hills about a hundred or a hundred and fifty feet high, which rise from the harbor so gently, that a heavily laden wagon can be drawn over without an extra team. There are no deep gullies, and no sand. About half a mile back from the landing lies a beautiful and very fertile plain, called "Vallejo Valley," several miles wide and extending from the lower part of the harbor, upon which it opens, northward to Suscol. I have never seen a city provided with such a magnificent place for country residences as this.

Fourthly.—The town must be the main trading point of a large agricultural population. Suscol is now one of the best-tilled parts of the state.

Fifthly.—Vallejo has peculiar railroad facilities. She can have railroads running almost the entire distance on level plains to Petaluma, Santa Rosa, Healdsburg, Sonoma, Napa, Fairfield, Marysville, Sacramento, Stockton, Amador Valley, Oakland, San José, etc.

Sixthly.—She has a climate similar to that of San Francisco, where ice never forms and summer clothes are never worn—precisely that degree of constant coolness most favorable to continuous labor, both mental and physical.

Seventhly.—She has the U. S. Navy-Yard on Mare Island at her doors, and must supply the laborers to do the work there. This navy-yard must always be of great importance, and the government works alone will build up a considerable town. Puget Sound is the only other point on the American coast of the Pacific where a navy-yard can be established, and a long time may pass before one will be required there.

Eighthly.—There is a possibility that the difficulties in the navigation of the Sacramento River, below Sacramento City, will make it necessary to connect that place and the valley north of it with the deep water, by a railroad to Vallejo.

It is estimated that forty thousand tons of dirt are carried down into the Sacramento River every day from the mines, and the effect to fill up the bed of the river is proved by past experience. The time must come when goods must be carried from the bay to the central and northern mines by land, and the sooner the preparations are made for this approaching change, the better. That land shipment, when commenced, must go through Vallejo.

Ninthly.—In San Francisco great damage is done to the wharves by the shipworm; in Vallejo the water is so fresh that wharves and boats are secure against that scourge.

§ 286. *Visalia.*—Visalia is situated in the "Four Creek country," about fifteen miles northeastward from Tulare Lake. The "Four Creek country" is formed by Cahuilla Creek, which, after leaving the Sierra Nevada, spreads out into a number of channels, and these again subdivide, and moistening a considerable district of rich soil, render it very productive. Visalia has a population of one thousand five hundred, and is the largest town in the San Joaquin basin southward of Stockton. It has now the trade of a radius of nearly a hundred miles, and having a central situation will no doubt maintain its relative importance. The town was overflowed with water several feet deep by the great freshet of January, 1862.

§ 287. *Red Bluff.*—Red Bluff, at the head of navigation on the Sacramento River, one hundred and seventy-five miles above Sacramento City, is a place of about one thousand inhabitants. It is the point at which the merchandise is landed from the Sacramento steamers for Shasta, Siskiyou and Trinity counties, with an aggregate population of 17,114. Sometimes the river is so low that the boats cannot reach Red Bluff, but this is a rare event. There have been attempts to remove obstructions above Red Bluff so that the head of navigation should be at Cottonwood, but these attempts have hitherto been vain.

§ 288. *Martinez.*—Martinez is situated on the southern side of the Straits of Carquinez, in a little valley. The population is about six hundred. It is a quiet, pleasant village, where trade is dull and life slow. It is connected with the world by a steam ferry-boat, running to Benicia. It would probably be a much more prosperous place if the title to the land on which it is situated were clear, but it is claimed under three grants, all of which have been confirmed, and nobody knows which one will take it. The town has a peculiar climate, with the same cool temperature as at Benicia, very much like that of San Francisco, but fogs are rare, and the winds are broken by hills. It is a climate similar to that of Oakland, without the wind, and is therefore very favorable to fruit. The sea-breeze pouring through the Straits of Carquinez, keeps off the frost, and some unexplained cause throws all the fog over to the northern side of the strait, thus giving the fruit on the southern shore an abundance of sunshine. The town was founded about the time the gold was discovered.

§ 289. *Pacheco.*—The town of Pacheco, or Pachecoville, is one of the newest in the state, having been founded late in

1858. It is built at the head of navigation of the Pacheco Slough, and is the shipping port of Pacheco, San Ramon, Diablo and Taylor valleys. The distance to Martinez is four miles, further than farmers like to haul their grain, when they can avoid it. To bring the shipping point nearer to them, Pacheco was built. The distance from Pacheco to the bay, in a straight line through the tule land, is about four miles; the distance by the slough is six miles. The slough is bare at low water; at high water it is navigable for sloops and schooners drawing six feet. In 1859 Pacheco shipped 180,000 sacks of grain. The population is about six hundred.

§ 290. *Suisun.*—Suisun, a village of about sixty houses, is on the western bank of the Suisun Slough in Solano county, about ten miles in a direct line from Suisun Bay and sixteen miles by the slough. The place was commenced on a little island, a couple of hundred yards in diameter, and no part of it more than a foot above the highest tide. It is surrounded by tules, or salt-water rushes, growing on land overflowed at every high tide and bare at low tide. Two roads lead from the dry land of the valley to the city—one of them a plank-road, now in a very dilapidated condition. Most of the streets are subject to overflow by spring tides, and the marks of the water can be seen upon them even when dry. A few lots have been raised above high tide by bringing earth from other places, and enclosures are made by digging ditches, in which the water is never more than two feet below the surface. The island, being in the tule, was not included in the Suisun grant, and it was claimed, in 1853, by two men who laid off the town, and who are now in litigation with each other about the unsold land. The place owes its importance to its advantages as the shipping point of the valley. The Suisun Slough is said to be the best slough in the state; that means, that it is wider and deeper than the sloughs through which vessels reach Stockton, Sonoma, Napa, San Antonio, Petaluma, Pacheco, and Alviso. Vessels drawing nine feet of water can enter the slough. The town is sixty-five miles from San Francisco, with which city there is regular steamboat communication. The town is one mile from dry land, on the edge whereof, immediately north of Suisun, lies Fairfield, which is the county seat, and has about five hundred inhabitants. Fairfield is on the route of the projected Benicia and Marysville railway.

§ 291. *Benicia.*—Benicia is situated on the northern side of the Straits of Carquinez. It was laid out in 1847, and for a time it aspired to be the great commercial city of the coast, which aspiration it did not abandon until as late as 1853. It was twice made the state capital, and twice deserted by the legislature. It has about one thousand inhabitants. The houses are scattered about so far from each other that the town is called, in sport, "The City of Magnificent Distances." A ferry-boat crosses the strait to Carquinez about six or eight times every day. The town site is composed of low bare hills. The climate is very windy.

§ 292. *Napa.*—Napa was laid off in 1848 by Nathan Coombs at the ford of Napa River, on the road from Benicia to Sonoma. In those days there were no bridges or ferries, and the position of the ford determined the location of the town. Now the ford is never used, but the investment of capital has made the town permanent. If mere natural advantages were to be taken into account, the town would be at Suscol, which is six miles nearer to the bay, and always accessible by small steamers, while at low tide the boats must stop several miles below Napa. The population of the place is about one thousand. The houses are of wood and brick, and neatly built.

§ 293. *Crescent City.*—Crescent City is a seaport of six hundred inhabitants, fifteen miles south of the Oregon line. The place was founded in 1853, with the expectation that because of its proximity to the mines of the Klamath and Rogue River basins, it would become an important commercial point for the imports of Southern Oregon and Northern California. Its founders, however, were disappointed in this expectation.

The people at the head of the Sacramento valley, knowing that an attempt was making to cut off a large part of their trade, went to work industriously and made a good wagon road to Yreka, and thus reduced the freights to that place very much. The country westward of Yreka is very rugged, and as the people of Crescent City had not the capital to make a wagon road, their goods had to be transported at much expense on mules; and Yreka and vicinity continue to make their imports and exports by way of the Sacramento River. Crescent City, therefore, remains a small place, but it supplies a district within a range of forty or fifty miles to the east and northeast. Trinidad is a small seaport which is the chief trading point of the miners in Klamath county.

§ 294. *Arcata.*—Arcata is a town of eight hundred inhabitants, at the northern end of Humboldt Bay. It was founded in 1850 with the expectation that the miners in the basin of the Trinity River would obtain all their imports through it; but the only means of conveying goods from Humboldt Bay was a bad mule trail through a rugged country infested by hostile Indians, so the trade continued to go by the way of the Sacramento River. It is probable, however, that the current of this trade will be changed within a year or two. A wagon road is being made to Weaverville, a distance of eighty miles, and the Indians have been driven away. At the present time fifteen hundred pack-mules are used in conveying goods from Arcata to the various parts of Humboldt, Klamath, and Trinity counties. The name of Arcata was adopted in 1860; previous to that time the town was called Union.

§ 295. *Anaheim.*—Anaheim is the only German town in the state. It was laid out by Germans, built up by Germans, and is populated and owned by Germans. But it will never have the foreign character which marks many German villages in the valley states of the Mississippi, where the English language is not known to any of the people. None of the Anaheimers have come direct from Germany; all of them have lived for some time among the Americans, and most of them speak English fluently. The English language will be the predominant tongue, although German will long be cherished. The number of residents is now three hundred, with a certainty that they must rapidly increase. Anaheim is a tract of land a mile wide by a mile and a half long, in the valley of the Santa Anna River, Los Angeles county. It was unoccupied, and supposed to be of little value in 1857, when it was bought for two dollars an acre by a German company of fifty members, mostly residing in San Francisco. They were incorporated as a joint-stock association. The land, containing one thousand one hundred and sixty-eight acres, was divided into fifty lots of twenty acres each, with a little town-plat in the middle, and convenient streets. The place was given in charge of a superintendent, who held his position two years, in which time he planted and cultivated eight acres of every lot with vines, and put willow hedges (nearly all the fences in Los Angeles county are of willow) all around the outer boundary of the tract, and along the principal streets inside. During a large part of the time he hired fifty laborers. The total expense for the two years was seventy thousand dollars, or one thousand four hundred dollars per lot of twenty acres, including eight acres of vine. The owner of a vineyard lot had a little town-lot of half an acre besides. In December, 1859, the property was divided by lot among the members, many of whom have now removed to the place and made their home there. There are nearly six hundred acres lying vacant, and the welfare of the vineyards requires that this land should be cultivated, for now it is covered with weeds and brush, and is the home of innumerable hares, squirrels, and gophers, which eat the vines, young trees, and grapes. When cultivated and irrigated, these pests will be drowned out and driven off, and the labor of the vine-grower will be much reduced upon the land now under tillage. When the whole tract shall be filled with bearing vines, it will produce twice as much wine as the town of Los Angeles does now, and nearly as much as that town will be able to produce when all its present vines shall be in full bearing—for about half of its million vines are still less than four years of age. Anaheim has some advantages over Los Angeles, in the regularity of its plan, and perhaps, also, in location (for it is nearer the ocean, and farther from the snowy mountains) and in the extent of rich land in its neighborhood, and in its location near the direct line of travel between San Pedro and San Bernardino, the latter being not only an important place for its own trade, but still more for its trade with Salt Lake. On the other hand, Los Angeles has the start, the capital, the houses, the merchants, and all those advantages which an old and prosperous town has over a little upstart of a village, and Anaheim must long, if not always, be a place of inferior importance, as compared with it.

§ 296. *Monterey, etc.*—Monterey, which previous to 1849, was the political capital and commercial centre of the territory, is now one of the least important towns of the state. Most of the houses are of adobes. The population is about one thousand.

Sonoma, which was founded in 1823, and was, up to 1850, the most important town north of the bay of San Pablo, has been gradually losing its trade and population; but the extensive production of wine in which its citizens are now engaged, may bring it up again. The houses are mostly of adobes. The tallest adobe house in the state, containing three high stories, is in Sonoma. It was commenced in 1835, and was never finished. After the walls had been erected, they were covered by a roof which projected three feet on all sides, and thus the building has remained. The walls are about three feet thick. The old buildings of the Mission of San Francisco Solano are in a very dilapidated condition. Sonoma is ten miles distant from San Pablo Bay, but the Sonoma Slough is navigable to the Embarcadero, which is only three miles from the town.

San Diego has a population of about eight hundred. The town once aspired to be the terminus of a railroad, which was to cross the continent about latitude 32°, but such aspirations are now pretty nearly abandoned. The harbor is good, but the land in San Diego county is not rich, and it may be said that the town has no brilliant destiny before it, at least not in the near future. The Mission of San Diego was the earliest in California and was founded in 1769.

In Santa Barbara and San Luis Obispo, nearly all the inhabitants are Spanish Californians, and nearly all the houses are of adobes. Both are small, dull places, founded in the last century, and the sites of old missions. The San Luis Obispo Mission was founded in 1772, that of Santa Barbara in 1782.

§ 277. *Mining Towns.*—Weaverville, Shasta, Oroville, Quincy, Auburn, Downieville, Mokelumne Hill, Jackson, and Mariposa, are large mining towns, and are the county-seats of various mining counties, but they possess no features of sufficient importance to entitle them to separate notice here. The mining towns of California, as a general rule, are built in cañons, with irregular crooked streets. The main business street runs through the middle of the cañon and is densely lined with stores, billiard rooms, liquor shops, restaurants, and inns. Many of those houses are of brick and fire-proof. The dwellings are scattered about irregularly; some are neatly built and are surrounded with pleasant gardens; the majority are miserable little shanties or log-cabins, with no yard, flowers or fruit-trees to give an appearance of home. The population of these towns is not permanent. One year they are here, next they are elsewhere. In 1854 Oroville was laid out; in 1857 it cast one thousand votes, and was the third town in the state; in 1860 its glory had departed, and at least a dozen towns have now a larger population and a larger trade.

§ 298. *General Remarks.*—In taking a general view of the towns of California, we perceive that all those of the most note are exclusively American. San Francisco, Sacramento, Marysville, Stockton, Nevada, Grass Valley, Placerville, Columbia, Petaluma, Yreka, Shasta, Sonora, Oakland, Vallejo, Santa Clara, Downieville, Weaverville, Mokelumne Hill, Auburn, Oroville, Jackson, Santa Cruz, Santa Rosa, Healdsburg,

Oscata, and Crescent City, are American in their origin and population. Not one of the old Spanish towns, save San José, has gained any thing in population since 1846. Los Angeles, Sonoma, Monterey, Santa Barbara, San Luis Obispo, and San Diego have just about the same number of inhabitants now as previous to the American conquest.

CHAPTER XIII.

TOPOGRAPHICAL NAMES.

§ 299. *Introductory.*—The topographical names of California differ much from those of other states in the Union, where there is a disagreeable repetition of familiar names. Our people have not attempted to immortalize Franklin, Jefferson, Madison, Adams, Henry, Randolph, Clay, Cass, Benton, Webster, Taylor, Fillmore, Polk, Pierce, or Buchanan, by affixing their tiresome patronymics to counties or towns. All our prominent places are designated by titles comparatively new to the English language, and strange to Americans.

The topographical names of the state are derived from three languages—Spanish, English, and Indian. Most of the names along the southern coast and about the bay of San Francisco—districts which were populated by the Spaniards long before the Americans came to the country—are Spanish. The larger rivers in the Sacramento basin were known to the Spaniards, and were named by them previous to 1846. The mining districts of the Sierra Nevada and the Klamath basin, and the coast north of 40°, were first explored and settled by the Americans, and therefore the names are of English origin. The Indian names are numerous.

§ 300. *Sacred Spanish Names.*—The Spanish names may be divided into the sacred and profane. The first Spanish settlers were Catholic missionaries, in whose almanac every day is named after some saint, and in whose faith the saints were but little below divinity. It was customary for them to keep the saints constantly in mind, and when they came to a strange place, to name it after the saint upon whose day they had reached it. Thus it is that nearly all the settlements made by or under the missionaries are sanctified.

The male saints have "San," the females "Santa" to precede their Christian names, as in English we have "St." Some uneducated Americans corrupt the "San" or "Santa" before certain Spanish names into "St.," and say "St. Francisco," "St. Lucas;" but the more intelligent Americans adhere strictly to the Spanish spelling and pronunciation of topographical names.

The Missions were all named from saints or sacred dogmas. There are San Miguel, San Gabriel, and San Rafael (from the three highest angels, Michael, Gabriel, and Raphael), San Juan Bautista and San Juan Capisteano (St. John the Baptist and St. John of Capisteano), San Francisco de Assisi and San Francisco de Solano, San Luis Rey and San Luis Obispo (St. Louis the king and St. Louis the bishop), San Carlos, Santa Clara, Santa Barbara, San José (St. Joseph), Santa Inez Virgen y Martyr (St. Inez the virgin and martyr), San Antonio de Padua (St. Anthony), San Fernando Rey (St. Ferdinand the king), San Buenaventura (St. Good Fortune), La Purisima Concepcion (the Most Pure Conception), Nuestra Señora de Soledad (our Lady of Solitude), and Santa Cruz (the Holy Cross).

Among the saints whose names are applied to places not missions, are San Pedro (Peter), San Pablo (Paul), San Mateo (Matthew), San Andres (Andrew), San Marcos (Mark), San Simeon, San Joaquin (Joachim), San Nicolas, San Clemente, San Lorenzo (Lawrence), San Leandro (Leander), San Pascual, San Ramon, San Felipe (Philip), San Cayetano (Cayetan), Santa Marta (Martha), Santa Maria, Santa Paula (Pauline), Santa Rosa, Santa Isabel, Santa Margarita, Santa Catalina, Santa Susana, Santa Lucia, and Santa Gertrudis. Other Spanish sacred names, not derived from saints, are Trinidad (Trinity), Sacramento (Sacrament), Jesus Maria (Jesus the Son of Mary), and Nuestra Señora La Reina de los Angeles (Our Lady the Queen of the Angels).

§ 301. *Profane Spanish Names.*—Among the Spanish profane names are Agua Fria (cold water), Agua Caliente (hot water, or warm spring), Vallecito (little valley), Esperanza (hope), Campo Seco (dry field), Garote, Hornitos (little ovens), Salinas (salt places), Alameda (a place of elms or cottonwood trees), Saucelito (a little clump of willows, more properly spelled Sauzalito), Bodega (a vault), Laguna Seca (dry lagoon), Cienaga (puddle), Merced (mercy), Buena Vista (good view), Contra Costa (the opposite coast, the shore opposite the bay of San Francisco), Del Norte (of the north), Plumas (feathers), Tulare (a place of tules), El Dorado (the golden land), Fresno (ash), Nevada (snowy), Sierra (mountain chain), Placer (gold diggings), Calaveras (skulls), Mariposa (butterfly), Alcatraz (pelican), Farallones (points of rock in the sea), Corte Madera (place where wood is cut), Monte (the mountain or forest), Loma Prieta (black hill), Monte Diablo (the devil's mountain), Montecito (little mountain or little forest), Alamo (elm or cottonwood tree), Alamo Mocho (the cropped cottonwood), Pájaro (bird), Coyote and Tejon (a badger). Some of these names have been changed by the Americans. The Spaniards say, el Rio de las Mariposas (the river of the butterfly), el Rio de las Calaveras, el Rio de los Pájaros, la Isla de las Alcatraces, la Bahia de San Francisco (the bay of San Francisco), La Mision de San Gabriel (the Mission of San Gabriel), el Rio de las Salinas. The Americans drop the common Spanish nouns of *rio*, *bahia*, and *mision*, and say Calaveras River, Salinas River, the Mission San Gabriel, etc. Though the plural form of Calaveras and Salinas has been preserved, the singular has been adopted for Pájaro River, Alcatraz Island, and Coyote Creek. Pájaro River was so named because of the great number of wild geese and ducks which were formerly seen in its valley. Cape Mendocino was named after the patronymic of an early Spanish navigator on this coast. Amador county and Amador valley were named after José M. Amador, who was formerly manager of the property of the Mission of San José, about 1835. He lived in Amador valley, and in 1848 he went with a number of

Indians to mine in what is now Amador county. Vallejo, Pacheco, Martinez, and Alvarado, are the names of prominent men among the Spanish Californians. Some Spanish names have been changed into English. The American River was formerly called *el Rio de los Americanos*, because the Americans entering California usually came down the banks of that stream. The Feather River was called *el Rio de las Plumas*, the river of feathers. The "Plumas," after having been abandoned as a designation for the river, was given to the county in which the river takes its rise. The Yuba River was called by the Spaniards, *el Rio de las Uvas* (the river of the grapes). The ignorant Spaniards wrote the main word *Ubas*, the *b* and *v* being often confounded in the Castilian tongue. The Americans gave the English pronunciation to the initial *u*, then wrote it "Yubas," as they pronounced it, and finally changed it to the singular. Angel Island was formerly called *la Isla de los Angeles*, and Mare Island was called *la Isla de las Yeguas*. The town of Benicia was laid off in 1846, and was first called "Francesca," one of the Christian names of the wife of M. G. Vallejo, on whose land the town was to be built; but in March, 1847, the name of the town of Yerba Buena was changed to San Francisco, and the projector of Benicia, Mr. Charles D. Semple, thought it necessary, for the purpose of avoiding confusion, to change the name of his city on paper, so he adopted "Benicia," another name of Mrs. Vallejo. The town of Sonora was so named because the majority of the first miners there were from Sonora. The New Almaden quicksilver mine, for some months after the nature of the ore was discovered, was called *la Mina de Santa Clara*. Its present name was derived from the great quicksilver mine of Almaden, in old Spain. The Enriqueta quicksilver mine was named after Enriqueta (Henrietta) Lawrence, the daughter of the managing owner of the rancho at the time the mine was discovered.

§ 302. *Indian Names.*—The Indian names in California are numerous. Among them are Siskiyou, Klamath, Shasta, Tehama, Colusi, Yolo, Napa, Sonoma, Mokelumne, Tuolumne,

Chowchilla, Cahuilla, Suscol, Suisun, Cosumnes, Temécula, Temascal, Jurupa, Petaluma, Tomales, Yreka, Ukiah, Cuyama, Cocomongo, Mayacmas, Bolbones, Guilicos, Huichica, and Hoopah. Most of these are the names of tribes of Indians. The Mokelumne, Tuolumne, Chowchilla, Cahuilla, and Cosumnes Rivers were called by the Spaniards el Rio de los Moquelumnes, el Rio de los Tuolumnes, etc. The second syllable of Moquelumne was changed by the Americans, to be spelled with a *k*, which has the same sound as *qu* before *e* in Spanish. Cahuilla is sometimes vulgarly spelled "Kaweah" by Americans, who thus represent the Spanish pronunciation as nearly as possible. Klamath and Shasta were formerly written "Tlamath" and "Tshastl." Sonoma, by some persons written "Zonoma" in early times, is an Indian word meaning "valley of the moon." Temascal means an Indian sweat-house. Solano is a Spanish word meaning the south wind, but Solano county was so called after the chief of the Suisun tribe of Indians. I have not been able to learn whether his name was given to him by the Spaniards, or was of Indian origin. Marin county was also named after an Indian chief. Yreka is a corruption of Wi-é-kah, which means whiteness, and is the Indian name of Mount Shasta, at the foot of which the town is situated.

§ 303. *American Names.*—Now we come to the American names. Towns are named after Jackson, Washington, Lafayette, and Stockton (the last was in command of the American navy on this coast during the Mexican war). Bigler, one of the governors of the state, has been immortalized by having his name affixed to a lake. The patronymics of Alexander Humboldt and J. A. Sutter are affixed to counties. Trinity River was so named because the white men who discovered it in the mountains, supposed it emptied into the bay of Trinidad, which had been discovered by the Spaniards several centuries ago. Marysville was first called Yubaville, and then named after Mrs. Mary Covilland, one of the founders of the place. Among the pioneer miners of Calaveras county were Murphy, Angel, and Carson, and they became the eponyms (to use a word coined by Mr. Grote) of the places where they stopped, first called Murphy's Camp, Angel's Camp, and Carson's Camp, now become permanent towns, which have discarded the "camp," and assumed the titles, "Murphy's," "Angel's," etc. It is better to drop the *s* and the apostrophe, as is sometimes done. "Yankee Jim's Camp"—the surname of "Jim" was never known to the general public—is now simply Yankee Jim. Messrs. Downie, Weaver, and Heald were the respective eponyms of Downieville, Weaverville, and Healdsburg; and Folsom was named after the owners of the rancho on which it was laid out. The knowledge or supposition of rich diggings is indicated by some of the names of towns. For instance, Ophir, Gold Hill, Quartzburg, Placerville, Oroville, Rich Bar, and Tin Cup. Placerville was, in 1849, called Hangtown, because it was the first place where any person was hanged by Lynch law. Oroville is a compound of *oro*, the Spanish word for gold, and *ville*, the French word for city. Tin Cup was so named because the first miners there found the placers so rich that they measured their gold in pint tin cups. Many of the bars and camps in the mining districts are named after the discoverers or first settlers. There are Scott's Bar, Long's Bar, Kelly's Bar, Kanaka Bar, Negro Bar, Chinese Camp, etc. Other places are named from the native places of the first settlers, as Mississippi Bar, Ohio Bar, Iowa Hill, Michigan Bluffs, Illinoistown, Alleghanytown, etc. Pine Log is so named because there was, in early times, at that place a pine log across the South Fork of the Stanislaus River, in such a position as to offer a very convenient crossing to miners. Some of the mining camps are named from tragic events which occurred there: thus there is a Murderer's Bar, a Dead Man's Bar, and a Dead Shot Flat. The following is a list of some curious names of mining localities:

Jim Crow Cañon,	Happy Valley,	Ground-Hog's Glory,
Red Dog,	Hell's Delight,	Bogus Thunder,
Jackass Gulch,	Devil's Basin,	Last Chance,
Ladies' Cañon,	Dead Wood,	Greenhorn Cañon,
Miller's Defeat,	Gouge Eye,	Shanghai Hill,
Loafer Hill,	Puko Ravine,	Shirttail Cañon,

Guano Hill,
Rattlesnake Bar,
Whiskey Bar,
Poverty Hill,
Greasers' Camp,
Christian Flat,
Rough and Ready,
Ragtown,
Sugar-Loaf Hill,
Poker Flat,
Wild-Cat Bar,
Dead Mule Cañon,
Wild Goose Flat,
Brandy Flat,
Gridiron Bar,
Hen-roost Camp,
Lousy Ravine,
Lazy Man's Cañon,
Logtown,
Git-up-and-git,
Gopher Flat,
Stud-horse Cañon,
Bob Ridley Flat,
One Eye,
Push-coach Hill,
Puppytown,
Mad Cañon,
Slap-Jack Bar,
Quack Hill,
Pepperbox Flat,
Nigger Hill,
Seventy-six,
Piety Hill,
Hog's Diggings,
Brandy Gulch,
Liberty Hill,
Love-Letter Camp,
Paradise,
Blue Belly Ravine,
Sluice Fork,
Shinbone Peak,
Seven-up Ravine,
Loafer's Retreat,
Humpback Slide,
Swellhead Diggings,
Cayote Hill,
Poodletown,
Yankee Doodle,
Horsetown,
Petticoat Slide,
Chucklehead Diggings,
Mount Zion,
Barefoot Diggings,
Plug-Head Gulch,
Skunk Gulch,
Coon Hollow,
Poor Man's Creek,
Humbug Cañon,
Bloomer Hill,
Grizzly Flat,
Rat-Trap Slide,
Pike Hill,
Port Wine,
Snow Point,
Nary Red,
Gas Hill,
Ladies' Valley,
Graveyard Cañon,
Gospel Gulch,
Chicken-Thief Flat,
Hungry Camp,
Mud Springs,
Skinflint,
American Hollow,
Gold Hill,
Pancake Ravine,
Centipede Hollow,
Nutcake Camp,
Seven-by-nine Valley,
Paint-Pot Hill.

Butte county was named from the *buttes* or high hills on its border. Cache Creek was probably so called because some trappers buried or *cached* something on its banks many years ago. *Butte* and *Cache* are words of French origin, introduced into the English language by trappers.

Anaheim is derived from Ana, the Spanish for Ann, and the German word *heim*, meaning home—and the compound means Anna's home. The Ana was suggested by the Santa Ana valley, in which Anaheim is built.

§ 304. *Etymology of "California."*—There is much doubt about the etymology of the word California. Some authors contend that it is a compound, derived from the Latin words *calida* (hot) and *fornax* (furnace). The name was first bestowed in the sixteenth century, by Spanish navigators, upon Lower California, the southern point of which reaches into the tropics; and being a dry, treeless, bare, and desolate country, it may well have appeared to them to be hot as a furnace. Some persons, however, suppose the word to be of Indian origin. The Spaniards and Mexicans called the peninsula *Baja California*, or Low California; also, *Antigua California*, or Old California; and the coast further north, *Alta California*, or High California; also, *Nueva California*, or New California. The two were called *Las Californias*, or the Californias. The state constitution was framed in 1849, and commences, "We, the people of California," etc. This, therefore, is *the* California, and the peninsula south of us is not meant or thought of, unless we use the adjective prefix, and say Lower California.

§ 305. *Pronunciation of Names.*—In the pronunciation of the names of Spanish and Indian origin, the letters have usually the Spanish sounds. *A* is like "a" in far; *e* like "a" in fare; *i* like "ee" in meet; *o* like "o" in go; *u* like "oo" in fool. *H* is silent; *j* and *g*, before *e* and *i*, have a sound similar to that of the English "h;" *s* never has the sound of *z*, but is always like "ss" in hiss. *Qu*, before *e* and *i*, is like "k." *Ll*, is "lli" in William; *ñ* is like "ni" in union. There are no diphthongs in Spanish. Every vowel is sounded separately. Words ending in a vowel in the singular, have the accent on the syllable next the last; those ending in a consonant, on the last. In case any vowel has an accent marked over it, then that vowel has the accent. The Spaniards of old Spain pronounce the *z* before all vowels, and the *c* before *e* and *i*, like "th" in thick; but the Mexicans give them the sound of *s*.

The errors which Americans most frequently commit in pronouncing Spanish words are, in giving to *a* the English sounds of "a" in fat and fate; giving to *s* the sound of "z;" to *j* and *g*, before *e* and *i*, the same sounds as in English; to *gu* the sound of the English "w;" and in putting the accent on the first syllable—English fashion. The following may serve as a further guide to the proper pronunciation of some of the names:

Spanish Names and Pronunciation.	Spanish Names and Pronunciation.
Diego—dee áy go.	José—ho sáy.
Suisun—soo ee sóon.	Jesus Maria—hay sóos mah reé ah.
Alameda—ah lah máy da.	Puta—póo tah.
Sierra—see ér ra.	Tejon—tay hóne.
Nevada—nay váh dah.	Farallones—fah rahl yó nes.
Mateo—mah táy o.	Gabriel—gah bree ále.
Monterey—mon ta ráy ee.	Rafael—rah fah ále.
Luis Obispo—loo éess o bées po.	Miguel—mee gále.
Los Angeles—loce án gel es.	Pájaro—páh hah ro.
Vallejo—val yáy ho.	Coyote—co yó tay.
Vallecito—val yay thée to.	Pacheco—pah cháy co.
Joaquin—ho ah kéen.	Cahuilla—cah oo eél ya.
Juan Bautista—hwan bah oo téestah.	Querétaro—kay ráy tah ro.

This table is not a perfect guide to pronunciation, but only an approximation.

Placer has been anglicized so much that it is commonly spoken with the accent on the first syllable. Mokelumne and Tuolumne have the accent on the antepenultimate and the vowel short. Siskiyou has the accent on the first syllable. Sutter is pronounced with the *u* like "oo" in foot. Mokelumne is often mispronounced Mac ál a my, and the Cosumnes River is not unfrequently called the Macosme. Folsom is pronounced like the adjective fulsome. Yosemite has four syllables with the accent on the antepenultimate (Yo sém i te).

CHAPTER XIV.

PRESENT AND FUTURE DEVELOPMENT OF THE STATE.

§ 306. *General Summary.*—Twelve chapters of this book have been filled with a detailed statement of the nature and characteristics of the resources, industry, and society of CALIFORNIA. In this chapter, I shall present a summary of their main features.

We have, then, before us a state, lying in the midst of the temperate zone, on the western coast of North America; bounded on one side by the Pacific Ocean, and on the other by a high range of mountains; reaching through nine degrees of longitude and three of latitude; with a coast-line eight hundred miles long, and a total area of about one hundred and sixty thousand square miles. The heart of the state is drained by two large rivers, which run from north and south, unite midway, and in their course to the sea form three large and deep bays, with secure and spacious harbors. On these bays and their tributaries, there are nearly one thousand miles of navigable streams now used by steamboats and sailing-vessels.

The climate near the ocean is the most equable in the world. At San Francisco, there is a difference of only seven degrees between the mean temperatures of summer and winter—the average of the latter season being 50° and of the former 57° Fahrenheit. Ice and snow are never seen in winter; and in summer the weather is so cool, that heavy woollen clothing is worn every day. There are not more than a dozen days in the year too warm for comfort at mid-day, and the oldest inhabitant cannot remember a night when blankets were not necessary for a comfortable sleep. The climate is just of that character most favorable to the constant mental and physical activity of men, and to the unvarying health and continuous growth of animals and plants. In the interior, the summers are much warmer than near the ocean; while in the mountains the winters are much colder. By travelling a few hundred miles, the Californian can find almost any temperature that he may desire—great warmth in winter, and icy coldness in summer.

The rocks of the state are chiefly granite and tertiary sandstone; the former occupying the high mountains, the latter the valleys. In former eras there were several, or perhaps many volcanoes in the range of the Sierra Nevada. Mount Shasta was one of them, and it now has hot springs on its summit, and sends up sulphurous vapors. On the western slope of the Sierra Nevada, about half way between the summit and the foot, are numerous beds of slate and veins of quartz. The same formations are found in the Klamath basin and in other parts of the state; and in nearly every case they are auriferous. There is scarcely a county which does not contain gold. The districts which contain enough gold to support a mining population, have an area of about ten thousand square miles. The gold-yield of the state is about forty-three million dollars annually—more than that of any other country, save the colony of Victoria, in Australia.

The number of men engaged in mining may be estimated at eighty thousand. Our placers and auriferous quartz veins are almost inexhaustible; there are great mountains of gold-bearing gravel which cannot be washed away for a century to come; and the quartz-lodes will last still longer.

The gold-mining of California is conducted in the most thorough and enterprising manner. Although the main principles of the sluice and the hydraulic washing were known and used, on a small scale, long before the discovery of gold in California, it was here that those modes of working were first perfected, applied on an extensive scale, and brought into universal use. Large rivers are turned out of their beds; mountains are pierced by tunnels; hills are washed away; and the rivers roll thick with mud to the sea through summer and winter.

The silver-mines of the state were discovered only a short time ago, and their value is not yet fully known; but that some of the ore is wonderfully rich, is established beyond a doubt. The silver districts are in the basin of Utah, at an elevation of five thousand feet or more above the level of the sea, in the midst of a desert country.

In quicksilver, California is the richest country in the world. There are extensive beds of sulphur, asphaltum, and plumbago, and large lakes and springs impregnated with borax.

The natural scenery of California is varied and grand. The Yosemite valley is a chasm ten miles long, two miles wide, and three thousand feet deep, in the heart of the Sierra Nevada, without its equal in the world for sublime and picturesque scenery. It has a dozen great cascades, the highest of which has a fall of thirteen hundred feet. The Mammoth Trees are the largest known growths of the vegetable kingdom. There are likewise in the state mud-volcanoes, natural bridges, many caves, and numerous hot and mineral springs, some of which throw out great columns of steam.

The animals and plants of California are peculiar to this coast. The finest group of coniferous trees in the world is that of this state. The mammoth tree, the redwood, the sugar-pine, the red fir, the yellow fir, and the *Thuja gigantea*, all reach the wonderful height of three hundred feet; the mammoth tree grows to be thirty feet in diameter, the redwood twenty, and the others from eight to twelve.

The grizzly bear is the largest and strongest indigenous animal of the continent; and the Californian vulture is, next to the condor, the largest bird that flies. The sea near our coast teems with halibut, turbot, mackerel, herring, sardines, anchovies, and smelts; while sturgeon and salmon are abundant in our rivers.

There are 40,000,000 acres of tillable land in the state, but not more than 1,000,000 acres are now cultivated. In 1860, the aggregate product of grains and roots of annual growth amounted to 14,470,000 bushels, being an average of twenty-four bushels to the acre cultivated, and of thirty-eight bushels to each inhabitant of the state. The crop of barley was the largest, measuring 5,700,000 bushels; that of wheat, 5,000,000 bushels; oats and potatoes, each, 1,500,000 bushels; and maize, 500,000 bushels. The barley forms thirty-nine per cent. of the 14,470,000 bushels; wheat, thirty-four per cent.; oats and potatoes, each ten per cent.; maize, three per cent.; and beans, peas, sweet potatoes, buckwheat, and rye, one-half of one per cent. each.

Farmers in California have many advantages over men of the same occupation in other parts of the United States. The winter is never so cold as to interrupt their work, and there are no storms of rain and hail to destroy their grain and hay. They need no barns. Barley thrives better than in any other part of the world. The soil and climate are also particularly favorable to the growth of wheat, which unites the valuable qualities of whiteness, dryness, and glutinousness, to a greater degree than any other wheat in the world. Our average crops are also larger than in any other place where manure is not used extensively. The yield of hops is large, and the facilities for drying them, so as to preserve their strength, are better than in any other land where they are cultivated. Our kitchen vegetables grow to an unparalleled size. Nowhere else have pumpkins been seen to reach two hundred and fifty pounds in weight each, beets one hundred and twenty pounds, white turnips twenty-six pounds, solid-headed cabbages seventy-five pounds, carrots ten pounds, water-melons sixty-five pounds, onions forty-seven ounces, Irish potatoes seven pounds, sweet potatoes fifteen pounds, and so forth. Some cabbages and beets have spontaneously become perennials here, continuing to grow from year to year, and remaining green throughout winter and summer; and many of our kitchen vegetables might be converted into perennials by preventing them from going to seed.

The abundance, excellence, and variety of our fruit aston ish the stranger, though he may have come from the markets of London or New York, which draw tribute from whole hemispheres. No market on the globe surpasses ours in variety, and yet it is not ten years since we began to import fruit-trees direct from the Eastern states and Europe. Our mild winters permit the trees to grow during nine or ten months in the year, and they grow more rapidly, and reach maturity more speedily, than in any other country where they are so healthy, and bear so abundantly. The pear and apple trees which were planted by the missionaries thirty or forty years ago, are still in perfect health, and some of them produce as much as a ton of fruit to the tree every year. The apple and pear seem to have found here their most congenial clime. There are no worms in our apples; no curculios in our plums or cherries; no Hessian fly or weevil in our wheat. The olive and the fig grow luxuriantly beside the apple and the pear. We can produce olives better than any of the olive-producing regions of the Mediterranean, because we have none of those storms of thunder and hail and rain, which frequently destroy the crops in southern Europe and Asia Minor. The vine produces more abundantly than in any part of Europe, and the crop has never failed or been destroyed here, as often happens there. A yield of one thousand gallons of wine to the acre is as frequent, proportionately, in California, as of four hundred in France or Germany. Our gardens are, in time, to be the most beautiful in the world, resplendent with conifers and deciduous trees, with the flowers of the temperate zone, and the luxuriant plants of the tropics. The shrubs which in New York remain small, and live only under shelter, as delicate exotics, are naturalized in San Francisco, grow almost to tree-like size, remain green throughout the year, and bloom during most of the months. The rosebush is covered with flowers from January to December.

Domestic herbiverous animals live and increase without shelter, and without cultivated food. They reach their full growth a year earlier than in the Eastern states. The absence of extreme cold gives them a more rapid growth, and exemption from many diseases. Sheep produce more wool, are healthier, increase more rapidly, and are kept at far less cost in California than in any American state east of the Rocky Mountains. Bees increase more rapidly, and make more honey than there is any record of their doing elsewhere. Thunder and rain storms kill a large proportion of the silk-worms in Italy, France, Turkey, and China every year; in the valleys of California we never have any lightning, and no rain during the season when the silk-worms feed.

The wages of labor in California are higher than in any other part of the world. Mechanics' wages are generally from two dollars and fifty cents to four dollars per day; common laborers, from one dollar and seventy-five cents to two dollars and fifty cents per day; farm laborers, and men and maid servants, from twenty dollars to thirty dollars per month. Our imports and exports of treasure are larger in proportion to our population than those of any other state. Our chief city is favorably situated for commerce, and its harbor always contains vessels of the largest size from every sea. It has an undoubted supremacy in the commerce of the north Pacific. We have no paper money, and no current coin less than a dime.

The inhabitants of the state, numbering nearly four hundred thousand, represent in their nativities every American state, and every continent, and every country of Europe, and many of the countries of Asia and Africa. Our population is unsurpassed in intelligence, experience in travelling, and skill in the arts. Our society is liberal in tone, and free in intercourse.

With many drawbacks, which have been set forth clearly and unreservedly, California is still the richest part of the civilized world. It possesses most of the luxuries of Europe, and many of the advantages which the valley of the Ohio had forty years ago. It offers an open career to talents. In the few years of its history it has astonished the world, and its chief glories are still to come. The arts, the sciences, the refinements of life, are to find a favored home in California.

Why is it then that the permanent population of the state has not increased more rapidly? Why have so many of the early immigrants left her shores, never to return, by their departure depriving her of the greatest element of wealth? The great cause is the mismanagement of land-titles by the federal government, and the consequence is, that the people have been unable to obtain secure homes, and therefore have gone to the Eastern states, where they could find permanent residences. This mismanagement has prevailed both in the mineral and agricultural districts, and has produced incalculable evils.

§ 307. *Sale of Mineral Lands.*—The welfare of every civilized state requires a permanent population, a well-regulated society, a steady business, and a secure investment of capital proportionate to the industrial ability and production of the people. These requisites are indispensable to all national prosperity. Their want, if long continued, must inevitably be followed by national ruin. They are wanting in a large portion of California.

In the moral and social, as well as in the physical world, cause and effect are inseparably connected; adequate means never fail in leading to correspondent ends; prosperity or ruin comes not by mere chance, but is the necessary result of the adoption of good or evil counsel. The ill-regulated society and unsound condition of business in our state, are traceable mainly to the insecure tenure of our lands; and as a necessary means to attain social, commercial, and individual health, we must have perfect land-titles. I shall speak first of the mineral counties.

It is a necessary consequence of the want of secure land-titles in the mining districts, that the inhabitants should be unsettled. There is nothing to fix them in any one place, while many motives impel them to frequent removals; and the result is, that a considerable portion of the mining population is truly nomadic in character. Most of them have poor claims, or none at all; and they enact laws, or establish customs having the force of laws, that all claims shall be small, usually not more than one hundred feet square. These small claims are worked out in a month or two, or at most in a year or two, and then the miner must go. Perhaps he will find his next claim within ten miles, perhaps not within fifty. When he gets a claim he may not be able to work it out; he must not only occupy his claim, but he must work it. If he absent himself from it more than three days, during the season in which it can be worked, for other cause than sickness, it becomes forfeit to whomsoever will seize it. In no case can he who mines in the river-beds, banks, flats, or gulches consider his claim a home for life; in one case out of a thousand it may employ him for ten years. Quartz and tunnel claims are more lasting, and many of them will not be exhausted in a lifetime; but the miners employed in these are a small portion of the total number.

The miner is not only not tied to his claim by ownership, or the hope of long employment and lasting profit, but he is constantly tempted by other tracts which are open to him without price. He may consider himself owner of all the unoccupied land in the country. He can take and use any of it. No one has a better title than he. Every unoccupied gully, flat, hill-side, river-bar, river-bank, and quartz-vein is persistently trying to seduce him. He can scarcely take a pleasure-walk on a Sunday morning without seeing some place which invites him to come there and settle, to desert his old home and make a new one. And when there is nothing to protect him against such temptations, save his belief in the superior mineral wealth of his first location, that belief may often be changed by a very brief examination of the new place. He has no title to the spot where he dwells, no substantial improvements, no property of any kind save such as he can carry on his back at one load.

The world never saw such a people of travellers as the Californians. There are now about 350,000 white inhabitants in the state, and more than 250,000 others have gone "home" during the last twelve years, four-fifths of them never to return. Not one-fifth—probably not one-tenth—of the miners of 1849 are now in the state, and it would be a difficult, and perhaps an impossible task, to find a Californian mining town, one-twentieth of whose population has been permanent there since 1850.

In regard to the men leaving California, it must in fairness be stated that many of them are actuated by a desire to be with their families, and they see that it is much cheaper for them to go to New York than to have their families come to San Francisco; and there are cases where the families would make very great objection, even overlooking the cost of passage, against moving to a land so far from all their relatives. But, on the other hand, it must be considered also that all the men who leave the state, do so seeing and acknowledging, before they go, that in climate, mineral resources, the profits of labor and trade, the enterprise, intelligence, and generosity of the people, the independent spirit of the poor, the democratic spirit of the rich, and the frank friendliness of all, California is far superior to any other part of the American Union, while it has many advantages in other respects. Such an acknowledgment, coming from men leaving a state with which many of the most interesting associations of their lives are connected, implies a great evil somewhere. Although some of them go "home" because they cannot bring their families to California, yet this is not the fact in one-fourth of the cases; they go because they do not wish to live here, because they will not live here.

Another evil effect of the want of secure land-titles, and the consequent unsettled character of the population, is the want of good houses and substantial improvements of all kinds. The dwellings throughout the mines are, as a class, mere hovels, even in the oldest and most thickly-settled districts. In the towns it is necessary to have some substantial stores, as a protection to the valuable goods kept in them; but with these exceptions, and a few fine residences, even nominal "cities" are collections of shanties, scattered about with little regard to order, and fitted up with little provision for comfort.

The wandering character of the population, and the want of permanent and comfortable homes, render the mines an unsuitable place of residence for families. There are a few women in the mines, and of these few a considerable share are neither maids, wives, nor widows. The general proportion of adult men to adult women, throughout the mining districts, is probably not less than three to one, and to married women, four to one.

It sometimes happens that miners having wives in the eastern states have them come to live in the mines; but in a considerable proportion of cases this arrangement is not a permanent one. Anxious as the inexperienced wife may have been to live with her husband, and willing as she might be to share his privations, the result has often been that she found life in the mines unsuited for herself and her children. There are many good, virtuous, and intelligent women living in the mines, and perhaps as well contented there as they would be in any other part of the world; but there are not enough of them.

If there were no other evil than this scarcity of women traceable to the present tenure of the mineral lands, that one fact would be enough to settle the question that the mines must be sold. The family is no less essential to the good order of society and the prosperity of the state than it is to the happiness of the individual. A community of American families must have permanent homes; they must own the land in fee-simple; and there cannot be a large community of families in the mining districts of California unless the land there be sold.

The scarcity of women is again the first link in a great chain of evils. Some men in the mining counties would like to marry, but cannot find wives to their choice. They must either travel thousands of miles to get a wife abroad, or take some awkward girl just entering her teens, without education or experience in society, and entirely incompetent to take charge of kitchen or nursery. The scarcity of wives and married women converts many men into tempters, and they must cause much misery. And women, knowing that they are scarce, and therefore in demand, are incited to calculate the chances and the profits of fidelity and chastity as compared with infidelity and infamy. Family quarrels often ensue, and the state has a sad notoriety for the frequency of its separations and divorces. A trustworthy gentleman informs me that, during a visit to a mining town in a remote part of the state, about seven years ago, he was informed that there were in the town one hundred and twenty-seven women, forty-nine of whom, though married, were living with men not their husbands. The case is certainly without a parallel, in the state or elsewhere; but the condition of affairs in this respect has changed very much for the better since 1855.

The want of families, and the comparative scarcity of intelligent and good women, deprive the community of many of the most wholesome pleasures and ennobling influences which are found in other states. The man who has no wife or sweetheart to work for is improvident; and, unchecked by such public opinion as can reign only where well-regulated families are numerous and society permanent, he gives himself up to dissipation, feeling confident that none of his neighbors will cut his acquaintance on that account.

As the people are among strangers, and do not expect to remain among them long, reputation loses its value, and public opinion its power; and thus forces of great influence in preserving the good order of society elsewhere have comparatively little influence in the mines.

The scarcity of families and the consequent unstable state of society make servant-girls shy of the country, and the few here demand enormous wages—five and six times greater than in New York. This may at first sight appear to be a fact of little importance, but it has really driven thousands of fam-

19*

ilies from the state, and prevented thousands of others from coming.

These various social evils chafe and foment one another, and the consequence is, that the miners who have come to the state intending only to remain a few years are not likely to change their intention. It is of course the ambition of most men in the country to have homes of their own; to have wives and families, to be with them and to enjoy their society. Since they do not propose to become permanent citizens here, if married, they do not bring their families with them; if unmarried, they do not marry while here. The necessary effect of this state of affairs is, that there is an exceeding anxiety to get away from the country as soon as possible. A feverish excitement prevails through the whole people. Speculation has risen to an unexampled height. The game is, to make a fortune in a few months or to be bankrupt; and there are tens of thousands to play at it. Men complain that they cannot enjoy life in the mines; that life there is a mere brutal existence; and they become desperate in their anxiety to leave it, to go elsewhere, where peace and comfort, permanent homes and social order prevail; where numerous well-regulated families furnish agreeable company for the married, and where numerous accomplished young ladies furnish not less agreeable company to the unmarried. Most men in California do not live here to enjoy life, but to make money, so that they may enjoy life in some other country. Not that the people are parsimonious—far from it; but they are puffed up with extravagant expectations, or rather determinations. Unless they can earn very large wages, they will not work at all. The merchant will not be content with a regular business, paying ten times as much profit as he could make with a like capital in the Eastern states; he must go into wild speculations, and risk every thing upon a remote chance of making a sudden fortune. The frequency of insolvencies, particularly in the towns, is inexplicable, at first, to a man who comes here without understanding the peculiar condition of our society; and the same man, going through the mines, will be astonished to see that the much-abused Chinese are the only class who are always industrious. The miner will often do nothing for weeks and months, running up long bills for "boarding," while he waits for rain, or the completion of a ditch, or for something else to turn up. He is too high-minded to accept small pay, and would rather be idle—at the risk of the boarding-house keeper and store-keeper. His idleness is frequently called "prospecting;" he travels about hunting for a place to work; and this prospecting may be said to employ nearly a fourth part of the mining population. The consequence is, that a large portion of the miners are always moneyless, or provided with an exceedingly small amount of money. At other times they fall upon rich deposits, and then try to make up in dissipation for past privations. And so the mining population comes to be an improvident one—unsteady, fond of gambling and other wild amusements. The fact is that there is not in the whole world such another reckless, thriftless, extravagant, improvident population as in the mining districts of California.

Another evil effect of our present system of land tenure in the mineral districts, is to be found in the gradual lowering of the general character of the population in the mining counties. Most of the steady, prudent, economical men leave the state with more or less money, while the dissipated, thriftless fellows remain; the latter class increasing in numbers, the former decreasing every year. The only means of fixing and increasing the former class, and giving them the proper influence in our society, is to give them permanent homes; and this policy will at the same time drive away the wrecked specimens of humanity among us, and compel them to seek homes in the Cimmerian darkness beyond our borders.

It is one of the great evils of the tenant-at-will system, that there is little security for the investment of capital. Land should be the main stock of wealth and the main basis of credit, and the increase of its value with increasing population should be one of the main sources of riches in every new country; but of this kind of property the mining districts are deprived by unwise policy. As it is now, it is almost impossible to induce the capitalists of San Francisco to invest money in, or loan money on, mining enterprises; they have learned by bitter experience that there is little safety for money invested in canals and quartz-mills, where there is no title to the lands save possession, which might be lost by abandonment or forfeiture at any moment. The consequence is, that the permanent improvements in the mines are rare, in comparison with the number which there would be if the mineral lands were sold; and where money is borrowed to make such improvements, extravagant rates of interest are paid. As a result of the comparatively small amount of capital invested, and the lack of security for large investments of money in mining enterprises, there is little demand for labor, and the state is full of poor men anxious to get work, but unable to find anybody to employ them. The immigrants to a new country are generally poor men; and, unless the state of business is such that they can confidently expect to obtain profitable employment on their first arrival, there is little encouragement for them to come. In regard to the certainty of success upon first arrival, California offers less encouragement to the immigrant than many of the states in the Mississippi Valley, and will not offer more until the tenure of lands in the mines shall be changed.

The present policy drives away the money produced in the state. Why do we send $40,000,000 of gold away every year? Simply because we cannot give good security for it. We have nothing to give as security. We offer to pay twice as much interest as anybody else, and our offer would be gladly accepted, if there were a certainty that we would pay as we promise; but there is no certainty, no security. The $30,000,000 shipped by California to New York last year, would have drawn $6,000,000 yearly interest here, while it will draw only $3,000,000 there; and the $8,000,000 shipped to London would have drawn $1,600,000 here, while it will draw only $450,000 in England; but the owners of the $40,000,000 prefer the

$3,450,000, with the New York and London security for capital and interest, rather than the promise of $7,600,000, with the danger of losing both capital and interest in California. It is true, there is a natural drain of specie from countries where labor is high, to those where labor is low, because the latter import little and export much; and of course, in this respect, California must necessarily become tributary to China, the Atlantic states, and Europe; but, on the other hand, our relations with the great centres of capital are so intimate, that we can get all the money we want at California rates of interest if we will but give perfect security for it, and pay the interest without fail. It would be no light matter for us to owe a hundred millions, and pay California interest on it, to European capitalists; but it would still be better than to do without the money, without the improvements which it would build up, without the population it would attract, and without the fixed wealth it would create.

Again, the present system exercises a most prejudicial effect upon the finances of the state, and bears very unequally upon the citizens. The farming districts, where the inhabitants own the land, pay heavy land taxes; whereas mining claims pay no taxes at all. The result is, that the taxation upon the men in the valleys is about three times as heavy as upon those in the mountains. The miners generally have no homes, and no fixed property, and cannot be forced to pay taxes. Most of the mining counties are deeply in debt, and many are going deeper every year. The only way to equalize the taxation is to sell the mineral lands, and compel the miner to pay a tax upon his mine, as well as the farmer on his farm.

The proposed sale of the mineral lands is opposed by two arguments: first, that it will lead to monopoly; and, secondly, that gold mining in this state is so precarious, that miners could not afford to have permanent residences and support families.

These two arguments are antagonistic to each other; both cannot be sound; at least one of them must be fallacious. The "monopoly" argument presupposes not only that the persons employed for wages to work the mines will earn enough to support themselves and families, but also that the monopolists will make a large profit, otherwise their monopoly would not last long. The "precarious" argument presupposes that only a small portion of the mining land will continue for any considerable time to pay a living profit; and that therefore there is little encouragement for capitalists to invest their money in mining land. The "monopoly" argument presupposes the investment of large amounts of capital—the very thing which the mines most need; the "precarious" argument presupposes that mining will be more profitable for a man who runs about than for one who stays at or near one place.

All the great social evils which I have mentioned as prevailing in California, are traceable directly to the roving character of the people; render the population permanent and you necessarily cure the evils. It is admitted that our mines will not be exhausted, and that the number of miners in the whole state will not decrease much, if at all, during the next fifty years. It is entirely safe to predict that Siskiyou, Nevada, Shasta, Placer, El Dorado, Plumas, Sierra, Tuolumne, and Calaveras, will be mining counties in 1950. Now if the mining is to be continuous, why should not the miner be permanent? There is no necessity that he should be a nomad; on the contrary, his own pecuniary profit and the welfare of society require that he should have a fixed residence, and not until he gets that, can he be a valuable citizen.

But it is said the mining population cannot be permanent, because mining is a "precarious" business. Well, I should like to know what business would not be "precarious," if conducted as mining has been in this state during the last ten years. Here are one hundred thousand men, mostly without homes; not staying in any one place more than four months at a time, on an average; spending one day out of three in prospecting; refusing to work unless they can make big wages; running successively to Gold Lake, Gold Bluff, Kern River, Fraser River,

Mono Lake, and Cariboo;—how could any occupation be other than precarious, managed in such a manner? Of course mining can be made precarious, and these fellows who are always running about are the very ones to make it so. It will not be made more precarious by permanence. If the five thousand miners of El Dorado, and the four thousand miners of Tuolumne, will just stay where they are, instead of changing places with each other three times every year, they will not lose any thing on the score of the precariousness of their business. I venture to assert that gold mining in California, conducted prudently, is not an uncertain business at all. A careful man can, with a certainty, earn more than he could as a farmer on the prairies of Illinois, where farming is one of the least precarious occupations in the world. The permanent citizen can afford to mine prudently; the nomad comes here to make his "pile" in a few years; he has no wife with whom to live joyously, and, as a matter of course, his mode of mining is precarious.

But it is said the capitalists will monopolize the mineral lands; and yet there is not a week that the honest miners do not come to San Francisco to solicit capitalists to invest their capital in mining enterprises; and when such an investment is made to assist a canal or quartz-mill, all the miners in the vicinity are glad, and property rises in value. Why is there more danger of monopoly in mineral lands than in the agricultural lands? Are the former more sacred than the latter? Is it to be supposed that capitalists will buy up the mineral lands and then not work them, but let their money lie idle? Certainly not; capitalists would be in no hurry to invest largely at first in the mineral lands: and if they should they would employ large numbers of laborers, to the great benefit of the whole country. And the same honest miners who have such an abhorrence of "monopoly,"—are not three-fourths of them determined to leave this land of unmonopolized freedom to return to the Eastern states, where capital is king, and where there are no laws to prevent the rich men from monopolizing the whole country? The assertion that the sale of the mineral lands would offer dangerous advantages to capital, is as much as to say that the sale would be followed by the investment of capital, and a general rise in the value of property in the mines, and an increase in the amount of their production.

This "monopoly" argument has been used for years, and the miners have come to believe it without ever examining it or seeing its absurdity. Instead of capital driving poor men out of the mines, it would bring them in; it would create a demand for labor; and the ten thousand men who are now in the mines, anxious to obtain permanent employment, would then get what they have been seeking in vain during the last four years. If capitalists buy up mining lands, of course they will do it with the intention of digging for the gold, and to do that they must employ laborers. This kind of labor is not dishonorable; it is such labor as most of the mechanics in California, as well as elsewhere, are engaged in all their lives: that is, labor for a fixed salary. It is just such labor as is done now by a large portion of the quartz, and hydraulic, and tunnel miners, who consider themselves quite as independent, and their occupation as honorable, as if they were cabinless and claimless surface diggers. The labor for fixed wages will not be unprofitable; on the contrary, it will remove all precariousness from the workman's mode of life, and will give him a good and certain income, with which he will always be able to live comfortably. It is not improbable that wages would rise after a sale of the mineral lands. Of course, every purchaser would wish to open his claims at once, and workers would be in demand. The great danger, if the mineral land were offered for sale, would be, not that too much capital, but that not enough would come into the mines. Just in proportion to the amount of land sold would be the amount of benefit done to the state. If none were sold, the present state of affairs would continue, and the greatest enemies of the sale could not say that any harm had been done; if a little were sold, the change would be but small; if much were sold, there would be a great increase in the value of mining property and in the demand for labor. The result of a well-managed sale, or donation system, would be that the present miners, and not distant capitalists, would come into possession of the richest places, and that every man in the state could, at a trifling cost, obtain a claim that would furnish him with profitable employment for many years. There are certain places in the mines where the claims are mostly in quartz-veins or deep banks, which will require many years to work them out, and there the population is comparatively stable. Of these places, Grass Valley and North San Juan may be taken as examples. The traveller sees at once, on approaching them, that there are more comfortable homes, more families, and more peace and sobriety among the inhabitants, than in the majority of the mining towns. The difference is a very great and important one, and if it can be removed by elevating the other towns to the level of those two, the sooner the better.

The "monopoly" argument was used in Illinois, against the sale of the mineral lands there, and prevailed for a time; the consequence was, the population was made up of vagrants, and the dwellings were all shanties, and society was no society at all. Finally the lands were sold, and the result was a great benefit to the people and the mining districts, in every social and industrial respect.

It may be objected to the sale of the mineral lands, that "the wisdom of our ancestors" has determined that mines should always belong to the government, and be open to all persons willing to work them. The objection may be recognized as a good one when that policy is proved to be wise by evidence and argument—not till then. The reason of the ancient policy was, that most of the land was owned by ignorant and unenterprising people, chiefly nobles, who, if they had owned the minerals, would have allowed the natural wealth of the land to remain undeveloped. But that state of affairs does not and never can exist in California. On the contrary, nobody can so safely be trusted to get all the gold out of a tract of land as the fee-simple owner of it.

The federal government has refused to sell the mineral lands to the state, and the surveyor-general has instructed his deputies not to "sectionize" the land in the mineral districts, or within several miles of where any miners are at work. The truth is, that a large part of the land in the mining region contains so little gold that it never can pay the miner, but is well suited for agricultural and horticultural purposes. Californians confidently expect that some of the finest fruits and wines of our state—and that is as much as to say of the whole world—will be produced in the mining counties, within five years from the present time; and the government should pursue such a policy as will encourage the occupation and cultivation of all the land suitable for such purposes.

If the sale were once determined upon, undoubtedly difficulties would arise as to the manner of carrying it into execution; but these would be of little import, as compared with the evils caused by the present system. The whole mineral district should be surveyed at once, and sold in lots to persons who will live on, or work them, varying in size, according to location and supposed mineral wealth, from one hundred and sixty to eighty, forty, twenty, ten, five, two and a half, and one and a quarter acres. Perhaps it would be advisable to grant at first no lots where many miners may be at work within a small space. Large lots of ten, twenty, and forty acres, now unoccupied, and which would long remain unoccupied under the present system, would find abundant buyers should the government propose to grant the fee-simple.

The offer of the mineral lands of the state, comprising about 10,000,000 acres, for sale, would present one of the greatest opportunities in the world for large numbers to secure great and certain wealth at a small immediate outlay; and not only every man now in the country, but every one who has been here, would exert himself to the utmost to become the owner of a tract of land, the mines of which would probably clothe him and his sons, and his sons' sons, in wealth, and which, if it were barren in gold, would still keep him in comfort with its agricultural products. From the moment it is known that the mineral lands will be sold, California's regeneration will begin. Californians will then determine to make this their permanent home; money will be saved; and at the time of sale, every man will seek to become owner of a tract of mining land, which shall enrich himself and his children. After the sale, titles being secure, comfortable houses will be built, wives will be sent for, mining will be conducted economically and steadily, claims will be worked which now will not pay, our population will increase, and so will the yield of our mines; the capital produced here will be retained; other capital will come from abroad, to obtain secure investment on safe titles; poor men, coming from abroad, will always obtain employment, and thus can get a start; railroads and turnpikes will be made; land will rise in value; the state will obtain its revenue honorably by the taxation of capital; society will become permanent, and public opinion powerful; dissipation will diminish; and California, instead of being socially the worst, will become the best state in the Union.

The question of the sale of the mineral lands is then the question of the future of the state. The advocates of the measure may not succeed this year, or the next; but they will, they must, succeed finally. The fight is between the permanent interest of California on one side, and, on the other, the temporary interests of some roving miners, who care nothing for the state, save to get its gold and then leave it. As for all the unmarried Americans (whatever their occupation) who may come hither to spend a few years—to carry away our gold if they are successful, or to remain with us as human wrecks if they fail—all these are no better for California than so many Chinamen; I call them "white Chinamen." They will not become permanent citizens; the yellow Chinamen cannot; so there is not much difference between them. If there is any legal, constitutional, and just measure by which we can drive the white and the yellow Chinamen out of the country together, and obtain white, permanent Californians in their stead, I, for one, shall be heartily in favor of it. "California for permanent Californians," is the proper motto for every faithful citizen of this state. We must have a political war; the permanent Californians must conquer the rovers, and compel them to settle down or leave. The great question is not whether we shall produce much gold or little; it is whether we shall have social and industrial order or disorder, which is equivalent to the question of the permanency or vagrancy of the population. I am confident in the belief that the sale of the mineral lands would cause a considerable increase of our gold yield; but no matter how great a decrease might ensue, state policy requires that the sale should be made, in any case. The gold now dug does little benefit to California; it slips through, like water through a sieve; and serves only to attract the vagrants who visit the state merely to despoil it. All the money under heaven will not pay for maintaining a system under which three-fourths of the people of a large district are vagrants, that is, rovers, and where six-sevenths are men.

It is not unreasonable to assume, that if the present system of mining titles be maintained, there will be very slow change for the better in the vagrancy of the miners and the inequality of the sexes during the next ten years; and I do not hesitate to say that, rather than the present state of affairs should continue, the state government should take effective measures to put a sudden end to all gold mining in the state, by declaring it a felony, and making it punishable by severe penalties, so that thereafter the people of California would turn their attention to such pursuits as farming, horticulture, and stock-raising, which contribute to the lasting profit of the state as well as the temporary profit of individuals.

California may be compared to a maiden who has been reared to love the paths of purity and peace, but who has been introduced of late into a corrupt society, and is now surrounded by men who wish to dishonor her—enjoy her for a short time, and after gratifying the base impulses of the moment, to have no further thought about her welfare, but to desert her forever, careless whether their desertion prove her ruin or not. From these men her virtuous soul turns with indignation and abhorrence. She welcomes no suitor save him who comes offering his whole heart in a life-long union, under solemn promise that she alone shall be loved and cherished by him.

Now let us turn to the manner in which the land-titles in the agricultural districts have been managed.

§ 308. *Mexican Grants.*—Upper California, when conquered by the Americans in 1846, contained about five thousand Mexican inhabitants, who, with their fathers and grandfathers, had lived here sixty or seventy years. Their chief occupation and the main source of their wealth were furnished by their herds of kine, horses, and sheep. Most of them dwelt in the country, upon ranches which had been granted to them for purposes of pasturage by the Mexican government. They held their lands under written titles, supposed to be, in most cases, legally perfect under the laws of Mexico. The government of that country never questioned or denied the validity of such grants as those held by the Californians. The grants were made to suit the habits and wants of the people. The Californians owned large herds, which were never fed on cultivated food, never kept in fields, nor placed under shelter. In a country where an almost unbroken drought reigns from May to November, and where cattle get no food, save wild and indigenous grasses, much more land is required to sustain a cow, than in those lands where careful cultivation and frequent rains provide a regular and certain abundance of food through the year. A fertile soil, like that of a large portion of the Mississippi valley, will sustain five or six head of cattle to the acre; but here three acres of uncultivated fertile land are necessary for the support of one cow. Herds of thousands of kine were not uncommon in California under the Mexican dominion. To accommodate these cattle, great tracts of land were necessary. The public land was granted not by the acre, as in the American states, but by the square league (containing four thousand four hundred and thirty-eight acres), which was "the unit of measurement" in granting public lands outside of the towns. The government granted away its lands willingly, and without compensation; no pay was required; the only condition of the grant was, that the grantee should occupy the land, build a house on it, and put several hundred head of cattle on it. Whenever he promised to comply with these conditions, he could get a grant of any piece of public land, of eleven square leagues or less, for which he might petition. It was a grand Mexican homestead law; and the chief complaint made about it was by the government, that the number of applicants for grants was not greater. The grants were not made according to the American land system, which would have been entirely unsuited to the wants and habits of the Mexican people. The public lands in California were never surveyed. I do not know whether a Mexican surveyor was ever seen in California; I feel confident that no ranch was ever surveyed, and its boundaries described, with bearings and distances, previous to 1846. The descriptions of the land granted were very vague. In most cases a certain number of leagues were given, within well known natural land-marks, which might include a district of fifty or a hundred miles square. In such case, the grantee could locate his ranch at any place within the limits. Sometimes a grant of so many leagues was made, at a place to which a name had been affixed by the Indians or Californians, and then the ranch included that spot; sometimes a ranch was described as bounded on one side by a range of mountains, on another by a river, and on other sides by ranches of older date. The Californians did not quarrel about their boundaries. If A's cattle crossed to B's ranch, for better pasture in the summer, B's would probably go to A's at another season. The herds were not closely kept. The cattle roamed about almost in a wild state, often unseen of man for months. So wild were they, that though they knew very well that a man on horseback was a superior animal and their master, yet they considered a man on foot as a base and ferocious beast, and attacked him as they would attack a wolf. Their owner knew his property only by the brand placed on them when they were calves. From the time when the redhot iron burned into their flesh, they roamed untouched by the hands of man, until fate decreed that they should be slaughtered to furnish fresh meat for their master's household, or hide and tallow for foreign commerce. Evidently this people, with such habits and such occupations, did not need to have their lands precisely described. Most of the titles were legally valid under the Mexican law. There was no motive to commit fraud, because land was of little value, and great tracts of rich soil were, up to the time of the American conquest, open to every petitioner. In most cases the actual occupation took place previous to 1840, and had never been interrupted. This occupation, the most conclusive proof of good faith, and an equitable title in itself, was notorious, and susceptible of proof by hundreds of witnesses. The paper titles were mostly of indubitable genuineness, written by the hands of well-known officials, bearing regular numbers, referred to in public lists of land-titles, and mentioned in government documents of various kinds. The proof of the genuineness of the title-papers, the good faith of the claimants, and the equitable validity of the claims, in nine cases out of ten, was abundant, and, to any man at all acquainted with the subject, indubitable. It was then evidently the duty of the government of the United States to provide for the summary examination of the documents, and in every case, where genuine title-papers were found with ancient occupation, to order a survey for the establishment of boundaries, giving to the claimant at least a *prima facie* recognition of title, subject, perhaps, to investigation in the courts, if any person should see fit to assail the validity of the grant. But the federal government pursued a policy very different from this plain duty. It delayed action through 1848, 1849, and 1850; and first, in 1851, passed an act nominally to "settle" private land claims in California, but really to unsettle them and the whole country, and keep them unsettled. That act provided for the organization of a court, or land commission, to try these claims ; declared every grant of land in California to be legally void, though it might be equitably good; and provided that every equitably good claim should be lost to the owner, unless he should sue the United States in that court, and gain the suit there or on appeal; and that there should be an appeal to the United States District Court, and thence to the United States Supreme Court. In all these courts the claimant was to be opposed—that is, persecuted—by a law agent appointed by the United States, with instructions to contest every claim to the utmost. The land commission organized in San Francisco, on the first of January, 1852, and continued its sessions until the third of March, 1855, when it expired by limitation. It had received eight hundred and thirteen petitions. The owner of land, under grant from Mexico, was compelled to petition the government of the United States for the privilege of keeping it. Of these eight hundred and thirteen petitions, some were for lands which had never been occupied; in some cases there were two or three petitions from different persons, claiming the same piece of land under the same original grant. In some cases the original grantee had sold out a large ranch to a number of Americans, each of whom presented a petition for his piece; and, in perhaps twenty-five or thirty cases, the title papers were forged ; leaving about six hundred original ranches, which had been held under indubitably genuine written title and notorious occupation.

Thus there were eight hundred and thirteen important lawsuits, involving the titles to ten million acres,—nearly all the private lands in the state,—to be tried in one court. This tribunal had three judges,—good lawyers, and industrious, honest men. No serious complaint has ever been made against any of them. They did what they could. When, at the end of three years, the time came for them to close their court, they had dispatched all the cases. The trials had been fair, the hearings deliberate and public, the opposition on the part of the United States law agents stubborn. All the law agents were competent men, and no one can justly complain that the interests of the United States were neglected by any one of them. The claimants had been kept in litigation three years; they had been compelled to bring numerous witnesses from remote parts of the state, to pay for interpreters, to fee lawyers, at rates unheard of before in the world, to dance attendance upon the court, and to leave their homes and their business for months at a time; but this was not enough. In every case where the land commission confirmed a claim, the United States government ordered an appeal to be taken to the United States District Court. This was nominally an appeal, but really an order for a new trial. Every question of fact and law was opened anew. Witnesses were again examined; the whole case was tried as in the original proceeding. There are two United States District Courts; one for the northern and another for the southern part of the state; each being the appellate court for all the lands within its own jurisdiction. Each of these two courts had other business besides land suits; and in the northern district, where the most important cases lay, the court had almost as much admiralty business alone as the judges of federal districts in the Atlantic states have to manage. Both these Californian district judges were good men. In these courts, too, the "interests of the United States" were protected by able and industrious lawyers, instructed to oppose the Mexican land claims to the utmost. Seven years have elapsed since the first case was appealed from the land commission, and there are now a number of cases still undecided in the District Courts: but in most of the cases decided, the claims of the Mexican grant-holders were confirmed a second time. The federal government, still not satisfied to let the claimants enter their lands, ordered appeals to the United States Supreme Court at Washington. This order was not accompanied by any proper provision to pay the clerks for making out the transcripts; and as the appeal could never be decided, and the claimant never get a perfect title, until the transcript should be sent up, and as the transcript never could go up until the clerk had received his fees, so the claimant was often compelled to pay the expenses of the transcript, amounting in some cases to several hundred dollars. This was an expense which custom and law impose upon the appellant, but in these cases the United States made no provision for repaying the respondent, although he was compelled to advance the money. After the appeals had been taken to the court of the last resort, the United States Attorney-General ordered the appeals to be dismissed in about four hundred cases, and in about forty cases the United States Supreme Court have given judgment in favor of the claimants, making four hundred and forty claims finally confirmed. About one hundred and forty claims have been abandoned by the claimants or finally rejected by the courts, and this estimate would leave two hundred and thirty cases still before the courts for adjudication upon their merits.

I have said that four hundred and forty cases have been finally confirmed, but final confirmation is not equivalent to final settlement. Up to 1859, it was supposed that when judgment on appeal had been rendered in private land claim, by the United States Supreme Court, in favor of the claimant, the litigation between him and the federal government, so far as that title was concerned, was at an end. But a new law was passed, requiring the surveys of the Californian ranches to be subject to review by the United States DistrictCourts. The exact boundaries of the claim could only be determined by a survey; and in large ranches, where the boundaries were not clearly defined, the location of the ranch became a matter of very great importance, often involving values of tens and even hundreds of thousands of dollars.

The consequence of the new law was, that four hundred and twenty out of the four hundred and forty finally confirmed claims, are thrown into the courts again; their settlement is postponed for an indefinite time; the owners are burdened with new litigation, with indefinite deferment of their hopes, with increased costs; and the country is again cheated out of quiet titles, permanent settlers, permanent improvements, and all those blessings of inestimable value which come only with numerous fixed and happy homes, and the best regulated social order.

While the government has thus, during twelve years, not simply refused to confirm the land-titles granted by Mexico, but made bitter and unceasing war upon them, and compelled the claimants to bear the expense of the warfare, these claimants have had to suffer from the assaults of other and still more dangerous and vexatious enemies—the squatters; who, while ostensibly left without countenance by the law, were really often engaged in an offensive and defensive alliance with the officers of the government. The squatters took the land, occupied it, drove away the owner's cattle, cut down his trees, fenced in his springs, paid him no rent, paid no taxes, by their influence forced him to pay the taxes on the land they were occupying, and assessed the taxes at most exorbitant rates. This system was not rare, but frequent—it was practised on not one, but a hundred ranches. And then, with the money derived from the land thus obtained, they paid lawyers to appear in the name of the United States, contest the owner's title, and delay a decision; and, after decision, to get up a contest about the survey and delay a settlement of the boundaries. I do not mean to say that every Mexican claim is good, or every squatter wrong; my purpose in this article is only to complain of the vast injustice done to the owners of honest and legally valid claims, which are the great majority of all presented to the courts.

It is fourteen years since Americans became the rulers of California, and land-titles are no nearer a settlement than they should have been twelve years ago, if a proper system had been adopted. The great question about the boundaries, which should have been the main subject of action, is now just where it was then. The claimants have sold two-fifths of their land to pay the expenses of litigation—that is said to be a modest estimate by those familiar with the subject—and they are not yet done. They have been despoiled of two-fifths of their land, deprived of the possession of a large portion of the remainder, and prevented from selling it while they saw its value, in many cases, decreasing steadily with the decay of business consequent on the exhaustion of the richest placer mines.

The injury done to the country by the delay in the settlement of the land-titles is, to a considerable extent, irreparable. That delay has caused us to lose, or has prevented our gaining, a population of a million citizens, of the most valuable class. Two hundred thousand men have left our state forever—half of them because they could not get permanent homes here—and they prevented as many more from coming, who would have come if they could have had certain land-titles. Not less than fifty thousand men have left us because of the unsteadiness of business and the lack of employment, caused by want of unquestioned ownership of the soil. Thus I estimate that the delay in settling our land-titles has cost us two hundred and fifty thousand men, representing a total population of one million persons. The golden flood, the grand rush of business, the unexampled prosperity which passed over the state from 1849 to 1853, has passed away forever; it is too late to repair the damage; fifty years of peace and justice cannot place California where she now would have been, had justice and sound policy been adopted twelve years ago.

Thus I have explained the reasons which caused the desertion of California by many of the best men who have ever visited her shore. Fortunately, every thing in California is gradually becoming more stable; titles in the agricultural districts are gradually being settled; and it is now almost established beyond a doubt, that within a few years the federal government must sell a considerable portion of the land in the mining counties, at least those counties not occupied by miners. The mining, the agriculture, the commerce, the population, and the wealth must continue to increase, and her name shall be glorious in the records of industry and on the pages of universal history.

APPENDIX.

THE PROPORTIONS OF THE SEXES IN CALIFORNIA.

COUNTIES.	WHITES.			Aggregate population.
	Male.	Female.	Total.	
Alameda	5,489	3,059	8,548	8,927
Amador	6,151	2,101	8,252	10,930
Butte	7,770	1,967	9,737	12,106
Calaveras	10,088	2,458	12,546	16,299
Colusa	1,543	622	2,165	2,274
Contra Costa	3,395	1,790	5,185	5,328
Del Norte	1,050	291	1,341	1,993
El Dorado	11,844	3,671	15,515	20,562
Fresno	774	225	999	4,605
Humboldt	1,721	777	2,498	2,694
Klamath	1,077	143	1,220	1,803
Los Angeles	5,712	3,509	9,221	11,333
Mariposa	3,385	918	4,303	6,243
Marin	2,339	758	3,097	3,334
Mendocino	2,037	868	2,905	3,967
Merced	800	314	1,114	1,141
Monterey	2,708	1,597	4,305	4,739
Napa	3,445	2,003	5,448	5,521
Nevada	11,457	2,681	14,138	16,446
Placer	8,507	2,312	10,819	13,270
Plumas	3,284	567	3,851	4,363
Sacramento	14,738	6,954	21,692	24,142
Santa Barbara	1,816	1,362	3,178	3,543
San Bernardino	1,482	1,022	2,504	5,551
Santa Clara	7,426	4,399	11,825	11,912
Santa Cruz	3,148	1,764	4,912	4,914
San Diego	850	399	1,249	4,324
San Francisco	33,990	21,636	55,626	56,802
San Joaquin	6,131	3,178	9,309	9,435
San Luis Obispo	1,098	672	1,770	1,782
San Mateo	2,211	935	3,146	3,214
Shasta	3,295	1,023	4,318	4,360
Sierra	9,793	1,537	11,330	11,387
Siskiyou	6,252	1,306	7,558	7,629
Solano	4,681	2,446	7,127	7,169
Sonoma	7,425	4,357	11,782	11,867
Stanislaus	1,606	594	2,200	2,245
Sutter	2,390	970	3,360	3,390
Tehama	2,997	1,005	4,002	4,044
Trinity	4,469	639	5,108	5,125
Tulare	3,456	1,159	4,615	4,638
Tuolumne	12,575	3,488	16,063	16,229
Yolo	3,196	1,493	4,689	4,716
Yuba	10,255	3,180	13,435	13,668
Total	239,856	98,149	338,005	379,994

The preceding table shows the number of white males, white females, white persons, and persons of all colors, in each county of California, as reported by the United States census for 1860.

From this table, it appears that only 28 per cent., or about one-fourth of the white inhabitants, are females. There may be 100,000 white minors, 50,000 boys and 50,000 girls, in the state. Subtract these figures from the totals of males and females, and we ascertain that there are 183,856 white men, and 48,149 white women, or 238,005 white adults of both sexes in California. The white women number, then, only 20 per cent. of the adult white population. There are four white men to one white woman. If we subtract 48,149 from 183,856, we ascertain that the white men have an excess of 135,707, as compared with the white women. There are so many bachelors in the state—three white men out of four are unmarried. These figures are full of significance. They tell much of the social condition of California. They should serve as an advertisement throughout the world, of the scarcity of, and demand for women in California. All through the rich gold-producing counties, four men out of five have no wives, and at least two out of five would like to have wives, if they could get them. These disconsolate bachelors are many of them poor, and unable to support families in luxury, but they might be good husbands, and provide their wives with all the necessaries of life. Let the old and young maids of Christendom take note of this great fact; three-fifths of the adult white population of California are men without wives.

This disproportion between the sexes is greatest in the mining counties, and least in the agricultural districts. Santa Barbara has 42 per cent. of females; San Bernardino 40 per cent.; Los Angeles and San Francisco, each 38 per cent.; Sonoma and Santa Clara, each 36 per cent.; Alameda and Santa Cruz, each 35 per cent.; Contra Costa and Solano, each 34; Sacramento, 32; San Mateo, 29; Yuba, El Dorado, and Shasta, each 23; Tuolumne and Mariposa, each 21; Butte, 19; Nevada and Calaveras, each 18; Siskiyou, 17; Sierra, 13; Plumas and Trinity, each 12, and Klamath 11. Only eleven white females in Klamath county to eighty-nine white males! If we exclude the children, and count only the adults, the disproportion between the sexes, in all these counties, is still greater.

EARTHQUAKES.

In the chapter on Climate, I have given a list of some of the most notable earthquakes which have been felt in California since the establishment of the missions. The following, however, were omitted inadvertently.

On the 11th Oct., 1800, six severe shocks were felt at San Juan Bautista, and every house was shattered and rendered uninhabitable. The same earthquake was felt with much severity at San José.

On the 21st June, 1808, twenty-one shocks were felt at San Francisco, and the few houses then existing were seriously injured.

THE END.

www.ingramcontent.com/pod-product-compliance
Lightning Source LLC
LaVergne TN
LVHW020928110826
845150LV00004B/802